AF697186

GEOGRAPHICAL DIMENSIONS OF ENERGY

The GeoJournal Library

Series Editor: WOLF TIETZE

Editorial Board:

John E. Bardach, Honolulu
Pierre Biays, Lille
Asit Biswas, Oxford
Bruce Currey, Bedford Park
J. Demek, Brno
Yehuda Gradus, Beer-Sheva
Arild Holt-Jensen, Bergen
Huang Ping-wei, Beijing
Johannes Karte, Bonn
P. M. Kelly, Norwich
C. Kergomard, Lille
C. Gregory Knight, University Park
M. P. Lapping, Guelph
W. Lauer, Bonn
Lee Shu-tan, Nanjing
Richard F. Logan, Los Angeles
Walther Manshard, Freiburg i. Br.
German Müller, Heidelberg
Paul Müller, Saarbrücken
Hiroshi Sasaki, Tsukuba
Akira Suwa, Tokyo
Jörn Thiede, Kiel
H. Th. Verstappen, Enschede
E. M. Yates, London
M. M. Yoshino, Tsukuba
Alexander Zaporozec, Madison

Geographical Dimensions of Energy

Edited by

Frank J. Calzonetti
Department of Geology and Geography and Regional Research Institute, West Virginia University

and

Barry D. Solomon
Federal Energy Regulatory Commission, Washington, D.C.

D. Reidel Publishing Company

A MEMBER OF THE KLUWER ACADEMIC PUBLISHERS GROUP

Dordrecht / Boston / Lancaster

Library of Congress Cataloging in Publication Data

CIP

Main entry under title:

Geographical dimensions of energy.

(GeoJournal library)
Includes index.
1. Power resources–Addresses, essays, lectures. 2. Energy development–Addresses, essays, lectures. 3. Electric power-plants–Location–Addresses, essays, lectures. 4. Environmental protection–Addresses, essays, lectures. I. Calzonetti, Frank J. II. Solomon, Barry D. (Barry David), 1955– . III. Series.
TJ163.24.G46 1985 333.79 85-10887
ISBN 90-277-2061-4

Published by D. Reidel Publishing Company
P.O. Box 17, 3300 AA Dordrecht, Holland

Sold and distributed in the U.S.A. and Canada
by Kluwer Academic Publishers,
190 Old Derby Street, Hingham, MA 02043, U.S.A.

In all other countries, sold and distributed
by Kluwer Academic Publishers Group,
P.O. Box 322, 3300 AH Dordrecht, Holland

TJ
163.24
.G46
1985

All Rights Reserved
© 1985 by D. Reidel Publishing Company, Dordrecht, Holland
No part of the material protected by this copyright notice may be reproduced or utilized in any form or by any means, electronic or mechanical, including photocopying, recording or by any information storage and retrieval system, without written permission from the copyright owner

Printed in The Netherlands

CONTENTS

PREFACE

This book is the first attempt to provide a comprehensive volume on the topic of energy geography. The book reviews research on energy geography, contains original refereed articles on energy and provides a chapter on future research directions in this subfield of geography. The book provides an overview of the research activity underway, and highlights the role of geographers and regional scientists in the study of energy topics. A diverse group of scholars contribute to the manuscript and the chapters show how different research methodologies and approaches are employed in investigating a range of relevant energy problems.

This book is intended for upper-level undergraduate students and graduate students in geography, regional science and related disciplines. It will also be useful to professional geographers who would like to know more about the energy geography subfield. Furthermore, the chapters will be interesting to applied geographers and energy policymakers who can see the value of the spatial perspective in addressing energy problems.

This book came about largely through the efforts of the Energy Specialty Group of the Association of American Geographers. The purpose of the specialty group is to improve communication and interaction among energy geographers and their peers, to increase the contribution of geographers in energy research, to increase the visibility and credibility of the work of geographers in energy research, and to assist in the development of energy course offerings and energy-related curricula which incorporate the training and perspective of geography. All proceeds from the sale of this book will be donated to the AAG Energy Specialty Group.

Participants of past Energy Specialty Group meetings noted that there really is no comprehensive volume on energy geography, and that a book on the topic should be written. This book is in response to this discussion. Many members of the energy specialty group were involved in the early planning of this book. Special attention should be given to Mike Pasqualetti, Jerome Dobson, Russell Lee, Susan Macey and Tom Wilbanks who reviewed early efforts to formulate a volume on this topic. Each contribution in the book has been reviewed by an external referee. We wish to thank the following individuals who provided such important service: Tim Allison; Diane Cameron; Susan Cutter; Jerry Dobson; Pat Harker; Ed Hillsman; Dee Knifong; Andy Krmenec; Lans Lundquist; William Miernyk; Byron Miller; Jeffrey Osleeb; John Robinson; Graham Rowles; Greg Sayre; John Sorenson; Derek Spooner; Cara Williams; and Diana Wion.

It would not have been possible to complete this volume without the support of the Regional Research Institute at West Virginia University. Most of the chapters were typed at the Institute. Greg Sayre, Diana Wion and Dwayne Porter provided great assistance in proofing the manuscript. Greg Sayre, Diana Wion and Cara Williams prepared the index. We are most grateful to Beth Noel, Carla Uphold and Dianna Martin for their high standards, patience and good humor through endless revisions of the manuscript.

March 1985
Regional Research Institute
West Virginia University
Morgantown, West Virginia

Frank J. Calzonetti
Chair, AAG Energy Specialty Group

INTRODUCTION

B.D. Solomon
Federal Energy Regulatory Commission
Washington, D.C. 20426

F.J. Calzonetti
Department of Geology and Geography, and
Regional Research Institute
West Virginia University
Morgantown, WV 26506

The purpose of this book is to illustrate the richness of research in "energy geography", a field that investigates the spatial implications of energy. While most of the contributors in energy geography are trained as human geographers there has been quite a large amount of research on the geographical aspects of energy problems undertaken by regional scientists economists, behavioral scientists, physical scientists and many others. While this breadth increases the range of topics available for study, it is often confusing to outsiders who may argue that energy geography does not have an exclusive research domain, or that it does not contribute to important public policy formulation.

One objective of this volume is to clearly demonstrate the importance and relevance of energy geography. The book shows that energy geography can be delimited as a discrete field of scientific inquiry and that it can shed light on important public policy questions. The chapters illustrate the wide range of perspectives that researchers utilize to increase our understanding of energy issues. While this volume is far from exhaustive and does not represent all of the good research underway, it does provide a sample of the types of research that is being conducted. All of the chapters are original contributions that have been externally reviewed by experts in the field.

In defining the field of energy geography, it is important to ask "where can one find the work of such people?" Energy geographers work in universities and colleges, research organizations, government, and the private sector. Publishing outlets include academic journals and books as well as professional research reports. While academic journals which publish articles in this area have included all of the major geography and regional science journals, lately there has been considerable interest in disseminating such research results in cross-disciplinary energy policy journals, as well as in prestigious, widely read magazines such as Science, Scientific American, Environment and

F. J. Calzonetti and B. D. Solomon (eds.), Geographical Dimensions of Energy, 1–6.
© 1985 by D. Reidel Publishing Company.

Bulletin of the Atomic Scientists. The three leading energy policy journals in which geographers have published the results of their energy research are Energy Policy, Energy: The International Journal and Energy Systems and Policy. Two of these latter journals have had geographers on the editorial review boards since their beginnings in the early 1970s. Peter Odell with Energy Policy, and Bruce Hannon with Energy Systems and Policy.

While the modest (if not widely recognized) professional success of energy geography as a field may be linked to its broad range of ideas and approaches to energy studies, the latter tends to obscure what the field truly constitutes, as well as what knowledge it might potentially contribute. To date, no one has yet successfully defined the field although several scholars have grappled with this important question. There have been numerous works by geographers on specific energy topics (e.g. renewable energy resources; power plant siting; offshore energy) or energy in selected counties (e.g., energy in Britian; energy in the USSR). There are only a few previous works which are more comprehensive assessments of the field of energy geography. Manners (1964, 1971) provided one of the earliest and most comprehensive volumes which treated energy geography as a specialized area of economic geography. Transportation and energy flows are emphasized in The Geography of Energy. After acknowledging that the distribution of energy resources is one of the major forces shaping the geography of energy, Manners carefully details the geographical aspects of energy markets, energy transportation, and political factors that influence energy development, use and energy facility siting. This work was a masterpiece written before information on energy production, transport and use was widely available.

In 1971, Nathaniel Guyol's Energy in the Perspective of Geography was published as a volume in the Foundations of Economic Geography series. Guyol takes an approach different from Manners. His approach is more international in scope than Manner's volume, and Guyol goes into great detail summarizing the value of energy accounting. He uses the Netherlands as his example. The volume mainly is a description of energy production and consumption tables and is not as successful in relating energy geography to the field of economic geography as is Manner's book. Data summarized by Guyol is now readily available in U.S. Energy Information Administration publications, but in the early 1970s these data were not widely available and a book summarizing energy accounts was of great value.

Three years after the 1973 oil crisis Earl Cook's Man, Energy, Society was published. This was an introductory text designed for energy courses in geography and related disciplines. Cook did not focus on the economic geography of energy as did Manners and Guyol but attempted to provide a comprehensive volume which considered all aspects of the importance of energy to society. The book includes sections on energy resource endowment, energy technology, energy use in agriculture, as well as sections dealing with energy resource depletion, imports of energy systems, and trends in energy use. The book is descriptive and does not attempt to summarize research by energy geographers nor is it concerned with the relationship of energy geography to research efforts in the discipline.

There has also been a few articles which summarize the field of energy geography. Fitzsimmons and Walton (1978) summarize books that provide an overview of energy problems and issues and discuss publications which provide energy data. Pasqualetti (1979) reviewed the area of energy research by physical geographers, and listed research topics in geomorphology, climatology, biogeography, water resources and environmental issues. Pasqualetti shows that there is much room for physical geographers to contribute in the study of energy geography. Wilbanks (1982) reviews the geographer's role in applied energy studies. He argues that energy geographers must be "energy experts" in order to gain respect outside the discipline. Wilbanks then provides a summary of research in applied energy geography.

There has also yet to be developed a fully satisfactory taxonomy of the field of energy geography. While several organizational frameworks are possible, each one has inherent flaws, in that one type or another study of the geography of energy might not fit into the framework. This actuality is again due to the broadness of the field. Nonetheless, each of the following organizational schemes has been considered for presenting the material in the book.

The first taxonomy could be called the "generalist" approach. This approach classifies energy geography research into five major areas: patterns of energy development; patterns of energy use; energy policy and planning; energy and regional development; and locational decision making. While comprehensive, these categories also lose some appeal due to their broadness, since their elements are not altogether obvious, nor do all the topics fall under the broad rubric of human geography.

A second taxonomic approach considers energy geography research by political scale. This would include urban, rural, regional, local-state, nation-state, international, and possibly global, and natural resource area. However, neglected research areas under this scheme include energy transport and flows, and research techniques such as spatial or multiregional modeling, and geographic information systems.

A third approach focuses on the historical development of geographers' contributions to energy studies. Research categories include, in chronological order: resource development; energy facility siting; environmental impact studies; socioeconomics and energy use patterns; energy policy; information systems; and waste management. Unfortunately, once again, some of these areas are not the sole province of human geography, and such a taxonomy may thus be counterproductive.

The fourth approach uses the third one as a point of departure, and focuses (yet still imperfectly so) on more strictly geographical topics. This approach has been adopted for the book, which is thus divided into the following four major sections: energy resource development; power plant siting and land use; energy use behavioral patterns; and multiregional and environmental issues. Geographical techniques, while _not_ fully covered by the book, can fit into one or more of the included sections. A few other topics, inevitably, might also fall into two (or more) of the section categories.

An additional approach to organizing the field (and the book) might focus solely on energy resources or fuels (i.e., coal, oil and gas, etc.). While this would restrict the coverage to energy production-side issues, the majority of past research efforts in energy geography could be classified in this manner as well. Therefore, it is instructive (as well as convenient) to provide a thematic literature review of previous research along these lines, to yield further and much more detailed background to the field of energy geography.

We begin in Part I with thematic literature reviews of major geographical studies on the primary energy resources. Clearly, most of this work has focused on coal and nuclear fission power, yet geographers have made important contributions to the study of other fuels as well. Ira Sheskin makes this clear in his review on oil and gas. Frank Calzonetti illustrates the long-standing attention geographers have given to the study of coal, while Martin Pasqualetti highlights the changing foci of nuclear power research. The final reviews by Stephen Sawyer on renewable energy and Barry Solomon on energy conservation indicate the recency of geographers' work in these areas, and the strong overlap of the subject matter with related fields.

Part II focuses on some critical economic, behavioral and planning issues for production of the major energy resources. Ian Manners begins by comparing Scottish and American experience in handling developmental impacts of offshore oil and gas. The next chapter presents a Thunen land use model for evaluating biomass energy resources, developed by Donald Jones and John Krummel. The model is useful for determining the competitive location of alcohol fuel production facilities. Also taking a spatial perspective, Jeffrey Richetto then discusses the practical issues faced by developers in marketing unconventional energy sources, in this case coalbed methane gas in Alabama (US). Finally, James Johnson summarizes his research on evacuation behavior from nuclear power plant accidents, and highlights spatial factors which will be instructive to planners and government policymakers.

Part III is concerned with the power plant siting and land use nexus. It begins with a discussion by Sherman Feher on the changing patterns of electric power facility locations in the US, which gives special attention to structural change and renewable energy fuels. The next chapter by Benjamin Hobbs reviews the literature on modeling imperfect spatial energy markets, and focuses on transport factors. This is followed by two companion studies on electric power controversies in Great Britain. First, John Fernie and Stan Openshaw demonstrate that population densities near nuclear power plants are much greater in the UK than in the US, and argue forcefully for stricter standards in the former. Second, Susan Owens reviews the role of citizen participation and conflict in 68 power plant applications in the UK. While ending on a pessimistic note, she raises some important questions about the functioning of democratic institutions in the siting arena. Lastly, Philip Pryde compares the land requirements for a wide range of solar electricity options, noting the great variability of the "land factor" for future solar facilities.

Part IV turns to energy consumption issues, with a heavy focus on energy conservation behavior. Marilyn Brown leads off with an examination of the determinants of home energy use, based on an urban case study in Illinois (US). Peter Nijkamp and Douwe Tiemersma next describe a widely aclaimed energy model that was developed in Europe, underscoring its spatial modules. They then proceed to apply it to various scenarios for Holland. The next chapter by Allen White investigates the conservation investment behavior of home buyers in times of rising energy prices, with a case study conducted in a suburb of Boston. The last chapter by Susan Macey addresses home energy consumption of elderly people, giving particular attention again to conservation investments. This study highlights a case study done in Lexington, Kentucky.

Part V addresses multiregional and environmental issues in an international context. Special attention here is given to current public policy issues. James Zucchetto compares and contrasts the energy intensities (including energy from the natural environment) of numerous national economies worldwide, and also over time. Stephen Lonergan follows with an entropy interpretation of energy-based regional growth and development in Canada. The next chapter by Frank Calzonetti and Robert Hanham models the relationships between changing fossil fuel prices and state government revenues in the US. Two related studies then address air quality problems of coal combustion. Thomas Georgianna, Barry Solomon and Kingsley Haynes, first apply time series methods to determine if the federal Clean Air Act, as well as major mineworker strikes, have had differential regional effects in the US. Mark Berkman and Peter Blair next use an input-output model to assess the regional impacts of possible "acid rain" controls in the same nation. Finally, Rolf Sternberg closes by holistically considering the socioeconomic and environmental changes accompanying massive hydroelectric energy development in the Brazilian Amazon.

Part VI completes the book with a chapter by Thomas Wilbanks on the prospects for the field of energy geography, outlining some of the more important research areas which remain to be pursued.

References

Fitzsimmons, A.K. and K.J. Walton. 1978. Toward a Geography of Energy: A Review of Base Sources and Literature. Journal of Geography 77: 61-66.

Pasqualetti, M.J. 1979. Energy. Geographical Survey 8: 35-40.

Cook, E. 1976. Man, Energy, Society. San Francisco: W.H. Freeman and Company.

Guyol, N.B. 1971. _Energy in the Perspective of Geography_. Englewood Cliffs, NJ: Prentice-Hall Inc.

Manners, G. 1971. _The Geography of Energy_. London: Hutchinson University Library.

Wilbanks, T.J. 1982. Location and Energy Policy. In _Applied Geography. Selected Perspectives_ ed. J.W. Frazier, pp. 219-232. Englewood Cliffs, NJ: Prentice-Hall, Inc.

PART I

Research Contributions in Energy Geography

CHAPTER 1

COAL

Frank J. Calzonetti
Department of Geology and Geography and
Regional Research Institute
West Virginia University
Morgantown, WV 26506

ABSTRACT. There has been much written on the geography of coal. The fuel exerted great influence on industrial development and settlement in the past, and was a major topic of investigation among economic and human geographers. The revival in the prospects for coal stimulated a new interest in the topic among modern geographers.

1. INTRODUCTION

Geographers have written more on coal than any other energy form. The fuel was the mainstay of modern civilizations as the discipline of geography was emerging and coal remained important throughout most of the twentieth century. Indeed, coal was the principal source of energy in the US until the second world war and in the UK coal provided 90 percent of the total energy needs in 1950. Geographer's early contributions were descriptive and most articles on the topic were surveys of the characteristics, history and potential of coal mining districts. The topic of coal geography became less vital in the 1950s and 1960s as oil and gas consumption rose and nuclear power offered great promise. The 1973 oil crisis led to a reevaluation of coal as many countries hurt by rising energy import bills had vast domestic coal reserves. This urgency aroused a new interest in coal among geographers particular in the US. Coal had enormous potential and could be converted into electricity, gaseous fuels and liquid fuels. Shifting to coal, however, would be difficult and would require dealing with numerous logistical, environmental, economic, and political problems. Most of the problems were inherently geographical and extremely contentious. Geographers are contributing on many fronts, but their efforts fall into three major areas: 1) studies of the coal resource delivery system; 2) impacts of facilities; and 3) regional growth and decline. First the contributions of early studies on the geography of coal are discussed.

F. J. Calzonetti and B. D. Solomon (eds.), Geographical Dimensions of Energy, 9–18.
© 1985 by D. Reidel Publishing Company.

There was striking similarity in the organization of early studies on the geography of coal. They usually began with a description of the geology of the region and coal output trends, traced the historical development of the region, then discussed markets, transport and prospects for the future (Birch 1934; Bainbridge 1936; Elkins 1953, Monkhouse 1955; Deasy and Griess 1956; Pounds 1958; and Hefford and Smith 1967). Some efforts were made to survey the industry for an entire nation, such as Harding's (1946) account of the US coal mining industry and Belden and Salter's (1935) paper on coal resources in China. In some studies, the "center of coal production" was traced across a particular coal mining region (Murphy and Spittal 1944; 1945). Little effort was made to explain the factors leading to shifts in the location of production. Indeed, Deasy and Griess (1967) explicitly stated that it was difficult, if not impossible, to identify the factors that lead to the growth and decline in coal mining activity. They preferred to forecast growth and decline on the basis of trend lines of coal production in US counties. An inspection of their conclusions (Wion 1984) is a lesson in the inaccuracy of trend line extrapolation.

Other studies, however, acknowledged that there indeed existed intraregional factors that led to decline in certain areas and growth in others (Bainbridge 1949; Roepke 1955). Rawstron (1951; 1966), for instance, investigated the location and utilization of power plants in Great Britain and acknowledged the relationship between coal-fired power plants and coal field markets. Rawstron identified restrictive factors (such as water availability) which affected power plant location. The actual development of a coal field was also investigated by geographers. Wilson (1968) presented a simple model to explain the development of a hypothetical coal field which stemmed from his investigation of mining activity in New South Wales. Wrigley (1961) showed a close linkage between coal output, industrial growth and population change in the early phase of industrial development in France, Belgium and Germany.

2. THE COAL RESOURCE DELIVERY SYSTEM

The coal resource delivery system refers to coal production, coal preparation and the transportation of coal to final destinations where it is converted into other energy forms or directly consumed. In many countries coal production is highly localized and the volumes transported are not that large making it possible to investigate problems in the coal resource delivery system by more straightforward approaches. For instance, in the UK, annual coal output is about equal to that of the US state of Kentucky. The nature of the UK coal industry has been characterized by Manners (1964, 1981), North and Spooner (1982) and Spooner (1981b). In the US, coal reserves are widely distributed, often located distant from markets, and coal transportation is a major element of the final price of coal. In addition, coal transportation bottlenecks have arisen. With the emergence of quantitative geography, geographers began to use new tools in the analyses of coal problems. A

Great Lakes study by King *et al.* (1971) demonstrated the importance of modeling coal flows, and coal transportation modeling is now a major topic of investigation by geographers. Given the complexity of the coal resource delivery system in the US, modeling serves to simplify the system and allow one to isolate key problems of geographical interest. For instance, a number of studies were completed by Osleeb and Ratick (e.g. 1985) which focused on various aspects of coal shipment. They directed their attention to transshipment problems involving railways and ports and have extended their effort to evaluate the ability of US ports and railroads to handle increased US coal exports. Another study of a smaller region was conducted by Elmes (1985) as part of the Pennsylvania Coal Study (Mach and Knight 1978). In this case, coal shipments to utilities were modelled.

While many studies focused exclusively on coal transportation, others attempted to model the entire coal resource delivery system, including the location of energy facilities (Haynes *et al.* 1983; Hauser 1971). In the 1970s there were a number of large national models springing from attempts by the US government to become self-sufficient in energy, but naive assumptions regarding regions and transportation networks often limited their usefulness (see Schanz *et al.* 1979). More recently, the spatial resolution of models has improved, increasing their value as planning tools. The Brookhaven Regional Energy Facility Siting model (Maier *et al.* 1979) is an example of a multicommodity, transshipment model. It is based upon county-level data which attempts to evaluate specific energy facility sites.

The problem of siting coal facilities has been a topic of much interest by geographers (Calzonetti and Eckert 1981). Large power plants and synthetic fuel facilities require large inputs of coal and water, generate physical wastes and cause local social and economic disruption (Rubin and Solomon 1983). Sites for energy facilities must recognize the economic and engineering objectives of the energy developer in addition to the impacts of energy facilities and the ability of a site to absorb these impacts. Site screening procedures provide an approach useful in identifying potential sites for energy facilities. Dobson (1979) showed how this approach could be used for planning new power plant sites in the state of Maryland, but the lack of need for new facilities precluded the implementation of the strategy. The goal of such tasks is to identify sites which are acceptable according to economic, environmental, engineering, political, and social attributes. Church and Bell (1981) illustrate how different models result in different rankings of locations for energy facilities. Problems include data decision rules used, variables considered, and cell scoring procedures. Methods to deal with multiobjective decisionmaking in facility siting have been discussed in detail by Keeney (1980) and surveyed by Solomon and Haynes (1984).

3. IMPACTS OF COAL FACILITIES

Coal produces undesirable physical and socioeconomic impacts at every stage in the coal resource delivery system. Interest in the land impacts of coal mining go back at least to Miller's (1952) study of strip mining in Pennsylvania and Doerr and Guernsey's (1956) paper which summarized the affects of underground and surface coal mining. At that time they noted the difficulties in assessing the extent of damage because of the lack of accurate record keeping. Interest in the environmental impacts of coal mining rose in the late 1960s and 1970s as new areas became targets of mining operations. The massive expansion of coal production, largely through surface mining operation was the focus of much study in the US (Calzonetti et al. 1980; Luten 1980). There also arose conlfict between agricultural and coal interests, mainly in the western states (Georgianna and Haynes 1981). The disposal of waste from power plants even became a problem in many areas (Calzonetti and Elmes 1981). In the UK, subsidence and colliery waste disposal continued to be a problem in coal-mining areas, and the expansion of mining into agricultural regions was very contentious (North and Spooner 1982).

In addition to the land impacts of large energy projects, the scale of energy activity led to socioeconomic impacts of small western towns (Muschett 1978; Ostresh 1981). The contributions made by geographers investigating the impact of energy development on small communities has increased. The way in which mining operations affected settlement of workers has also been a topic of study. Although the socioeconomic impacts of energy development in the western US is considerable, public attention may be focused upon the most severe cases (Science and Public Policy Program 1981; Leistritz and Murdock 1981). Large coal projects have also been proposed in the eastern US. Solomon and Rubin (1985) provide an econometric method for estimating the impacts of large-scale synthetic fuel facilities in the eastern Interior US coal field (see also Solomon 1985).

The major impacts of coal mining are localized, but power plants produce atmospheric pollutants that travel great distances. Carbon dioxide emissions, for example, are a global problem, and acid rain is an interregional and international problem which is quickly receiving the attention of geographers. In North America and Europe, acid rain is a sensitive political problem pitting coal producing regions against regions impacted by acid rain. In the US, midwestern and Appalachian coal interests view impending acid rain legislation as a direct threat to jobs, but those in New England and Canada view acid rain as a threat to their environment. Some coal interests claim that Canadians are overstating their concern over acid rain in an effort to export surplus electricity to the US (Friedman and McMahon 1984). The impact of air pollution controls on the economy has been considered (see Rose 1983), and in this book Blair and Berkman show the regional economic impacts of acid rain legislation in the US.

4. REGIONAL GROWTH AND DECLINE

Coal has been a mixed blessing to most regions. When markets are strong, coal provides jobs and revenues in a regional economy. During recessionary times, coal regions are often hard hit. In many cases there is long-term decline in coal regions resulting from such factors as a loss of markets, changing markets, seam exhaustion, and declining productivity. The growth and decline of mining regions has been intensely studied by geographers and regional economists.

Spooner (1981a) discussed mineral resource development in the context of the evolution of the space economy, in which the early stage of initial growth and expansion will ultimately lead to the decline of coal and regional depression. Deasy and Griess (1965) investigated the Northeast Pennsylvania anthracite region, one of the first US coal-producing regions to experience long-term decline. They observed that the decline persisted, even though the bituminous coal regions were growing and the state's economy was expanding.

The regional disparity in the fortunes of coal-producing regions is quite sharp in the UK. There has been a long-term decline in the peripheral coal regions by expansion of the central coal fields. The National Coal Board is investing in productive pits and expanding into new areas and reducing investment in less productive peripheral coal fields (Spooner 1981a). The 1984 British coal strike is testimony to the political explosiveness of this policy as miners attempt to retain jobs in the declining coal regions.

In the US, despite abortive attempts, there is no national coal policy. However, the regional variation in coal quality and the regional impacts of legislation and unionization has resulted in shifts in production from one region to another as discussed by Clements (1977), as well as Georgiana, Solomon and Haynes in this volume. In 1971, for instance, Wyoming produced 8 million tons of coal per year. By 1982, Wyoming's output stood at 108 million tons. Western coal is lower in sulfur content than eastern or interior coal. Western coal expanded partially because the Clean Air Act encouraged power plant operators to use low sulfur coal to meet air quality standards. After intense efforts by eastern politicians, the Clean Air Act Amendments of 1977 were enacted which require <u>all</u> new units to install scrubbers, irrespective of the sulfur content of the coal used. This should reduce the advantage of western coal over time (see also Calzonetti 1985).

Miernyk (1982, 1984) has perhaps provided the most evidence relating the impacts of the coal industry on regional economies. He argues that the rising value of coal since the late 1960s has been instrumental in causing higher personal incomes and declining unemployment in coal-producing regions in the US. He notes that rising energy prices have tipped the regional terms of trade within the US to favor energy surplus states at the expense of energy deficit states. This topic is developed further by the contribution in this book by Calzonetti and Hanham.

Geographers and regional scientists are making broad contributions to the study of coal and it appears that the topic will continue to be an important area of investigation. Most of the work appears to have been from the perspective of economic geography. Despite the major environmental impacts of coal production, transport and use physical geographers have not fully taken advantage of research opportunities. It is expected that political geography will find the topic of great interest as coal investment and use tends to put region against region in a highly volatile public policy arena.

6. ACKNOWLEDGMENTS

I am grateful to Derek Spooner, Barry Solomon, Diana Wion and Dwayne Porter for their helpful suggestions in preparing this paper.

7. REFERENCES

Bainbridge, T.H. 1936. The West Cumberland coalfield. Economic Geography 12: 167-74.

______. 1949. Population changes over the West Cumberland coalfield. Coalfield. Economic Geography 25: 128-33.

Belden, W. and Salter, M. 1935. The coal resources of China. Economic Geography 11: 304-06.

Birch, T.W. 1934. Development and decline of Coalbrookdale coalfield. Geography 19: 114-26.

Calzonetti, F.J., M.E. Eckert and Malecki, E.J. 1980. Siting energy facilities in the USA: Policiies for the Western states. Energy Policy 7: 138-52.

______ and Eckert, M.E. 1981. Finding a place for energy: Siting coal conversion facilities. Washington, D.C.: Association of American Geographers.

______ and Elmes, G.A. 1981. Metal recovery from power plant fly ash: An ecological approach to coal utilization. GeoJournal Supp. 3: 59-70.

______. 1985. Regional trends in coal use for electricity generation: Implications for the US and UK. In Geographic Issues in Mineral Resource Development, eds. H.E. Johansen, O.P. Mathews and G. Rudzitis. forthcoming.

Church, R.L. and Bell, T.L. 1981. A Comparison of two baseline screening approaches to regional energy facility siting. GeoJournal

Suppl. (3): 17-36.

Clements, D.W. 1977. Recent trends in the geography of coal. Annals of the Association of American Geographers 67: 109-25.

Deasy, G.F. and Griess, P.R. 1956. Geographical significance of recent changes in mining in the bituminous coal fields of Pennsylvania. Economic Geography 33: 283-98.

______. 1965. Effects of a declining mining economy on the Pennsylvania anthracite region. Annals of the American Geographers 55: 239-59.

______. 1967. Local and regional differences in long term bituminous coal production prospects in eastern United States. Annals of the Association of American Geographers 57: 519-533.

Dobson, J.E. 1979. A regional screening procedures for land use suitability analysis. The Geographical Review 69: 224-34.

Doerr, A. and Guernsey, J.S. 1956. Man as a Geomorphological agent: The example of coal mining. Annals of the Association of American Geographers 46: 197-210.

Elkins, T.H. 1953. The brown coal industry of Germany. Geography 38: 18-29.

Elmes, G.A. 1985. Modeling spatial interaction patterns of utility coal in Pennsylvania. Annals of the Association of American Geographers 25: in press.

Friedman, J.M. and McMahon, M.S. 1984. The Silent Alliance. Chicago: Regnery Gateway.

Georgianna, T.D. and Haynes, K.E. 1981. Competition for water resources: Coal and agriculture in the Yellowstone River Basin. Economic Geography 57: 225-37.

Harding, G.E. 1946. American coal production and use. Economic Geography 22: 46-53.

Hauser, D.P. 1971. System costs and the location of new generating plants in England and Wales. Transactions of the Institute of British Geographers 54: 101-21.

Haynes, K.E., Phillips, F.Y. and Solomon, B.D. 1983. A coal industry distribution planning model under environmental constriants. Economic Geography 59: 52-65.

Hefford, R.K. and Smith, D.L. 1967. The Leigh Creek Coalfield, South Australia: Resource development in a difficult environment. Annals of the Association of American Geographers 57: 503-18.

Keeney, R.L. 1980. Siting Energy Facilities. New York: Academic Press.

King, L.J., Casetti, E., Odland, J. and Semple, R.K. 1971. Optimal transportation patterns of coal in the Great Lakes region. Economic Geography 47: 401-13.

Leistritz, L.F. and Murdock, S.H. 1981. The socioeconomic impact of resource development: Methods for assessment. Boulder, CO: Westview Press.

Luten, D.B. 1980. Western coal mining. Landscape 24: 1-2.

Mach, D.L. and Knight, C.G. 1978. Alternative scenarios for Pennsylvania's coal economy. Applied Geography Conference 1, Binghamton NY: SUNY-Binghamton, 299-306.

Manners, G.M. 1964. The geography of energy. New York: Hilary House Publishers.

______. 1981. Coal in Britain. London: George Allen and Unwin.

Meier, P.M., Hobbs, B.F. McCoy, M. and Stern, R. 1979. The Brookhaven regional energy facility siting model (REFS): Model development and application. Upton, NY: Brookhaven National Laboratory, BNL-51006.

Miernyk, W.H. 1982. The differential effects of rising energy prices on regional income and employment. In High Energy Costs; Assessing the Burden, H.H. Landsberg (ed), Baltimore: The Johns Hopkins University Press, pp. 297-330.

______. 1984. Energy and Regional Development. In Energy Costs Urban Development and Housing, eds. A. Downs and K.L. Bradbury pp. 226-86. Washington D.C.: The Brookings Institution.

Miller, E.W. 1952. Penn Township-An example of local governmental control of strip mining in Pennsylvania. Economic Geography 28: 256-60.

Monkhouse, F.J. 1955. The South Limburg Coal Field, Economic Geography 31: 126-37.

Murphy, R.E. and Spittal, H.E. 1944. A new production map of the Appalachian bituminus coal region. Annals of the Association of American Geographers 34: 164-72.

______. 1945. Movements of the center of coal mining in the Appalachian plateau. The Geographical Review 35: 624-33.

Muschett, F.D. 1978. Geographic Scales of Economic and Environmental Impacts from Energy Facilities: The Colstrip Example. Applied Geography Conference 1, Binghamton, NY: SUNY-Binghamton, 307-17.

North, J. and Spooner, D.J. 1982. The Yorkshire, Nottinghamshire, and Derbyshire Coalfield: The focus of the Coal Board's investment strategy. The Geographical Journal 148: 22-37.

Osleeb, J.P. and Ratick, S.J. 1983. The Impacts of Coal Conversion of the Ports of New England. Economic Geography 59: 35-51.

Ostresh, L.M. Jr. et al. 1981. An empirical assessment of the social impact of rapid growth in counties of the Old West Region. Paper presented at the 77th Annual Meeting of the Association of American Geographers, Los Angeles, CA.

Pounds, J.G. 1958. The spread of mining in the coal basin of Upper Silesia and Northern Moravia. Annals of the Association of American Geographers 48: 149-163.

Rawstron, E.M. 1951. The distribution and location of steam-drive power stations in Great Britian. Geography 36: 249-62.

______. 1966. Electric power generation, In Nottingham and its Region ed. K.C. Edwards. Prepared for the meeting of the British Association for the Advancement of Science. Nottingham: 310-14.

Roepke, H.G. 1955. Changing patterns of coal production in the Eastern Interior field. Economic Geography 31: 234-47.

Rose, A. 1983. Modeling the macroeconomic impact of air pollution abatement. Journal of Regional Science 23: 441-59.

Rubin, B.M. and Solomon, B.D. 1983. Economic and fiscal impact analysis of energy development projects: Coal liquefaction in the Illinois Basin. Socio-Economic Planning Science 17: 11-20.

Schanz, J.J. Jr., J.W. Sawyer Jr. and Perry, H. 1979. The future regional distribution of national synthetic fuel capacity. Washington, D.C.: Resources for the Future.

Science and Public Policy Program. 1981. Energy From the West.

Norman, OK: University of Oklahoma Press.

Solomon,, B.D. and Rubin, B.M. 1985. Environmental linkages in regional econometric models: An analysis of coal development in western Kentucky. Land Economics 61: 43-57.

______ and Haynes, K.E. 1984. A survey and critique of multiobjective power plant siting decision rules. Socio-Economic Planning Sciences 18: 71-79.

______. 1985. The socioeconomic impacts of a regional synthetic fuels industry: An integrated econometric analysis. London Papers in Regional Science 15: forthcoming.

Spooner, D.J. 1981a. Mining and regional development. Oxford: Oxford University Press.

______. 1981b. The geography of coal's second coming. Geography 66: 29-41.

Wilson, M.G.A. 1968. Changing patterns of pit location on the New South Wales Coalfields. Annals of the Association of American Geographers 58: 78-90.

Wion, D. 1984. Growth and decline of the West Virginia coal industry. Paper presented at the Mid-Atlantic meeting of the Association of American Geographers, West Chester, PA.

Wrigley, E.A. 1961. Industrial growth and population change. Cambridge: Cambridge University Press.

CHAPTER 2

PETROLEUM AND NATURAL GAS

Ira M. Sheskin
Department of Geography
P.O. Box 248152
University of Miami
Coral Gables, Florida 33124

ABSTRACT. The geography of oil and gas has long been an important topic in economic geography. Interest in oil and gas grew in the 1970s and geograhers are investigating problems and issues in oil and gas distribution, production, transport and use. Regional geography forms an important segment of the literature.

1. CONTRIBUTIONS BY GEOGRAPHERS

It is appropriate to discuss the literature on the geography of petroleum and natural gas together, because these fossil fuels are in many ways "joint" products. Before drilling, a geologist usually cannot predict whether oil, gas, or both will be found in a given field. Petroleum and natural gas comprise approximately 43 and 25 percent of the United States' energy consumption, respectively; petroleum forms a major component of international trade. In addition, most oil and gas is produced in a few basins and many areas that consume large quantities of these fuels lack a resource base entirely. Thus, the production and distribution of these two fuels forms fertile ground for geographical analysis.

A number of economic geography textbooks provide significant overviews of petroleum and natural gas. One of the earliest works was Odell's (1963) economic geography of the oil industry. Alexander and Gibson (1979, 186-201) provide an extensive historical geography of petroleum and natural gas, as do Thoman and Corbin (1974, 332-53). An interesting discussion of international differences in oil and gas production is provided by Jumper, Bell, and Ralston (1980, 307-20) and Wheeler and Muller (1981, 238-49; 261-62) examine the production and consumption of both commodities.

Because of the importance of energy to the Middle Eastern economies, it is not surprising to find that the geographic literature on the Middle East pays particular attention to oil and its economic impact. Bill (1971) examined the impact of petroleum on planning in the Middle

F. J. Calzonetti and B. D. Solomon (eds.), Geographical Dimensions of Energy, 19–25.
© 1985 by D. Reidel Publishing Company.

East. Beaumont, Blake, and Wagstaff (1976, 268-94) provide a detailed description and maps of the major oil fields of the region. In addition, they devote the entire chapter on Libya to the impact of oil on that nation's economy. Fisher (1978) also provides an introduction to Middle Eastern oil reserves. In a follow-up study, Melamid (1973) analyzed the role of the geographical distribution of workers' residences on the development of oil fields in Iran. Melamid (1975) also examined the distribution system for petroleum products in Iran for the purpose of identifying economic regions. Sutton (1979) discussed the natural gas industry in Algeria. Maechling (1982) examined the security risks to energy production and trade in the Persian Gulf. Finally, Odell (1984) discussed various options available to OPEC in the 1980's.

Another research stream has concentrated on the oil and natural gas deposits in the Europe's North Sea. K. Chapman (1976) and Fernie (1976) provide general overviews of North Sea oil. Odell and Rosing (1974) developed a simulation model of the development of North Sea oil. Odell (1977) examined the North Sea's potential for natural gas in relation to the Western European gas market. P. Chapman (1977) discussed the utilization of the resource in the petrochemical industry. Smith et al. (1976) and Davis (1977) examined the impact on Scotland. Bailey (1978) discussed the division of the resources among the littoral states. Bohn (1981) examined Great Britain's North Sea oil. Hansen (1983) investigated the impact of oil on regional policy in Norway. G. Manners (1984) presented some of the costs and benefits of North Sea oil, with particular emphasis on employment. I. Manners (1978; 1982) has investigated environmental planning issues surrounding North Sea oil development, and has conducted similar studies for the Texas and Louisiana Gulf Coast of the U.S. (Manners 1983; see also his chapter in this book).

Non-geographers have taken the lead in developing various mathematical programming formulations to investigate the distribution of natural gas. These studies were mathematically robust, but geographically simplistic. Lawrence (1973) employed a quadratic programming approach to derive a set of optimal North American natural gas flows and prices, with the specific objective of identifying when additional capacity would be needed in a given pipeline corridor. Waverman (1973) employed the transportation problem to examine the effects of Canadian government policy on the cost-efficiency of the North American natural gas delivery system. Both the Lawrence and the Waverman models suffered the drawback of employing spatial units (such as US Census divisions and states) so large as to weaken specific conclusions. Brooks (1975), utilizing 35 supply and 49 demand regions, improved the spatial resolution of distribution models with a mathematical programming formulation of American natural gas flows. Although this facilitated analysis at a spatially-disaggregate level, the model suffered problems with respect to the estimation of pipeline capacities and transmission costs, thus rendering its policy usefulness suspect.

Osleeb and Sheskin have published a series of articles in the geographic literature on the distribution of natural gas, and carefully acknowledged spatial issues. They (1975) derived probable 1990 United States market areas for Canadian natural gas, employing varying sets of supply assumptions made by the former Federal Power Commission. Osleeb and Sheskin (1977) also employed the transportation problem to predict US natural gas flows between 17 North American supply and 56 demand regions, and to identify the areas of the United States which would be most prone to deficits in times of natural gas shortages. This model also helped to identify sites for the production of liquefied and synthetic natural gas.

Sheskin (1977) used a modified version of the out-of-kilter algorithm to model explicitly the natural gas pipeline network. This model improved upon the previous research efforts in that: 1) supply and demand regions were disaggregated into cohesive units, based upon pipeline network structure; 2) the pipeline network was modelled explicitly and sound estimates of pipeline capacity derived; and 3) transmission costs were calculated accounting for the critical nonlinear relationship between transport cost per unit and flow volume in a pipeline. Using this model to evaluate the alternative proposed routes to deliver Alaskan natural gas to the lower 48 states, Sheskin (1977) concluded that the gas should be delivered to the Midwest rather than California (see also Foster 1977). Sheskin and Osleeb (1982) used this same model to examine the possible impact of Mexican natural gas on the US market, and concluded that serious pipeline capacity problems may inhibit the use of Mexican supplies as a _supplementary_ source of natural gas. They also identified US areas most likely to benefit from Mexican imports. Lejtman and Weyant (1981) examined alternative Mexican export policies for its recent _oil_ discoveries, and Goldstein (1981) analyzed the political regulation of the Mexican oil resource.

Other optimization studies have been completed for oil and gas problems. For example, also using a mathematical programming approach, Osleeb (1979) evaluated the strategic oil storage program of the US Department of Energy (see also, Hansen 1979). Earlier, Odell, Rosing, and Beke-Vogelarr (1976) developed an "optimal" pipeline network for the U.K. sector of the North Sea.

A number of other interesting studies on the geography of oil and gas deserve mention. Melamid (1968) examined the Nigerian petroleun industry, and Olu-Abiodum (1974) examined the locational effects of the Nigerian Civil War on this industry. Odell (1970) examined international trade in oil and aspects of world power. Jones (1976) examined the question of sovereignty over oil in Antarctica. Stafford (1978) presented a discussion of the impact of offshore oil development in Florida. Lins (1979) discussed the effect of the exploration, development, and production of oil and natural gas on the land use of the Kenai Peninsula, Alaska. Hoffman (1979) presented various estimates of the size of oil and gas reserves in the USSR and Eastern Europe. Pacione (1979) provided an overview of Italian natural gas and Ma (1980) did the same for Chinese oil. Sheskin (1980) presented a broad overview of the supply, distribution, and demand for natural gas in the United States. Wallach (1980) discussed the West Side oil fields of

California. Rudzitis (1982) discussed the resolution of a controversy between the development of petroleum reserves and the shrimping industry. A small amount of work has been done on the development of synthetic oil resources, such as from shales and tar sands, as exemplified by Weaver (1977), Schein (1978) and Ratnieks (1977).

While the above review is far from exhaustive, it is clear that a substantial volume of literature does exist on the geography of petroleum and natural gas. Regionally, much of the work is concentrated around North America (particularly with respect to the flows of natural gas), the Middle East, and Europe's North Sea. Two topical areas of importance have been the examination of natural gas flows, and the impact of oil money on the economy of states. Given the importance of these fossil fuels to the world economy, and the spatial aspects of the industries, the geography of oil and gas and related synthetic fuels is an area in which geographers can make further significant contributions.

2. REFERENCES

Alexander, J.W. and Gibson, L.J. 1979. Economic geography. Second Edition. Englewood Cliffs, NJ: Prentice-Hall.

Bailey, R. 1978. Unequal share in the North Sea. Energy Policy 6: 320-31.

Beaumont, P., Blake, G.H. and Wagstaff, J.M. 1976. The Middle East. New York: John Wiley & Sons.

Bill, J.A. 1971. The challange of change, petroleum and planning in the Middle East. Focus. New York: American Geographical Society.

Bohn, J.A. 1981. Britain's North Sea oil. The Kansas Geographer 16: 27-37.

Brooks, R. 1975. Allocation of natural gas in times of shortage: A mathematical model of the production, transmission, and demand for natural gas under Federal Power Commission regulation. Unpublished PhD dissertation, Massachusetts Institute of Technology, Department of Operations Research.

Chapman, K. 1976. North Sea oil and gas: A geographical perspective. Problems in Modern Geography Series. London: David & Charles.

Chapman, P. 1977. The utilization of North Sea hydrocarbons as raw materials for petro-chemical manufacture. Geoforum 8: 155-68.

Davis, J.F. 1977. Bane or boon? North Sea oil and gas development in Northeast Scotland. Great Plains-Rocky Mountain Geographical Journal 6: 204-11.

Fernie, J. 1976. The development of North Sea oil and gas resources. Scottish Geographical Magazine 93: 21-31.

Fisher, W.B. 1978. The Middle East. Seventh Edition. London: Methuen.

Foster, H.D. 1977. A pipeline route for Alaska gas. The Geographical Magazine 50: 237-40.

Goldstein W. 1981. Political regulation of Mexico's oil boom: The policy choices of wealth. Energy Policy 9: 251-65.

Hansen, J.C. 1983. Regional policy in an oil economy: The case of Norway. Geoforum 14: 353-61.

Hansen, R.G. 1979. Site selection and analysis for the strategic petroleum reserve. Applied Geography Conferences 2: 71-80.

Hoffman, G.W. 1979. Energy projections--oil, natural gas, and coal in the USSR and Eastern Europe. Energy Policy 7: 232-41.

Jones, P. 1976. Whose oil resources?--The question of Antarctica sovereignty. Geography 61: 167-68.

Jumper, S.R., Bell T.L. and Ralston, B.A. 1980. Economic growth and disparities, a world view. Englewood Cliffs, New Jersey: Prentice-Prentice.

Lawrence, A. 1973. Pricing and planning in the US natural gas industry: An econometric and programming study. Unpublished PhD dissertation, State University of New York at Buffalo, Department of Economics.

Lejtman, T.M. and Weyant, J.P. 1981. Managing an oil bonanza: An analysis of alternative Mexican export policies. Energy Policy 9: 186-97.

Lins, Jr., H.F. 1979. Energy development at Kenai, Alaska. Annals of the Association of American Geographers 69: 289-303.

Ma, L.J.C. 1980. Oil from the wells of China. Geographical Review 70: 99-130.

Maechling, E. 1982. Security risks to energy production and trade: the problems of the Middle East. Energy Policy 10: 120-30.

Manners, G. 1984. North Sea oil: Benefits, costs, and uncertainties. Geoforum 15: 15-31.

Manners, I.R. 1978. Offshore oil: An overview. Policy Study No. 11. Austin, Texas: Center for Energy Studies, The University of Texas.

__________. 1982. North Sea oil and environmental planning: The United Kingdom expierience. Austin, Texas: University of Texas Press.

__________. 1983. Offshore energy development and the Texas Gulf Coast. Proceedings of the 2nd International Conference on Oil and the Environment, pp. 3-11. Halifax, Nova Scotia: Technical University of Nova Scotia.

Melamid, A. 1968. The geography of the Nigerian petroleum industry. Economic Geography 44: 37-56.

__________. 1973. Satellization in Iranian crude oil production. Geographical Review 63: 27-43.

__________. 1975. Petroleum product distribution and the evolution of economic regions in Iran. Geographical Review 65: 510-25.

Odell, P.R. 1963. An economic geography of oil. London: Bell.

__________. 1970. Oil and world power. Middlesex, England: Penguin Books.

__________. 1977. The potential for natural gas from the North Sea in relation to Western Europe's market for gas by the Mid-1980's. Geoforum. 8: 155-68.

__________. 1984. Outlook for the international oil market and options for OPEC. Energy Policy 12: 5-12.

__________, and Rosing K.E. 1974. The North Sea oil province: A simulation model of development. Energy Policy 2: 316-29.

__________, Rosing, K.E. and Beke-Vogelaar, H. 1976. Optimizing the oil pipeline system in the UK sector of the North Sea. Energy Policy. 4: 50-55.

Olu-Abioudun, J. 1974. Locational effects of the civil war on the Nigerian petroleum industry. Geographical Review 64: 253-63.

Osleeb, J.P. 1979. An evaluation of the strategic storage program of the US Department of Energy. The Professional Geographer 31: 393-99.

__________, and Sheskin, I.M. 1975. Current and future prospects for Canadian-US natural gas trade. In Proceedings of the Middle States Division, Association of American Geographers, eds. J.B. Garver, Jr. and J.H. Munson, 9: 76-79.

_______________. 1977. Natural gas: A geographical perspective. Geographical Review 67: 71-85.

Pacione, M. 1979. Natural gas in Italy. Geography 64: 211-15.

Ratnieks, H. 1977. Oil shale and the USSR. The Geographical Magazine 52: 46.

Rudzitis, G. 1982. Resolution of an oil-shrimp environmental conflict. Geographical Review 72: 190-99.

Schein, D.L. 1978. Resources and confict: Oil shale and national parks. Transition 8 (fall): 1-8.

Sheskin, I.M. 1977. Alaskan natural gas: Which route to market? The Professional Geographer 30: 180-89.

__________. 1980. A geographic approach to the study of natural gas. Journal of Geography 79: 86-99.

__________, and Osleeb, J.P. 1982. Mexican natural gas: Implications for the US market. Energy Policy 10: 27-41.

Smith, H.D., et al. 1976. Scotland and offshore oil: The developing impact. Scottish Geographical Magazine 92: 75-91.

Stafford, J.W. 1978. The impact of offshore oil exploration on Florida. The Mississippi Geographer 6: 38-43.

Sutton, K. 1979. Natural gas in Algeria. Geography 64: 115-19.

Thoman, R.S. and Corbin, P.B. 1974. The geography of economic activity. Third Edition. New York: McGraw-Hill.

Wallach, B. 1980. The West Side oil fields of California. Geographical Review 70: 50-59.

Waverman, L. 1973. Natural gas and national policy: A linear programming model of North American natural gas flows. Toronto: University of Toronto Press.

Weaver, G.D. 1977. Comparative analysis of oil shales and tar sands. Great Plains - Rocky Mountain Geographical Journal 6: 359-65.

Wheeler, J.O. and Muller, P.O. 1981. Economic geography. New York: John Wiley & Sons.

CHAPTER 3

NUCLEAR ENERGY

Martin Pasqualetti
Department of Geography
Arizona State University
Tempe, Arizona 85287

ABSTRACT, There has been much recent interest in nuclear power by geographers, reflecting the attention focused upon it by general public. Public attitudes, socioeconomic impacts, risk, equity considerations and waste transport and disposal are primary areas of attention.

1. INTRODUCTION

The contribution which geographers have made to the study of nuclear power has been rapidly increasing and diversifying in recent years, reflecting public and governmental interests as well as reacting to specific publicized incidents like that at Three Mile Island. Much of the increased interest has been basically geographical in flavor, quite suddenly stimulating the pursuit of variant research themes and the continuation of traditional ones with a new emphasis. A review of the geographical contribution to the study of nuclear power reveals five topics have emerged as present foci of consideration: public attitudes, socioeconomic impacts, risk, equity considerations, and waste transport and disposal. The evolution of these present themes is identifiable by tracing research emphases over the years.

The earliest geographical paper on nuclear energy appeared in 1951 on the topic economic geography (Jones, 1951). Jones' paper, plus another 13 (8% of the total 173 contribution identified between 1945-1984), constitute all that was published between 1950-1969. There was no identifiable topical focus during this period, although the use and impact of nuclear explosions, (a theme which was receiving considerable media and public attention at the time), attracted more emphasis than any other topic (Fosberg 1963; Hess 1962; Thomas 1969). Other topics included the general applications and descriptions of nuclear energy (Eggleston 1965; Mowll 1958a; Pryde 1983); the role of nuclear energy in energy balances (Hoffman 1957); its use to generate electricity (Hart 1956); and siting. This was a period of nuclear

F. J. Calzonetti and B. D. Solomon (eds.), Geographical Dimensions of Energy, 27–40.
© 1985 by D. Reidel Publishing Company.

history when the promise of cheap and abundant electricity from uranium was generally accepted. It was also a time when nuclear energy was not particularly controversial, either within public or private forums. This is apparant in the small volume and generally mundane subject matter of the early papers.

2. RECENT CONTRIBUTIONS

The focus of attention began to sharpen during the 1970s as environmental impact statements were required in the US after the passage of the National Environmental Policy Act in 1969 and in the UK after the Windscale Inquiry of 1977 (Hall 1977; 1978; Hancock 1977). Radiation received early consideration from Mason (1972a) and Cook (1971; 1975), but has received no attention from geographers since. Three topics which started to attract steady interest in the 1970s have carried over into the 1980s. The first topic, nuclear facility siting, garnered attention as the entire energy industry started looking more carefully at the problem of power plant location (Mason 1971; Anderson 1971; Richetto 1978; Semple and Richeto 1976; 1979; Solomon 1982). Near the end of the decade, nuclear waste products and their disposition attracted attention from scientists when it became apparent that disposal was becoming a widespread public concern as well as a major impediment to the expansion of the nuclear industry (e.g. Eyre 1978; Hare 1978; Kasperson 1977). The 1970s also witnessed some of the earliest work by geographers on such topics as the technological risk of nuclear power (Kates 1977; 1978) public attitudes (Hall 1979), and evacuation from nuclear disasters (Wolpert 1977). As national demands for appraisal and estimation of environmental impact expanded from the natural sphere to include human effects, the late 1970s also found the earliest socioeconomic work (e.g., Brunn, Johnson, and Zeigler 1979; O'Riordan 1979). Many of the papers of the latter part of the decade were in early response to the Three Mile Island accident.

The early 1980s saw a continued upswing in geographic attention to nuclear energy. One of the major thrusts was a sharp expansion of the interest in socioeconomics begun in the late 1970s. Like much of the attention received by nuclear energy, this was also partly a carryover from the work on fossil plants and fuels. In some cases there was a direct exchange of ideas among several disciplines, particularly at places such as the Energy Division at Oak Ridge National Laboratory where scientists from several disciplines worked side by side on similar topics. The socioeconomic emphasis constitutes a major volume of nuclear work conducted by geographers thus far in the present decade (Bergmann and Pijawka 1981; Chalmers, et al. 1982: Greenberg, Krueckeberg, and Kaltman 1984; Kasperson 1982; Metz 1978; 1982a, Pijawka 1982; 1984; Sorensen 1984). Risk, initially of interest because of the innate hazard of radioactive materials, gathered strong momentum as part of emergency preparedness by the start of the 1980s, largely as a direct response to the TMI accident (Cutter 1984;

Kasperson, 1982; Kates, Hohenemser, and Kasperson 1983; O'Riordan 1984a; Pasqualetti and Pijawka 1984; Sorensen 1984; Wolpert 1980; Zeigler, Brunn, and Johnson 1981). The TMI experience demonstrated the inadequacy of emergency preparations at nuclear power plants, and now that such plans are required of all nuclear power stations, geographers are becoming more involved in their preparation and assessment (e.g. Johnson and Zeigler, 1984).

Recently, there has been a resurgence of geographic interest in power plant siting, both in the US (Dobson 1984; Guiness 1980; Richetto 1984) and abroad (e.g. Openshaw 1980; 1982a; Openshaw and Taylor 1981). Partly this has been a reaction to stiffer environmental constraints and partly it has been a response to the emergency planning issues raised by TMI. As utility companies have found it increasingly difficult to locate approvable sites, alternative siting schemes have been forwarded (e.g. Baker, et al. 1980; Solomon, Haynes, and Krmenec 1980). However, geographers have added a new dimension in their siting studies which not includes the considertion of waste disposal routes and repositories, a theme of emphasis which began with two articles authored by Hare (1978; Hare and Aikin 1980) and continued into the 1980s (e.g. Hill et al. 1982). Attention given to waste issues has become quickly specialized, especially on the related aspects of socioeconomics (Kasperson 1980), legal matters (Metz 1982a; 1982b), values (Kasperson and Morrison; 1983), and most recently equity (Kasperson 1983a; 1983b; Kasperson and Rubin 1983; and Kasperson, Derr, and Kates 1984).

Several other topics which gained initial attention in the 1970s attracted further research attention in the 1980s. One was a continued interest in the general topic of attitudes (e.g. Kasperson et al. 1980; Zelinsky 1984) and several of the possible subdivisions which first received attention at TMI, the more general topics of nuclear waste (Brunn, Johnson, McGirr 1980), as well as history (Cook 1982), and scale (Wilbanks 1984). Likewise, the work on accidents diversified to concepts of risk cognition and evacuation behavior (Cutter and Barnes 1982), while socioeconomic interest progressed from issues of housing and services to matters of health (Kasperson and Lundbland 1983). Attention to a new topic, decommissioning, is just beginning (Pasqualetti 1984b; Solomon 1982, 1984).

Regardless of the topic emphasized, whether siting, socioeconomics, waste, or risk, all have been given attention as they relate to a public involvement and interest in nuclear energy. Although there was an accident in 1957 at Windscale, involvement began in earnest after the fire at Brown's Ferry (Alabama) nuclear power plant in 1975 and was broadened during several state initiatives which reached U.S. voters in November of 1976. It was rekindled, expanded and intensified less than two and one-half years later with the accident at Three Mile Island Unit 2. The issues concerning nuclear power have been kept before the public ever since because of plant-specific controversies such as the attempt to halt construction of the Seabrook plant in New Hampshire, the licensing difficulties (related to siting in a seismically active area) associated with the Diablo Canyon plant in California (Pijawka 1982), the financial collapse of several reactor projects in the state of Washington, the public vote against operating a new nuclear plant in Austria, and the lengthy Sizewell inquiry (O'Riordan 1984b) into the use of pressurized water reactors along the Suffolk coast in England.

While a review of the literature shows current research themes and new trends, it also reveals subjects which have steadily received little attention, such as geographical distributions (only Guinnes 1980; Ilbery 1981a; Mounfield 1961), planning (only James 1979; Pearch 1979), and economics (only Jones 1951). Moreover, despite the fact that the political process is ingrained in many aspects of nuclear power, only five citations took the theme of geopolitics directly (MacDonald 1980; O'Riordan 1978; Scott 1980). Environmental impact was also represented infrequently (e.g. O'Riordan 1979; Pasqualetti 1983b; Richetto 1980; 1984) although geographers have been heavy contributors in the broad area of environmental studies.[1] Fuel resources, neither in terms of distribution, transportation nor use have received much attention Cook 1980; Guyol 1958; Luten 1979; McDougall 1972). On other themes of seemingly key geographic interest and application, geographers have contributed even less. For example, geographers have not been involved in the study of released radiation or in the modeling of the patterns of radioactive particle dispersal or deposition. Nor have they concerned themsevles with the many issues associated with land use (e.g., Metz, et al. 1984), policy (e.g. Pasqualetti 1983a), or the nuclear fuel cycle (Cook 1980; Pasqualetti 1983b).

The contribution made on non-U.S. topics has been a relatively small 6 percent of the total (Ash 1977; 1979; Boyle and Robinson 1981a; 1981b; 1982; Fernie 1980; Fernie and Openshaw 1984; Greenhalgh 1975; Ilbery 1981a; 1981b; Soumagne 1973). Moreover, much of the attention given in print on non-U.S. themes has been by American geographers (e.g. Pryde 1979a; 1979b; Pryde and Pryde 1974). The emphasis on the US situation is partly explained by differences in political and social systems, media attention, the availability of data, and governmental responsiveness to public inquiries.

3. CONCLUSION

At the core of fundamental work on nuclear energy themes is a 'baseline' of contribution which has remained numerically small but steady over the years (e.g. on siting), and which has been punctuated by sharp increases in activity on specific topics when they move into the public eye. As nuclear incidents have retained a high public visibility, geographers have remained responsive in their research emphases, especially in areas of traditional disciplinary strengths. However, the thematic diversification by geographers most apparent in their recent nuclear work, testifies to a strong and varied background of applicable expertise and interests which is not as demónstrable in other disciplines.

In all, the geographic study of nuclear energy has been more generally noticed and valued than the work of geographers on other aspects of energy. Explanation for this stems from the unique nature of the radiation hazard, in varying real and perceived degrees, and the resultant interest in all aspects of nuclear energy, particularly geogaphical aspects of risk, evacuation, siting, attitudes, and equity. Now that many of these problems have received some analysis, however, the numerical contribution of geography, as that of many other disciplines, is beginning to ebb back to nominal levels along with research funding. As one meaningful step to slow this contraction and to retain some of the newfound disciplinary momentum and respectability, geographers could apply themselves to questions pertinent to such nuclear topics as decommissioning and related government energy policy.

4. FOOTNOTES

1. A random review of dissertation topics of the authors of nuclear papers, however, revealed they often had interest in a topic with applicability to nuclear power such as hazards or environmental impact.

5. REFERENCES

Anderson, M.G. 1971. A computer based model for nuclear power station site selection. Area 3: 35-41.

Ash, M. 1977. Energy and form: The Windscale file. Town and country planning November: 469-73.

______. 1979. The meaning of Windscale. Town and country planning June: 295-98.

Baker, E.L., West, S.G., Moss, D.J., and Weyant, J.M. 1980. Impact of offshore nuclear power plants: Forecasting visits to nearby beaches. Environment and Behavior 12: 367-407.

Bergmann, P., and Pijawka, K.D. 1981. The socioeconomic impacts of nuclear generating stations: An analysis of the Rancho Seco and Peach Bottom facilities. GeoJournal Supplementary Issue 3: 5-16.

Boyle, M.J., and Robinson, M.E. 1981a. French nuclear energy policy. Geography 66: 300-03.

______. 1981b. Plogoff says no to nuclear power. Geographical Magazine 53: 681-83.

______. 1982. A further note on French nuclear energy. Geography 67:

148-49.

______, Johnson, J.R., Jr., and McGirr, B.J. 1980. Locational conflict and attitudes regarding the burial of nuclear wastes. East Lakes Geographer 15: 24-40.

______, S., Johnson, J.R., Jr. and Ziegler, D. 1979. Final report in social survey of Three Mile Island area residents. East Lansing, MI Michigan State University.

Chalmers, J.D., Pijawka, K.D., Branch, D., Bergmann, P., Flynn, J., and Flynn, C. 1982. Socioeconomic impacts of nuclear generating stations: summary report on the NRC post-licensing studies. NUREG/CR-2750. U.S. Nuclear Regulatory Commission.

Cook, E. 1971. Ionizing radiation. In Environment: Resources, pollution and society ed. W.W. Murdoch, pp. 254-278. Stamford, Connecticut: Sinauer Associates.

______. 1975. Ionizing radiation. In Environment (2nd ed.), ed. W.W. Murdoch, pp. 297-323. Sunderland (Massachusetts), Sinauer Associate

______. 1980. The role of history in the regulation of the nuclear-fuel cycle. In Uranium mine waste disposal, ed. C.O. Browner, pp. 5-10. New York: Society of Mining Engineers of AIME.

______. 1982. The role of history in acceptance of nuclear power. Social Science Quarterly 63: 3-15. Reprinted in Nuclear power: assessing and managing hazardous technology, ed. M.J. Pasqualetti an K.D. Pijawka, Ch. 2. Boulder, Colorado: Westview Press 1984.

Cutter, S. 1984. Residential proximity and cognition of risk at Three Mile Island: Implications for evacuation planning. In Nuclear power: Assessing and managing hazardous technology, ed. M.J. Pasqualetti and K.D. Pijawka, pp. Boulder, Colorado: Westview Press 1984.

______ and Barnes, K. 1982. Evacuation behavior and Three Mile Island. Disasters: The International Journal of Disaster Studies and Practice 6: 116-24.

Dobson, J.E. 1984. Introduction to Section II-Spatial Assessments and Impact Mitigation. In Nuclear power: Assessing and managing hazardous technology, ed. M.J. Pasqualetti and K.D. Pijawka, pp. . Boulder, Colorado: Westview Press.

Eggleston, W. 1965. The nuclear age in Canada. Canadian Geographical Journal 71: 182-91.

Eyre, S.R. 1978. Nuclear power and nuclear waste. Progress in Physical Geography 2: 151-56.

Fernie, J. 1980. Nuclear power: Salvation or damnation? In A geography of energy in the United Kingdon, pp. 77-104. London: Longman.

______. and S. Openshaw. 1984. Decision-making and safety issues of nuclear power development in the United Kingdom. In Nuclear power: Assessing and managing hazardous technology, ed. M.J. Pasqualetti and K.D. Pijawka, Ch. 3. Boulder, Colorado: Westview Press.

Fosberg, F.R. 1963. Effects of nuclear explosions on vegetation. Geographical Review 53: 136-37.

Greenberg, M.K., Krueckeberg, D. and Kaltman, M. 1984. Population trends around nuclear power plants. In Nuclear power: Assessing and Managing hazardous technology, ed. M.J. Pasqualetti, and K.D. Pijawka, Boulder, Ch. 8. Colorado: Westview Press.

Greenhalgh, D.H. 1975. The nuclear power industry in Europe. Energia Nucleare 22: 320-5.

Guiness, P. 1979. Nuclear power and the American energy crisis. Geography 64: 12-16.

______. 1980. The changing location of power plants in California. Geography 65: 217-20.

Guyol, N. 1958. Nuclear fuels in the pattern of world energy supplies. Annals of the Association of American Geographers 48 266-267 (abs.)

Hall, D. 1977. Windscale: the TCPA case. Town and country planning. December: 517-19.

______. 1978. Windscale inquiry report. Town and country planning. May: 246-50.

______. 1979. The nuclear fallacy. Town and country planning. August: 148-49.

Hancock, T. 1977. The Windscale inquiry. Town and country planning. October: 451-54.

Hare, F.K. 1978. Nuclear waste and the hydrologic cycle. Bulletin of the American Meterological Society 59: 267-268.

______ and Aikin, A.M. 1980. Nuclear waste disposal: Technology and environmental hazards. In Nuclear energy and the environment, ed.

E.E. El Hinnawi, pp. 168-199. Oxford: Pergamon. Reprinted in Nuclear power: Assessing and managing hazardous technology, ed. M.J. Pasqualetti and K.D. Pijawka, Boulder, Ch. 13. Colorado: Westview Press, 1984.

Hart, J.F. 1956. Nucleo-electricity versus carbo-electricity in Britain. Geographical Review 46: 116-18.

Hess, W.N. 1962. New horizons in resource development: The role of nuclear explosions. Geographical Review 52: 1-24, January.

Hill, D., Pierce, B.L. Metz, W.C. et al. 1982. Management of high-level waste repository siting. Science 218: 859-864.

Hoffman, G.W. 1957. The role of nuclear power in Europe's future energy balance. Annals of the Association of American Geographers 47: 15-40.

Hohenemser, C., R.E. Kasperson, and R.W. Kates. 1977. The distrust of nuclear power. Science 196: 25-34. Reprinted with a postscript in R.W. Kates, C. Hohenemser, and J.X. Kasperson, eds. Perilous progress: Technology as hazard, pp. . Cambridge, Mass.: Oelgeschlager, Gunn & Hain, 1983.

Ilbery, B.W. 1981a. The diffusion of nuclear power in Europe. Geography 66: 297-300.

______. 1981b. Nuclear power in western Europe. Tijdschrift voor Econ. En. Soc. Geografie 72: 242-50.

James, P. 1979. Nuclear energy and planning. Town and country planning August.

Johnson, Jr., J.H. and Zeigler, D. 1984. A spatial analysis of evacuation intentions: Implications for radiological emergency response planning. In Nuclear power: Assessing and managing hazardous technology, ed. M.J. Pasqualetti and K.D. Pijawka, pp . Boulder, Colorado: Westview Press.

Jones, S.B. 1951. The economic geography of atomic energy. Economic Geography 27: 268-74.

Kasperson, R.E. 1977. Nuclear waste management and the public: Considerations for public policy. Testimony presented to the California Energy Resources Conservation and Development Commission, 26 May 1977. (Available from Center for Technology, Environment, and Development, Clark University).

______. 1980. Anticipating the socioeconomic impacts of nuclear waste facilities upon rural communities. In U.S. Senate Committee on Agriculture, Nutrition, and Forestry, Subcommittee on Rural Development. Hearings on the socioeconomic effects of a nuclear waste storage site on rural areas and small communities, 96th Congress, 2nd Session. Washington, D.C. pp. 57-63.

______. 1982. Societal response to Three Mile Island and the Kemeny Commission Report. In _The risk analysis controversy: An institutional perspective_, ed. H.C. Kunreuther and E.V. Ley, pp. 61-77. Berlin, West Germany: Springer Verlag.

______. 1983a. Editor. _Equity Issues in Radioactive Waste Management._ Cambridge, Mass.: Oelgeschalger, Gunn & Hain.

______. 1983b. Social issues in radioactive waste management: The national experience. In _Equity issues in radioactive waste manage ment._ R.E. Kasperson, ed. pp 24-65. Cambridge, Mass: Oelgeschalger, Gunn & Hain.

Kasperson, R.E. an Lundbland, J. 1982. Closing the protection gap: setting health standards for nuclear power workers. _Environment_ 24: 14-20, 33-38.

Kasperson, R.E. and Morrison, M. 1983. Value conflict in the nuclear controversy. In _The nuclear power controversy_, ed. H.T. Engelhardt and A. Caplan, New York Plenum Press (in press).

Kasperson, R.E. and Rubin, B. 1983. Siting a radioactive waste repository: What role for equity? In _Equity issues in radioactive waste management_, ed. R.E. Kasperson, pp.118-36 Cambridge, Mass: Oelgeschlager, Gunn & Hain.

Kasperson, R.E., with Derr, P., and Kates, R.W. 1984. Confronting equity in radioactive waste management: modest proposals for a socially just and acceptable program. In _Equity issues in radioactive waste management_, ed. R.E. Kasperson, pp. 331-68. Cambridge, Mass.: Oelgeschlager, Gunn & Hain.

Kasperson, R.E., Berk, G., Pijawka, K.D., Sharaf, A.B., and Wood, J. 1980. Public opposition to nuclear energy: Retrospect and prospect. National Academy of Sciences, Committee on Nuclear Energy and its Alternatives: Sociopolitical impacts of energy use and policy, ed. C.T. Unseld, D.E. Morrison, and D.L. Sills, eds. Revised and updated versions reprinted in _Science, technology, and human values_ 5 (Spring): 11-23.

Kasperson, R.E., Hohenemser, C.J., Kasperson, J.X. and Kates, R.W. 1982. Institutional responses to different perceptions of risks. In _Accident at Three Mile Island: The human dimensions_ ed. D. Sills, C.

Wolf, and V.B. Shelanski, pp. 39-46. Boulder, Colorado: Westview Press.

Kates, R.W. 1977. Assessing the assessors: The art and ideology or risk assessment. Ambio VI(5): 247-252.

______. 1978. Case study: Risk assessment of nuclear reactors: Three views of the reactor safety study. In Risk assessment of environmental Hazard, ed. R.W. Kates, pp. 87-97. Chichester, U.K.: John Wiley and Sons.

______, Hohenemser, C., and Kasperson, J.E., eds. 1983. Perilous progress: Technology as hazard. Cambridge, Mass: Oelgeschalger, Gunn & Hain.

Luten, D.B. 1979. Who owns the atom? Landscape 23: 1-2.

Mason, P.F. 1971. Some environmental considerations in the siting of nuclear power reactors along the California coast. Geography 56: 335-37.

______. 1972a. Rejoinder to comments on "Spatial variability of atmospheric radioactivity in the United States." Professional Geographer 24: 272-73.

______. 1972b. Some spatial implications of a massive industrial accident: The case of nuclear power plants. Professional Geographer 24: 233-36.

McDonald, J.R. 1980. Environmental hazard and political response: Nuclear sites as a focus of ecologic protest in France. Paper presented at the 76th Annual Meeting of the Association of American Geographers, Lousiville.

McDougall, H. 1972. Heavy water: The critical element in Canada's nuclear programme. Canadian Geographical Journal 85: 238-245.

Mathieson, R.S. 1980. Nuclear power in the Soviet Bloc. Annals, of the Association of American Geographers 760: 271-279, June.

Metz, W.C. 1978. Predicting the population influx which accompanies a nuclear power plant. Nuclear News 21: 48-51.

______. 1979. Eastern nuclear power plant information centers. Electrical World (Dec. 27).

______. 1980. Eastern nuclear power plant information centers. Electrical World 193: 52-55.

______. 1982a. Legal constraints to high-level radioactive waste repository siting. Impact Assessment Bulletin 1: 55-64.

______. 1982b. Legal constraints to siting. In Nuclear waste: Socioeconomic dimensions of long-term storage, ed. Steve Murdoch, F. Larry Leistritiz, and Rita R. Hamm, pp. 73-90. Denver: Westview Press.

Metz, W.C., Daum, M.L., Pearlman, K.T., and Waite, N.C. 1984. Nuclear energy and land use. In Nuclear power: Assessing and managing hazardous technology, ed. M.J. Pasqueletti and K.D. Pijawka, pp. Boulder, Colorado: Westview Press.

Mowll, J.U. 1958a. Applications of atomic energy. Geographical Review 48: 278-280.

______. 1958b. Prospects for nuclear power in Canadian mineral industries. Canadian Geographer 11: 34-42.

Openshaw, S. 1980. A geographical appraisal of nuclear reactor sites. Area 12: 287-290.

______. 1982a. The geography of reactor siting policies in the U.K. Transactions, Institute of British Geographers N.S. 7: 150-162.

______. 1982b. The siting of nuclear power stations and public safety in the UK. Regional Studies 16: 183-198.

______ and Taylor, P. 1981. U.K. reactor siting policy: Memorandum by the Political Ecology Research Group, in first report of the house of Commons Select Committee on Energy 1980-1: Minutes of evidence. 3: 990-1018. HMSO, London.

O'Riordan, T. 1978. Environmental issues: I environmental issues and the political agenda: Nuclear power. Progress in Physical Geography 2: 494-504.

______. 1979. Environmental issues: II Lessons from a nuclear acci dent. Progress in Physical Geography 3: 573-586,

______. 1984a. Introduction to section IV - Risk perception and emergency planning. In Nuclear power: Assessing and managing hazardous technology, ed. M.J. Pasqualetti, and K.D. Pijawka, pp. 239-246. Boulder, Colorado: Westview Press.

______. 1984b. The Sizewell B Inquiry and a National Energy Strategy. The Geographical Journal, July. 152: 171-182.

______. 1983a. Nuclear energy: The place of geography. Paper presented at the 79th Annual Meeting of the Association of American Geographers, Denver, Colorado.

______. 1983b. Nuclear power impacts: A convergence/divergence schema. Professional geographer and K.D. Pijawka.

______. 1984b. The decommissioning dilemma. Sierra, Sept/Oct. 69: 64-67.

______ and Pijawka, K.D. 1984. Nuclear power: Assessing and managing hazarouds technology, Boudler, Colorado: Westview Press.

Pearch, D. 1979. Planning and the Windscale inquiry. Town and country planning: April, pp. 9-10.

Pijawka, K.D. 1981. The socioeconomic impacts of nuclear generating stationa: Analysis of Rancho Seco and Pearch Bottom Facilities. GeoJournal Supp. 3: 5-15.

______. Pijawka, K.D. 1982. Public response to the Diablo Canyon nuclear generating plant. Energy, The International Journal 7: 667-80.

______. 1984. Public sector effects and social impact assessment of nuclear generating plants. In Nuclear power: Assessing and managing hazardous technology, eds. M.J. Pasqualetti, pp 171-188. Boulder, CO: Westview Press.

Pryde, P.R. 1978. Nuclear energy development in the Soviet Union. Soviet Geography: Review and Translation 19: 75-83.

______. 1979a. Nuclear power. In The Societ Energy System, ed. L. Dienes and T. Shabad, pp. 151-170. Washington, D.C.: U.H. Winston.

______. 1979b. Nuclear energy in the Soviet Union. Current History, 77: 115-18, 135.

______. 1983. Nonconventional energy resources. New York: John Wiley and Sons.

______ and Pryde, L.T. 1974. Soviet nuclear power. Environment 16: 26-34.

Ratnieks, H. 1979. Soviet nuclear energy without restraint. Geographical Magazine 52: 1,3,5.

Richetto, J.P. 1978. A locational strategy for the siting of electric energy facilities. Public Utilities Fortnightly 101: 29-33.

______. 1980. The environment as a factor for locating nuclear electrical facilities in the United States. Geografiska Annaler B 62: 39-46.

Richetto, J. 1984. Locating nuclear electric energy facilities in the United States: A note on structural relationships and the environment. In Nuclear power: Assessing and managing hazardous technology, ed. M.J. Pasqualetti and K.D. Pijawka, pp. 103-120. Boulder, Colorado: Westview Press.

Scott, M. 1980. Nuclear confrontation. Town and country planning: March, 91-73.

Semple, R.K. and Richetto, J.P. 1976. Locational trends of an experi mental public facility: The case of nuclear power plants. Professional Geographer 28: 249-253.

______. 1979. The location of electric generating facilities: Conflict, coalition and power. Regional Science Perspectives 9: 117-38.

Solomon, B.D. 1982. U.S. nuclear energy policy: Provision of funds for decommissioning. Energy Policy 10: 109-119.

______. 1984. Nuclear power plant decommissioning. In Nuclear power: Assessing and managing hazardous technology, ed. M.J. Pasqualetti and K.D. Pijawka, pp. 387-412. Boulder, Colorado: Westview Press.

______. Haynes, K.E. and Krmenec, A.J. 1980. A multiobjective power plant location model with hierarchical screening: Nuclear power in northern Indiana. Review of Regional Studies 10: 9-22.

Sorensen, J. 1984. Evaluating the effectivness of warning systems for nuclear power plant emergencies: Criteria and application. In Nuclear power: Assessing and managing hazardous technology ed. M.J. Pasqualetti and K.D. Pijawka, pp. 259-278. Boulder, Colorado: Westview Press.

Soumagne, J. 1973. Problems geographiques de l'Energre nucleaire en France. L'Information Geographique 37: 43-51.

Thomas, T.M. 1969. Subterranean nuclear explosions--future tools in mineral exploitation? Paper presented at 1968-9 Annual Conference of the Institute of British Geographers, London, (Abs).

Wilbanks, T. 1984. Scale and acceptability of nuclear energy. In Nuclear power: Assessing and managing hazardous technology, ed. M.J. Pasqualetti and K.D. Pijawka, pp. 9-50. Boulder, Colorado:

Westview Press.

Wolpert, J. 1977. Evacuation from the nuclear accident. In Geographical Horizons, ed. J. Odland and R.N. Taaffee, pp. 124-29. Dubuque, IA: Kendall/Hunt.

______. 1980. The dignity of risk. Transactions, Institute of British Geographers 5: 391-401.

Zeigler, D.J., Brunn, S.D., and Johnson, J.H. Jr. 1981. Evacuation from a nuclear technological disaster. The Geographical Review 71: 1-16.

Zelinsky, W. 1984. Introduction to section I - Acceptability and national response to the risks of nuclear power. In Nuclear power: Assessing and managing hazardous technology, ed. M.J. Pasqualetti and K.D. Pijawka, pp. 1-8. Boulder, Colorado: Westview Press.

CHAPTER 4

RENEWABLE ENERGY

Stephen W. Sawyer
Department of Geography
University of Maryland
College Park, Maryland 20742

ABSTRACT. Geographers have demonstrated widespread interest in renewable energy resources. This interest reflects both the general popularity of renewable energy and the special appeal that renewable energy's resource and spatial attributes have to members of the profession. These special characteristics include use of local and regional resources, use of flow resources on a sustained yield basis, greater local control and independence, reduction of international and intergenerational conflicts, reduced environmental disruption, and utilization of seasonal and daily climatic and biological patterns long studied by geographers.

1. CONTRIBUTIONS BY GEOGRAPHERS

Three general observations may be made regarding geographers' recent activities in this field. The first is the widespread interest noted above. This is evident in presentations at the annual meetings of the Association of American Geographers and regional geography associations. A second observation is that, despite numerous reports at professional meetings, geographers have formally published relatively little on the topic in books or in geography, energy, or related journals. This disparity exists, in part, because the content and editorial boards of geography journals are dominated by more traditional fields and the editorial boards of energy journals are populated by economists and engineers who are generally less receptive to themes addressed by geographers. Geographers have aggravated this situation by conducting research that has been overly descriptive or place specific rather than the quantitatively analytic or generally applicable. A third observation is that many geographers have channeled their renewable energy interests in direct implementation programs, particularly in developing countries. Most commonly, these individuals are members of interdisciplinary teams and are valued for their ability to deal with the technical aspects of renewable energy use while also having the social science skills and the site and situation sensitivity needed for

F. J. Calzonetti and B. D. Solomon (eds.), Geographical Dimensions of Energy, 41–45.
© 1985 by D. Reidel Publishing Company.

effective implementation. These individuals also contribute relatively little to the formal literature--even though many of them struggle with implementation research on a daily basis.

Renewable energy research has been pursued from four overlapping directions: as studies in regional resource availability, as more general energy resource assessments, in relation to land use planning, and in the context of policy analysis. The assessments of regional resource availability represent the most common theme. These efforts, both published and unpublished, represent new variations on geographers' traditional interest in regional analysis and inventory. Examples include assessments of geothermal energy's significant potential in the western U.S. (Pasqualetti 1979, Pryde 1977), the La Grande hydroelectric complex in Quebec (Hamley 1983), and reports on Soviet geothermal and solar development (Pryde 1979, 1984). Similar contributions for developing nations include numerous works by Smil (e.g. 1981, 1983a), team projects involving biomass use in Liberia (Ornsted, Hillsman, et al. 1982) and two forthcoming volumes on renewable energy resources and planning in East Africa.

A second theme focuses on the availability and use of renewable energy resources outside of a specific locational context. Best termed descriptive assessments or inventories, geographers have contributed to the literature on biomass (Smil 1983b, Smith 1981, Solomon 1980, 1981), geothermal energy (Pasqualetti 1980), active solar energy (Sawyer et al. 1984), and books on multiple renewable energy resources by Pryde (1983) and Sawyer (1985). More analytic modeling efforts that also fall into this category involve alternative methods of measuring resource availability and the comparison of renewable energy resource flows with demand requirements. One example of resource measurement is provided by Lougeay and Brazel's (1984) use of irradiation data to estimate solar insolation levels. Comparison of resource flows and energy demand is illustrated by Feldman and Wirtshafter's (1980) effort to model insolation availability during peak electricity demand periods.

Geographers' concern with land use planning and management is also evident in their renewable energy research. A major issue has been quantifying the space requirements of renewable energy technologies relative to one another and to other energy options. Pasqualetti and Miller (1984), Pryde (the volume) and Smil (1984) report on the large requirements of most renewable energy technologies, but also how numerous factors offset some of these needs (such as the ability to use the land for one or more conjunctive uses, the large life-cycle needs of coal, and the relative ease of returning most renewable energy sites to post-energy uses). Numerous variations on this theme exist. Examples include Ryerson's (1984) effort to model solar shadowing by buildings and natural topographic features to aid the design of solar access regulations in land use codes, Gruntfest and Eichen's (1984) study of public opposition to siting large renewable energy facilities in Vermont, and Pasqualetti's (1983, 1985) analysis of siting issues in geothermal energy development.

Policy analysis is a fourth theme evident in the literature. This topic is addressed implicitly in most of the publications cited above.

In addition, several efforts have focused directly on it. Examples include studies of the relationship of state renewable energy tax incentives and state energy, economic, and climatic conditions (Sawyer with Friedlander 1983), unsuccessful attempts to identify statistical associations between the value of state tax incentives and actual solar adoption rates (Sawyer and Wirtshafter 1985), and efforts to determine whether the policy perceptions of renewable energy end-users are similar to those of resource planners (Sawyer and Feldman 1981).

Future research by geographers may be expected in a wide range of areas where the skills and spatial perspectives of the profession provide the basis for valuable contributions. The need continues, for example, for regional resource assessments and studies on how greater renewable energy use can be incorporated into local and regional development programs. (To improve upon past efforts in this area, future ones must be more critical of the reliability and economics of the technologies.) Similarly, continuing analysis of actual patterns of renewable energy adoption is needed to better understand regional rates of use, innovation diffusion processes, and the effect of varying incentive programs. Greater information is needed on the environmental, socio-economic, and aesthetic impacts of renewable energy use. In particular, the growth of small scale, decentralized electricity production with renewable energy technologies will create a demand for more facility siting research and methods of minimizing adverse impacts and citizen opposition. Greater contributions by geographers may also be expected on the transfer of renewable energy technologies from pilot tests to widespread use in developing nations. The dismal record of this transfer to date in comparison to the desperate need and potential clearly calls out for the site and situation sensitivity that characterizes the profession.

2. REFERENCES

Feldman, S.L., and Wirtshafter, R.M. 1980. On the economics of solar energy. Lexington, MA: Lexington Books.

Gruntfest, E.C., and Eichen, M.A. 1984. Public opposition to large scale soft technology in Vermont: Four case studies. Journal of Environmental Systems 14: 137-146.

Hamley, W. 1983. Hydroelectric development in the James Bay region, Quebec. Geographical Review 73: 110-112.

Lougeay, R., and Brazel, A. 1984. A presentation of solar irradiation data suitable for solar energy applications. Solar Energy.

Ornsted, B.J., Hillsman, E.L., et al. 1982. The on-farm use of biomass fuels: Market penetration potential during normal and fuel emergency conditions. Oak Ridge, TN: Oak Ridge National Laboratory. ORNL-5894.

Pasqualetti, M.J. 1979. Geothermal energy. In Geothermal energy and regional development: The case of imperial county. eds. Rose, A. and Edmunds, S. pp.13-33. New York: Praeger.

_________. 1980. Geothermal energy and the environment: The global experience. Energy: The International Journal 5: 111-165.

_________. 1983. The site specific nature of geothermal energy: Its effects on land use planning. Natural Resources Journal 23: 795-841.

_________. 1985. Planning for the development of site-specific resources: The example of geothermal energy. The Professional Geographer. In press.

_________, and Miller, B.A. 1984. Land requirements for the solar and coal options. The Geographical Journal 150: 192-212.

Pryde, P.R. 1977. Geothermal energy potential in the western United States. Journal of Geography 76: 170-174.

_________. 1979. Geothermal energy in the Soviet Union. Soviet Geography: Review and Translation 20: 69-81.

_________. 1983. Nonconventional energy resources. New York: John Wiley and Sons.

_________. 1984. Soviet development of solar energy. Soviet Geography: Review and Translation 25: 24-33.

Ryerson, C.C. 1984. Mapping solar access in high relief areas for regional energy planning. The Professional Geographer 36: 345-352.

Sawyer, S.W. 1985. Renewable energy: progress, prospects. Washington, DC: Association of American Geographers.

_________, and Feldman, S.L. 1981. Technocracy versus reality: Perceptions in solar policy. Policy Sciences 13: 459-472.

_________, with Friedlander, S.C. 1983. State renewable energy tax incentives: Monetary values, correlations, policy questions. Energy Policy 11: 272-277.

_________, and Wirtshafter, R.M. 1985. Market stimulation by renewable energy tax incentives. Energy: The International Journal.

_________, Sorrentino, A., and Wirtshafter, R.M. 1984. Solar energy adoption patterns in the United States. Energy Systems and Policy 8: 161-182.

Smil, V. 1981. Can the poor countries afford biomass energies. In

Beyond the energy crisis, Vol. II. eds. Fazzolare, R.A., and Smith, C.B. Eds. pp. 677-686. Oxford: Pergamon Press,

__________. 1983a. Deforestation in China. Ambio 12: 226-231.

__________. 1983b. Biomass energies: Resources, links, constraints. New York: Plenum.

__________. 1984. On energy and land. American Scientist 72: 15-21.

Smith, N. 1981. Wood: An ancient fuel with a new future. Washington, DC: Worldwatch Institute.

Solomon, B.D. 1980. Gasohol, economic, and passenger transportation policy. Transportation Journal 20: 57-64.

__________. 1980. Agricultural energy: Debunking the net energy loss myth. The Environmental Professional 2: 292-295.

CHAPTER 5

ENERGY CONSERVATION

Barry D. Solomon
Federal Energy Regulatory Commission
Washington, D.C. 20426

ABSTRACT. There is great diversity in the work that geographers have written on energy conservation. Much of this work overlays with related disciplines, although many energy conservation topics are inherently spatial in nature. This research topic offers many opportunities for important geographical contributions.

1. INTRODUCTION

Energy conservation has been widely acclaimed as the major energy "resource" for the developed world, and an important option for less developed countries as well. Since conservation (or energy efficiency) is also well known to be embodied in numerous and disparate technologies, geographical research on its supply (displacement) potential has also been quite diverse. However, while a respectable and growing number of geographers have investigated issues in energy conservation, relatively few have used purely geographical approaches, and much of the significant work is recent in origin and has strong overlaps with that of colleagues in related disciplines. The latter phenomenon is perhaps more pervasive for conservation than for any other area of energy geography research. For example, work by geographers on efficient land use patterns blends with similar studies found in urban and regional planning, work on conservation in transportation might also be found in environmental psychology or sociology.

Given this extreme diversity of research orientations to energy conservation, the major contributions by geographers defy simple characterization. Nonetheless, most of the important work has addressed one of the four themes to be considered below: buildings, industry, transport and land use. In addition, several geographers have attempted to join the broader policy debates on energy conservation (Foster and Sewell 1976; Johnson 1978; Sorensen and Collins 1981; Hannon, et al. 1981; Hannon 1982; Sawyer 1984, 1985), which has assuredly increased the visibility of the discipline. Most

F. J. Calzonetti and B. D. Solomon (eds.), Geographical Dimensions of Energy, 47–54.
© 1985 by D. Reidel Publishing Company.

of the other research, while usually also resulting in policy implications, has focused on the narrower sectoral and locational issues to be discussed next.

2. BUILDINGS

Research on energy conservation in buildings has generally taken a behavioral approach, often similar to that found in innovation diffusion studies. However, with the exception of the Oak Ridge National Laboratory's broad-based, building energy use studies (see, for example, Blue, et al. 1979), virtually all of the relevant work has addressed strictly residential energy use patterns. Basic information on the influence of architectural design, orientation and location on home energy use and conservation can be found in Munro (1977) and McColl (1978), while Curtis, et al. (1984b, 1984c) lay out some fundamentals for understanding and increasing the rate of energy conservation in residences.

Recent work by Brown and colleagues examines several important practical issues related to conservation in residences (see also the chapters by Brown, Macey and White in this book). For example, in one study a stratified, random sample of residents in a central Illinois city was interviewed to determine the impacts of utility home energy use audits and federal energy conservation income tax credits (Brown and Macey 1985). It was found that conservation program adopters were usually better educated, wealthier, owned newer homes, and engaged in more conservation practices than non-adopters, and the adopters were also spatially concentrated in outer urban census tracts. While traditional mass media sources were usually relied on for program information (cf. Frazier 1976; Curtis, et al. 1984c), many people were not familiar with the conservation programs and their eligibility. In another study, government energy conservation programs that target low-income residents have been criticized for their ineffectiveness, and a more socially and spatially equitable aid allocation formula has been proposed (Cullen and Johnson 1984). Surprisingly, the adopters of the programs examined in Illinois increased home energy use over the three year study period (while non-adopters reduced energy use), which has important ramifications for the design and targeting of energy conservation policies.

A related, follow-up study by Brown and Rollison (1985) in the same study area focused on energy use behavior among low-income and elderly home-owners. Contrary to common belief, it was found that these residents were high energy users, particularly when an adjustment is made for building size. Furthermore, as also noted by Jackson (1980), low-income and elderly households use fewer than average residential conservation measures, especially insulation and other capital-intensive technologies. These segments of the residential population may thus also have a significant amount of "discretionary" energy use, beyond their minimal needs. To increase home energy efficiency, however, major barriers to increasing conservation activity must be overcome, which relate to information channels, values, beliefs, attitudes and social norms (cf. Brown and Macey 1983).

3. INDUSTRY

A small number of geographical studies have been conducted on energy conservation in industry, which is a neglected research area. Two studies of note are parallel site screening analyses and socioeconomic impact assessments for industrial cogeneration in Chicago, USA (Fowler, et al. 1984) and Leeds, UK (Beaumont and Keys 1981). The latter work emphasized the need to go beyond conventional financial project evaluations, due to critical socioeconomic and environmental issues raised by cogeneration. A popular conservation option in Europe, Japan and the USSR (and recently in the USA), cogeneration of process steam and electricity could, in many regions, eliminate the need for many new electric power plants. In a related paper by Rose (1984), a regional input-output (I-O) model is refined to determine the optimal level of conservation for a southern California natural gas utility company.

A few econometric modeling studies have determined energy conservation, efficiency, or intensity, either directly or implicitly, of various manufacturing sectors of a regional economy. For instance, the multiregional "MREEED" model of Lakshmanan (1983) estimates energy demand by end-use sector, as well as demand for energy and other factors in production processes. By generating own and cross price elasticities of demand, the model estimates energy intensity and substitution tendencies for U.S. regional manufacturing sectors. However, this approach fails to account for indirect energy use in production processes, that is ultimately embodied in final goods and services. The I-O approach can solve this problem, since every sector of the economy is interrelated. For instance, the method of embodied energy analysis has been widely used with national I-O tables in the U.S. to examine the net (direct plus indirect) energy intensity of all sectors of the economy. In one recent application, the potential for energy conservation through steel recycling is examined, which is found to be an uneconomical option (Hannon and Brodrick 1982). Related I-O energy analyses have been conducted for several nations of Western Europe, where focus is placed on the regional structure of energy intensities and flows (Beutel and Murdter 1982; Reutel 1983).

4. TRANSPORT

Geographers have conducted extensive research on energy conservation in transportation, often adapting methods from economics and civil engineering to examine state and local issues in fuel conservation and shortfalls, and vehicular and transport system energy efficiency. Virtually all of the studies use statistical modeling procedures now common to many other areas of geography.

State-level gasoline supply and demand in the U.S. has been heavily investigated econometrically be David Greene. After earlier studies established the importance of spatial aspects of gasoline demand or usage by the automobile (Soot and Sen 1979; Greene 1980), a closer look was given to the affect of the 1979 summer petroleum supply shortage on state gasoline demand (Green 1983; Green and Chen 1983). While the regional level of the shortage ranged from 0 to 8 percent of consumption, the shortfall was clearly unevenly distributed across the U.S. Several gasoline pricing and fuel rationing management policies were examined in this light. Due to differential state demand responses to the oil supply shortage, the most equitable solution was found to depend on a "correct" redistribution of available crude oil imports. Related work has examined the variety of problems faced by the U.S. federal government in determining state-level gasoline conservation targets (Greene and Walton 1981), and the fuel savings from the elimination of larger engine cars (McKenna 1977).

While the private automobile has understandably received considerable attention in the energy conservation literature, other modes have been considered as well. For example, Southworth and Janson (1982) have researched the energy use implications for Chicago of what is often the most fuel-efficient public transit option: the bus. Focus is given to model evaluation, particularly the link between alternative models in the transportation systems planning and management processes. Some work has also been done by a geographer on regional diesel fuel demand for heavy trucks in the U.S. (Greene 1984). The results show that fuel demand for trucks is very _insensitive_ to diesel fuel price changes, probably due to a growing demand for intercity freight haulage.

A few geographers have recently been involved with the assessment and planning of transportation efficiency improvements in less developed countries (see, for example, Beaumont 1982). A study of transportation energy usage patterns in Tunisia found diesel fuel (mainly for light trucks) to dominate over gasoline use, although conservation policy recommendations were made for a range of fuels and transport modes (Greene, _et al_. 1984a). A related study was conducted for Costa Rica, where recommendations for energy conservation focused somewhat more narrowly on highway transport (Green, _et al_. 1984b).

5. LAND USE

Research on energy efficient land use often overlaps with transportation studies, and two major perspectives can be identified. The first might be called positive analysis, where energy geography researchers have investigated the spatial patterns of cities under various assumptions about likely future energy prices (particularly gasoline prices), the value of work and travel time, and related factors. Examples include Dendrinos (1979), Romanos (1978), and

Beaumont and Keys (1982, ch. 5). Most notably, Keyes (1980); has argued that energy will have little effect on existing patterns of suburbanization, unless we experience massive energy price increases.

The second perspective is normative consideration of urban land use patterns. Assuming continually rising energy prices and supply constraints, Dutt and Costa (1977) called for the revival of the central business district, more mass transit, and generally more intense land use in the center city and existing suburban environs. Along the same lines, Sewell and Foster (1980) proposed reducing street widths, spacing street lighting, and making density less expensive. Curtis, et al. (1984a) cite examples of five types of efficient land use patterns in Canadian cities, while Hannon (1982) discusses the energy-efficient city in more general terms. Finally, Romanos and Hatmaker (1980) develop an optimization model to minimize transportation energy costs for new land uses in an existing urban area.

6. CONCLUSIONS

Geographers have increasingly found fruitful research avenues on energy conservation issues. More cases studies on energy use in buildings, transport, and efficient land use patterns will continue, often focusing on less developed countries where little work has been done to date. It will be important to emphasize the spatial components of this work, as well as the role of energy conservation in alternative political contexts. Another important research area is the simulation of the impacts of conservation policies in various end-uses and economic sectors, in the framework of regional econometric and I-O models in industrial studies, and gravity/entropy and optimization models in transport studies. Since existing models can be easily adapted for such simulation studies, this research area should become popular and extremely useful to policy-makers worldwide.

7. REFERENCES

Beaumont, J.R. 1982. The planning of urban transport in developing countries: towards an energy efficient future. In *Proceedings, second international conference on urban planning in developing countries*. Caracas.

_____, and Keys, P. 1981. Combined heat and power generation schemes: a feasibility study. *Environment and Planning A* 13: 623-34.

_____. 1982. *Future cities: spatial analysis of energy issues*. Chichester: Research Studies Press.

Beutel, J. 1983. Interregional analysis of energy flows. *Papers of* the Regional Science Association 53: 83-104.

_____, and Murdter, H. 1982. Input-output analysis of energy flows and the determination of optimal production activities. In Proceedings, third Hungarian conference on input-output techniques, pp. 225-50. Budapest.

Blue, J.L., et al. 1979. Building energy use data book, edition 2. ORNL-5552. Oak Ridge, TN: Oak Ridge National Laboratory.

Brown, M.A., and Macey, S.M. 1983. Understanding residential energy conservation through attitudes and beliefs. Environment and Planning A 15: 405-16.

_____. 1985. Evaluating the impact of two energy conservation programmes in a midwestern city. Applied Geography 5: 39-53.

Brown, M.A., and Rollison, P.A. 1985. The residential energy consumption of low-income and elderly households: How 'non-discretionary' is it? Energy Systems and Policy 9: in press.

Cullen, B.T., and Johnson, J.H. Jr. 1984. Energy assistance for the poor: An evaluation and alternative allocation procedure. Energy 9, 571-81.

Curtis, F.A., Lara, B., De Man, A., and Simpson-Housley, P. 1984a. Energy conservation and land use planning: A Canadian perspective. International Journal of Energy Research 8: 369-74.

Curtis, F.A., Simpson-Housley, P., and Youck, C. 1984b. Household energy conservation and locus of control: A research note. International Journal of Energy Research 8: 89-93.

Curtis, F.A., Simpson-Housley, P., and Drever, S. 1984c. Household energy conservation. Energy Policy 12: 452-55.

Dendrinos, D.S. 1979. Energy costs, the transport network, and urban form. Environment and Planning A 11: 655-64.

Dutt, A.K., and Costa, F.J. 1977. Spatial patterns and design policies for future American cities. Journal of Geogrpahy 76: 24-27.

Foster, H.D., and Sewell, W.R.D. 1976. Images of Canadian futures: The role of conservation and renewable energy. Report No. 13. Ottawa: Environment Canada.

Fowler, G.L., Baougher, A.H., and Janson, S.D. 1984. Evaluating sites for industrial cogeneration in Chicago. Energy 9: 97-111.

Frazier, J.W. 1976. Household responses to the energy crisis: An analysis of some coping behaviors of Akron households. Ph.D. dissertation, Department of Geography, Kent State University.

Greene, D.L. 1980. The spatial dimension of gasoline demand: An econometric analysis of state gasoline use. Geographical Survey 9: 19-28.

_____. 1983. Regional demand implications for gasoline supply shortages. In Systems and models for energy and environmental analysis, eds. T.R. Lakshmanan and P. Nijkamp, pp. 206-33. Aldershot: Grower Publishing Co.

_____. 1984. A derived demand model of regional highway diesel fuel use. Transportation Research B **18B**: 43-61.

_____, and Chen, C.K. 1983. A time series analysis of state gasoline demand, 1975-1980. The Professional Geographer 35: 40-51.

Greene, D.L., and Walton, G.H. 1981. Data and methodological problems in establishing state gasoline conservation targets. Transportation Research Record 815: 24-30.

Greene, D.L., Hu, P.S., and Rose, A.B. 1984a. Transportation energy use and efficiency in Tunisia. ORNL-6066. Oak Ridge, TN: Oak Ridge National Laboratory.

Greene, D.L., Kowalski, R., and Southworth, F. 1984b. Transportation energy conservation reconnaissance mission to Costa Rica June 3-16, 1984. Revised draft report, prepared for the U.S. Agency for International Development. Oak Ridge, TN: Oak Ridge National Laboratory.

Hannon, B. 1982. Energy and Land use. In Energy & land use, eds. R.W. Burchell and D. Listokin, pp. 189-213. Piscataway, NJ: Center for Urban Policy Research, Rutgers University.

_____, and Brodrick, J.R. 1982. Steel recycling and energy conservation. Science 216: 485-91.

Hannon, B., Herendeen, R.A., and Penner, P. 1981. An energy conservation tax: Impacts and policy implications. Energy Systems and Policy 5: 141-66.

Jackson, E.L. 1980. Perceptions of energy problems and the adoption of conservation practices in Edmonton and Calgary. Canadian Geographer 24: 114-30.

Johnson, W.A. 1978. Muddling toward frugality. San Francisco: Sierra Club Books.

Keyes, D.L. 1980. The influence of energy on future patterns of urban development. In The prospective city: economic, population, energy, and environmental developments, ed. A.P. Solomon, pp. 309-25. Cambridge, MA: MIT Press.

Lakshmanan, T.R. 1983. A multiregional model of the economy, environment, and energy demand in the United States. Economic Geography 59: 296-320.

McColl, R.W. 1978. Some basics for teaching and evaluating energy conservation in the home. Journal of Geography 77: 94-99.

McKenna, D.E. 1977. Fuel savings resulting from the elimination of eight cylinder and six cylinder automobiles. Transition 7: 9-10.

Munro, J.W. 1977. House orientation and space heating conservation. The California Geographer 17: 97-106.

Romanos, M.C. 1978. Energy-price effects on metropolitan spatial structure and form. Environment and Planning A 10: 93-104.

_____, and Hatmaker, M.L. 1980. Urban activity allocation under criteria of transportation energy efficiency. International Journal of Energy Research 4: 1-10.

Rose, A.Z. 1984. Evaluating the optimal level of energy conservation: the natural gas utility perspective. Working Paper No. 8, Mimeo. Regional Research Institute, West Virginia University, Morgantown.

Sawyer, S.W. 1984. State energy conditions and policy development. Public Administration Review 44: 205-14.

_____. 1985. Federal-state conservation programs: The states' assessment. Energy Policy 13: in press.

Sewell, W.R.D., and Foster, H.D. 1980. Energy conservation through land use planning, a synthesis of discussions at a symoposium held in Montreal, 26-28 March 1980. Working Paper No. 6, Lands Directorate, Environment Canada.

Soot, S., and Sen, A. 1979. Metropolitan work-trip energy consumption patterns. Traffic Quarterly 33: 275-95.

Sorensen, J.H., and Collins, N. 1981. Preliminary considerations for state energy emergency planning. ORNL/TM-7876. Oak Ridge, TN: Oak Ridge National Laboratory.

Southworth, F., and Janson, B.N. 1982. Energy use and emissions impact measurement in TSM. Transportation Engineering Journal 108: 328-42.

PART II

Resource Development Issues

CHAPTER 6

OFFSHORE OIL AND GAS DEVELOPMENT IN THE NORTH SEA AND THE GULF OF MEXICO

Ian R. Manners
Department of Geography
The University of Texas at Austin
Austin, Texas 78712-1098

ABSTRACT. In the last decade an increasing proportion of the world's crude oil supply has come from offshore fields. The growing importance of offshore oil has coincided with, and on occasion reinforced, the growth of environmentalism in industrialized nations. Against this background, this chapter analyzes and compares major policy initiatives in the United States and the United Kingdom intended to minimize the adverse social and environmental consequences of offshore drilling in the Gulf of Mexico and the North Sea.

1. INTRODUCTION

To the oil and gas industry "offshore is as magic a word today as Westward Ho was to the men of those violent days after America's Civil War in the 1860s" (Editorial in Offshore, May 8, 1978). This sense of optimism about the ocean's hydrocarbon potential is based on the experience of the past decade, in which exploration and development has been successfully extended into deeper, more challenging offshore environments. And while the pace of exploration slowed somewhat during the economic recession of 1981 and 1982, the proportion of the world's crude oil supply coming from offshore fields has continued to increase. By 1983 offshore production of crude oil had reached 14 million barrels per day, equivalent to 26 percent of the world's total crude oil output. Moreover, for all the risk and uncertainty that surrounds offshore oil exploration, the likelihood of discovering further large fields capable of influencing global energy flows appears much greater for the relatively unexplored oceans than for the well-drilled continents.

The growing importance of offshore oil provides an appropriate context for analyzing the policy response of industrialized nations to the growth of environmentalism, which O'Riordan (1981, vii) difines as the search for mediation between what reality appears to compel us to do and what we believe we should do to protect the long-term interests

F. J. Calzonetti and B. D. Solomon (eds.), Geographical Dimensions of Energy, 57–78.
© 1985 by D. Reidel Publishing Company.

of the entire community. In the United States in paricular, the Santa Barbara blowout in 1969 raised serious policy questions with respect to prospective uses of the marine environment and reinforced demands for a more rigorous accounting of the ecological costs of offshore energy development. This chapter, therefore, will identify the impacts that arise in the course of offshore oil and gas development and analyze the ways in which different communities have sought to reconcile social and environmental objectives with the development of energy resources. The analysis will draw heavily on the offshore experience in the Gulf of Mexico and the North Sea, but the insights derived are applicable elsewhere as other nations seek to resolve the conflicts and mitigate the impacts of offshore energy developemnt.

2. IMPACT OF OFFSHORE OIL AND GAS ACTIVITIES

What major impacts can be anticipated in the course of offshore oil and gas development? The range of potential impacts, their timing and duration are far better understood now than they were a decade ago; nevertheless, the magnitude and significance of these impacts remain as controversial as ever. In addition, while certain general impacts may be forecast prior to development, based on assumptions about the scale and timing of offshore activity, determination of local environmental and socioeconomic impacts must await completion of production plans and submission of proposals for site-specific facilities (Williams and Hom 1979).

During the exploration phase, the elements of risk and uncertainty dominate offshore operations. The focus of attention is essentially offshore, with much of the drilling activity likely to take place out of sight of land. With the exception of service bases for offshore supply vessels, onshore impacts during this phase are negligible. Indeed, should oil or gas be found in commercial quantities, local planning authorities are likely to discover that the level of onshore activity during the exploration phase has been deceptively low-key. Offshore the onset of exploration drilling is almost invariably accompanied by controversy over the safety of drilling operations and the chances of an oil spill. Concern over the incidence and potential impact of a spill is likely to be expressed most forcefully wherever the area leased for exploration includes valuable fishing grounds or borders a coastline of outstanding recreational and amenity value.

With the discovery and verification of commercially exploitable deposits, additional support facilities are required to bring the fields into production. As a result, the focus of most planning activity shifts to the coastal zone, although the safety of installations and the prevention of oil pollution during production and transportation remain major offshore concerns. In the United States, an additional concern has been the discharge of large quantities of drilling effluents (both drilling muds and drill cuttings) during field appraisal and development (National Research Council 1983).

Onshore issues during this development phase are likely to include the siting and location of such support facilities as platform fabrication yards and protection of the environmental and amenity resources of the coastal zone. It is, of course, impossible to foresee the precise nature, magnitude, or location of onshore development until the exploration phase of offshore activity has been completed and the existence of commercially viable fields has been established. From the perspective of the affected community, therefore, much of the uncertainty surrounding onshore development, at least in a planning context, is only removed as companies begin to formulate production plans.

The onshore facilities required to support offshore exploration and production include supply bases, platform fabrication yards, pipeline terminals, pumping sttions, and storage tanks; these facilities in turn create the opportunity for further downstream activities, including refining, gas separation, and petrochemical produciton. Each of these activities brings social, economic, and environmental changes of both an immediate and a long-term character. Local labor markets will be affected by the creation of new jobs and an influx of contruction workers; property values will be sharply increased; and housing, education, and social welfare programs will need to be expanded. Rapid economic and demographic growth will produce profound changes in both individual lifestyles and community structure, while simultaneously depleting the amenity resources of particular localities.

The major onshore impacts, both enviornmental and social, are summarized in Table 1. Not all of these impacts will be felt in full measures by all communities. Clearly, larger communities and those with existing infrasturcture are in a better position than others to cope with rapid offshore development. Smaller coastal communities with little previous experience of coping with rapid growth will be hardest hit by an energy-based growth boom. Such communities often lack the full-time public officials and planning staff or the budgeting and land use programs that their situations require. With a rapid population influx, housing and transportation facilities, as well as social services, are likely to fall short of community needs. In this situation there is likely to be growing concern over a deteriorating quality of life, uncertainty over what, where, and when development will occur, and anxiety over the cost and risk of public infrastructure investment. Unless effectively addressed, the adverse socioeconomic impacts experienced during the development and production phases are likely to have significant policy repercussions, feeding back to influence the attitudes of those in the local community who initially favored offshore development in anticipation of tangible economic benefits.

Table 1. Onshore Support Facilities and Associated Impacts

Phase	Major Support Facilities	Major Onshore Impacts/Issues: Environmental	Major Onshore Impacts/Issues: Social
1. Exploration	Supply bases	Use of existing all weather harbors and facilities; potential for conflict with existing uses (e.g., fishing)	Minimal impact on labor/housing markets but potential for diversion of labor from traditional maritime occupations
		May prove temporary if no commercial discoveries are made in immediate offshore area	
2. Development	Supply and service bases for offshore rigs	Strain on existing harbor space and facilities; probable need for harbor expansion including land reclamation and dredging	Expansion of engineering, transportation, communications, supply handling and storage, and administrative services; severe impact on local housing and labor markets
	Production platform construction	Locational requirements (sheltered, deepwater site) may restrict range of choice; potential for conflict in areas of high amenity value due to noise, visual intrusion, etc.	Severe impact particularly in remote, rural areas due to high but fluctuating labor requirements; work camps and other forms of temporary accommodation as well as character of work force may create social strains
	Pipe-laying	Minimized through careful route selection to avoid sensitive ecosystems; multiple landfalls may be avoided by means of coordination and joint-user schemes	Minimal and temporary
	Terminal and storage facilities (including tank farms, oil/gas separation, etc.)	Varies according to product characteristics and circumstances (e.g., oil requires no immediate treatment as gas does); potential for conflict if located in area of high amenity value or where land is already designated for alternative use (agriculture, forestry); risk of marine pollution greatest when surface transport and transshipment of oil occur	Dependent upon scale of facilities; severest impact during construction
3. Operation	Processing (refinery; natural gas liquids; ethylene processing and cracker plants; petrochemicals	Dependent upon scale but may include visual intrusion, air and water pollution, noise; safety issues; conflict with existing land use, etc.; constraints on siting may arise due to preexisting location of landfalls, storage, and other primary facilities	Dependent upon scale of facilities but severest during construction

Source: Manners (1978, 87).

3. NORTH SEA OIL AND THE SCOTTISH EXPERIENCE WITH IMPACT ASSESSMENT

The diverse problems that confront coastal communities during the different phases of offshore activity are well-illustrated in the North Sea and the Gulf of Mexico. The Gulf oL Mexico represents one of the most intensively developed offshore regions in the world. Production of oil and gas has been built up slowly over a period of nearly 40 years; moreover, this has occurred within a preexisting infrastructure (services, facilities, and institutions) developed to meet the needs of proven onshore fields, and at a time when environmental concerns were less well articulated (Manners 1983). The circumstances of exploration and development off the Scottish coastline have been very different. Here, the physical setting (deeper water and more extreme weather conditions) and the absence of onshore infrastructure posed a very different set of challenges to local communities and planners. In addition, given the dependency of the United Kingdom on overseas suppliers of oil, there was an understandable desire to bring North Sea oil fields on-stream as rapidly as possible.

In general, the experience of the United Kingdom suggests that (at lease in frontier areas) it is the establishment of essential onshore support facilities that causes the greatest social and environmental disruption. In Scotland, in particular, the development phase proved to be a time of reactive, ad hoc planning as communities struggled to come to terms with the number, size, and complexity of oil-related planning applications received following the discovery of oil. Virtually every account of North Sea oil development during the phase refers to the "overwhelming pressures" on local planning authorities (Baldwin and Baldwin 1975; Chapman 1976; Francis and Swan 1973; Mitchell 1976). Lacking the necessary staff and expertise to evaluate these applications adequately, local planning authorities were forced to make many decisions on a pro-forma basis. There was neither the time nor perhaps, in some cases, the inclination to undertake detailed project appraisal, to obtain independent verification of the data provided by developers, to investigate and evaluate alternative sites, to coordinate land use policies with other planning authorities, or to assess the application in the broader context of long-term market forces and national needs. With few exceptions, long-term planning goals were compromised as planners reacted to the immediate problem of coping with the short-term demands of offshore developers. Only as the pace of development began to slacken could local authorities begin to place greater emphasis on comprehensive forward planning. Yet many of the initial decisions on such issues as services bases, pipeline landfalls, and tank farms had served to narrow the range of choices open to planners and, thereby, to constrain subsequent development options.

Public concern over the possible effects of large-scale developments on environmental quality undoubtedly predated North Sea oil, but as Clark and Turnbull (1984, 135-37) have observed, the geographical concentration of development proposals along a coastline noted for its natural beauty and the likely impact of oil-related facilities on isolated communities characterized by distinctive lifestyles meant that such concerns were expressed with renewed vigor. Furthermore, the difficulties experienced by local authorities in dealing with oil-related development proposals indicated serious deficiencies in the British planning system, particularly with respect to documenting and taking into account the social and environmental effects of development proposals at the project submission stage.

Under the British planning system, all local authorities are required to prepare Structure and Local Plans. These detail the authority's policies and proposals with respect to development and other uses of the land and provide the context for approving or disapproving subsequent development applications. The general argument, particularly among central government planners, has been that such advance planning ensures that particular developments are sited in such a way as to minimize any adverse impact. Thus, development proposals submitted to a local authority are reviewed for conformity with the Structure/Local Plan, and are either approved or, if the proposed use is "nonconforming," reserved for further discussion with affected parties. With most applications the planning process is completed when the authority decides to grant unconditional permission, to grant permission subject to conditions, or to reject an application. It will be noted, however, that throughout the process a developer is under no obligation to prepare an environmental impact statement (EIS) unless specifically requested to by the local planning authority. Moreover, a developer wishing to obtain an indication of a local authority's intentions prior to incurring any major financial outlay (for example, before purchasing a site) may choose to seek "outline" planning permission. In such cases, the developer need only provide the most general information about a proposal; particulars relating to siting, appearance, access, and support services are left for subsequent consideration (known as "reserved matters") in the event that the outline application is approved. Once outline planning application has been given, the local authority is committed to allowing the development to take place, in one form or another, once any imposed conditions have been met by the applicant. Thus a situation exists whereby local authorities may have to consider a development proposal and assess its likely impact on the basis of very sketchy information.

The limitations of the British planning system were quickly revealed as the pace of North Sea activity quickened and as the focus shifted from exploration to field development. Local planning authorities along the Scottish coastline were overwhelmed with proposals for supply bases and platform construction yards. While supply bases could often be accommodated within existing harbors, the location of platform fabrication yards proved to be a much more troublesome issue (Manners 1982, 194-97). the best efforts of local planning authorities were frequently frustrated by inadequate and inaccurate information about specific proposals and by uncertainty over developers' intentions. In 1971, for example, Brown and Root submitted an application to build a platform fabrication yard at Nigg Bay on Cromarty Firth. In response to a questionnaire prepared by the Ross and Cromarty County Council, the contractor estimated that the construction force would number between 400 and 600 people, a figure that corresponded very closely to the size of the labor force then being laid off following completion of a nearby aluminum smelter. The council approved Brown and Root's application unconditionally. By 1974, with two platforms under construction at the yard, the local force numbered around 3000, placing an intolerable strain on the local housing market and service infrastructure and necessitating the purchase of two passenger ships to provide temporary accommodation. During construction of the Sullom Voe oil terminal in the Shetland Islands, the local council complained that "having assured (us) that a maximum of 1200 workers would be needed for the terminal developments, the oil industry are now indicating a figure of at least 2000 and possibly as much as 3000 at peak (1977/78)" (Shetland Islands Council 1976, 19). Even this proved to be an underestimate, as the construction force had grown to 5000 by 1982. Elsewhere, speculative applications by developers who submitted duplicate proposals in the hope of persuading at least one of the authorities involved to act expeditiously created tremendous uncertainty with respect to the real demand for land, housing, roads, and public services.

Recognizing the difficulties being experienced by local planning authorities, the Secretary of State for Scotland issued a circular requesting that the Scottish Office be notified of all major oil-related planning applications (Scottish Development Department 1974a). This was accompanied by a "technical advice note" in which local planning authorities were advised that such proposals required "rigorous appraisal" in order to ascertain the likely impact of the proposed development on the social, economic, and environmental character of the area (Scottish Development Department 1974b). This inititative by the Scottish Office appears to be the first acknowledgement by a central government department that major developments required detailed assessment over and above that normally provided for under the United Kingdom planning process (Clark and Turnbull 1984). Subsequently, the Scottish Development Department commissioned a group of geographers at the University of Aberdeen to prepare expertise in analyzing complex planning proposals in a methodical way" (U.K. Department of the Environment 1976; Clark _et al._ 1981.).

Interest in project appraisal for oil-related development was reinforced by the growing complexity and potential hazards of proposals for such downstream facilities as refineries and petrochemical plants. As a result, a number of environmental impact assessments have been completed for major oil-related projects (Scottish Development Department 1976; Manners 1982). In a number of instances, the studies have been initiated by the Scottish Development Department; in other cases, the local planning authority called in independent consultants; in yet other cases, studies were commissioned by the prospective developer. Many of these studies have been severely criticized in terms familiar to those who have worked with impact statements in the United States--descriptive, superficial, inadequate, incomplete. Others, however, have had a significant influence on subsequent planning decisions. The environmental impact assessment (EIA) on a proposed liquified natural gas (LNG) separation plant at Peterhead, for example, was instrumental in alerting the local authority to the hazards of the project as proposed and to the inadequacies of Peterhead Harbour in safely accommodating existing shipping movements, as well as the export of propane and butane from the LNG plant. It should be emphasized, however, that all such environmental assessments have been undertaken on an ad hoc, project-by-project basis.

Paradoxically, while the growing use of impact assessment in Scotland has provided a model for other local authorities, the Scottish experience appears to have reinforced the belief of the central government that EIA should <u>not</u> be made a mandatory planning requirement. Such an approach, it is argued, would impose an unnecessary and undesirable rigidity on the planning system. Rather, the scope, timing, and methodology of analysis should be tailored to suit the circumstances of individual proposals and to meet the informational needs of the decision-maker at different stages in the planning process (Scottish Development Department 1976, 3-4).

The preference for a more informal, less prescriptive approach, where the onus is on local authorities to determine whether, when, and how impact assessment is to be undertaken, has undoubtedly been influenced by the fear that a mandatory EIA procedure would delay the processing of planning applications. Given the government's desire to expedite the development of North Sea oil, any such delay was unacceptable. This view is explicitly expressed in a 1975 government working paper: "standardised methods (of impact assessment) may inhibit the achievement of the appropriate depth of analysis and the <u>rapid response required</u>" (Scottish Development Deaprtment 1975, 10; emphasis added). In this regard, there existed in the United Kingdom a rather negative image of EIA as a slow, bureaucratic, litigious process, derived in large part from the early experience of implementing the 1969 National Environmental Policy Act in the United States (Anderson 1973).

The overriding concern with expediting North Sea oil development is most clearly demonstrated (a) in the enactment by Parliament of special legislation exempting certain types of oil development from normal planning procedures, and (b) in the exercise of ministerial discretionary powers to "call in" planning decisions deemed to involve the national interest. The Offshore Petroleum Development (Scotland) Act of 1975, for example, which empowered the Secretary of State for Scotland to acquire land for "key oil-related activities" through compulsory purchase, provides a sharp contrast to the consistency provisions of the 1972 Coastal Zone Mangement Act in the United States. Its enactment was motivated primarily by fears in Whitehall that production platforms would not be available in the numbers or within the time frame desired if normal planning procedures were followed. However, as the British government was soon to discover, attempts to predict the pace of oil development and the demand for platforms are fraught with difficulty given the uncertainties involved (Mackay and MacKay 1975). And, in practice, many of the yards for which planning permission was granted remained uncleared or unused while other yards suffered severely from the uneven flow of orders, with all that this implies in terms of economic and social hardships for the local labor force (Chapman 1976, 168).

The exercise of "call in" powers, i.e., the removal of a planning application from local consideration for decision by the Secretary of State for Scotland, demonstrates the amount of administrative discretion enjoyed by decision-makers in the United Kingdom (O'Riordan 1981, 232-33). If a development proposal is called in, the Secretary may hold a public inquiry at which evidence is taken from all interested parties. But the Secretary is not statutorily bound by the findings of the inquiry, and is completely at liberty to set aside the hearing officer's recommendations. For example, the report of the public inquiry into a proposed refinery and oil storage facility on Cromarty Firth concluded that the national need for development at this site did not appear to be sufficiently great to override social and environmental objections to the proposal--yet the hearing officer's recommendations were rejected by the Scottish Secretary and the proposal approved.

The present situation, therefore, is that while the British government endorses the concept of impact assessment and accepts the need for project appraisal in the case of major development proposals, it is content to leave the decision on whether or not an EIA should be prepared to local planning authorities and to the developer. While the information generated in this manner may prove helpful in the context of evaluating specific proposals, there is, as noted by Roberts and Roberts (1984, 102), a real danger that impact assessment will become nothing more than a technical, pro forma exercise that is undertaken after many of the important policy decisions have been made.

Despite the absence of any statutory requirement for impact assessment, local and regional authorities in Scotland have been increasingly creative in their efforts to combine environmental assessment with forward planning. The genesis of this approach can be traced to the Coastal Planning Guideline for oil- and gas-related development published by the Scottish Development Department in 1974 (Figure 1). These guidelines, which defined <u>preferred development zones</u> and <u>preferred conservation zones</u>, were intended to supplement local land use controls and to encourage the siting of "new developments in a way that makes the best use of existing labour an infrastructure" while providing greater protection to those areas containing features of "national environmental, scenic, and ecologic importance" (Scottish Development Department 1974c, 6). The guidelines provided a sensible basis for a long-term coastal planning strategy, but their appearance so long after the onset of exploration off the

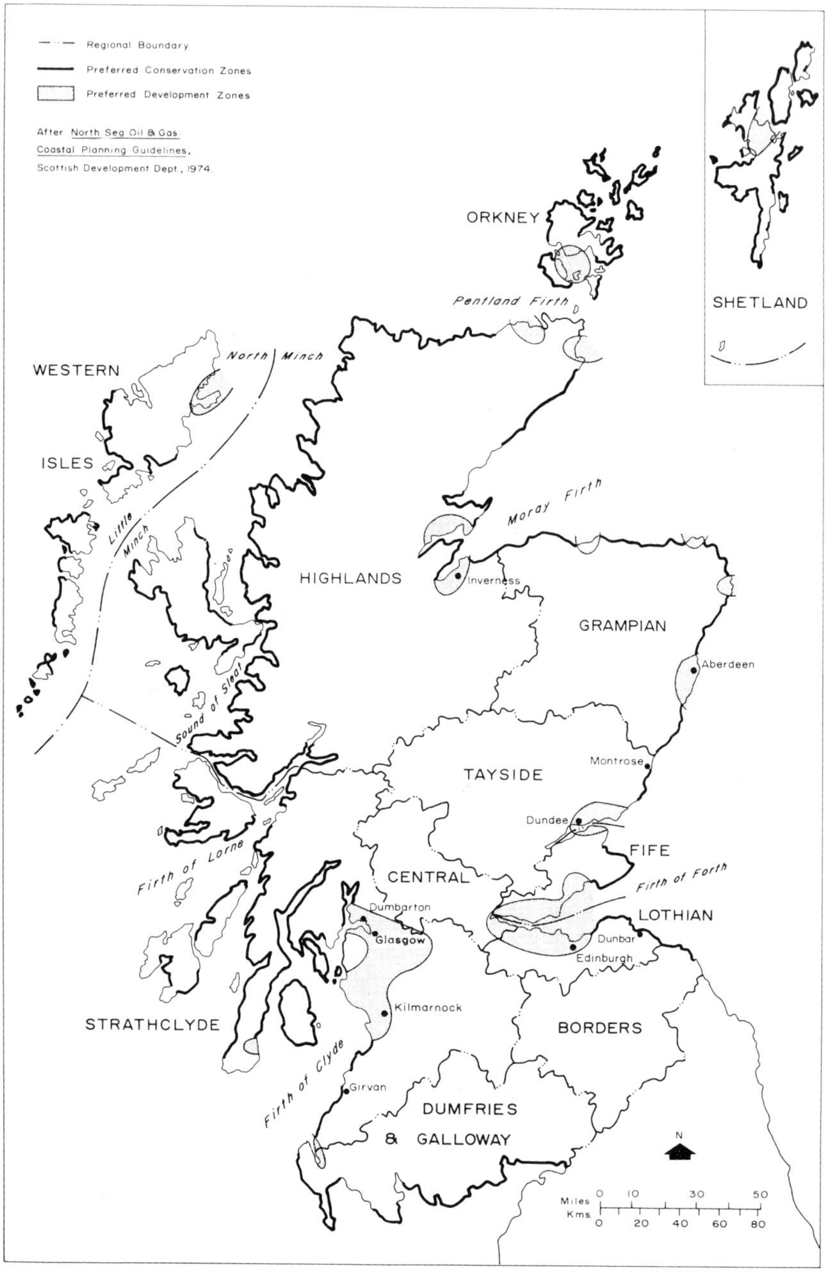

Figure 1. Coastal Planning Guidelines for Oil- and Gas-Related Development in Scotland.

Scottish coastline considerably diminished their value. Indeed, when the preferred development zones are compared with those areas where development was already taking place, it is difficult to avoid the conclusion that the guidelines were a post-factum rationalization for what had already occurred rather than a framework for forward planning (Manners 1982, 217).

A more effective application of this approach to land use planning has evolved in response to the National Planning Guidelines for petrochemical developments (Scottish Development Department 1977). At one level the guidelines provided local authorities with a much-improved information base, outlining the likely demands for labor, land, and services that would be generated by such plants, as well as the safety and pollution issues that would need to be considered when dealing with a planning application. At the same time, the guidelines clearly reflected the national interest in facilitating petrochemical development, noting that "no opportunity for desirable petrochemical development should be lost through lack of preparedness ... if local and structure plans identify sites as being suitable for petrochemical development, companies can hope to satisfy their requirements and establish themselves faster and more easily" (Scottish Development Department 1977, 2). Recognizing the reactive, ad hoc nature of land use decisions during the early years of onshore development, many regional authorities moved quickly to identify potential sites and to revise their structure plans and development control policies accordingly (Turnbull 1979). Although the specific approach and site selection criteria varied from one regional authority to another, the overriding concern has been to evaluate the potentialities and limitations of particular areas for petrochemical activity, thereby eliminating unsuitable sites (on grounds of environmental amenities, social costs, safety, or infrastructure needs) ahead of specific development proposals. The petrochemical contingency plan prepared by the Strathclyde Regional Council, for example, reviewed prospective sites on the basis of physical and engineering criteria and assessed the major social and environmental impacts that could be anticipated at each site under alternative development scenarios (Strathclyde Regional Council 1978). The final plan identified those sites within the region that could accommodate different levels of petrochemical development with least social and environmental disruption. In effect, without prejudging individual proposals that remain subject to detailed project appraisal, combining impact assessment with forward planning in this manner offers a way of siting facilities that are considered to be in the national interest, while simultaneously minimizing long-term costs to the local community.

4. IMPACT ASSESSMENT AND MITIGATION IN THE GULF OF MEXICO

As noted earlier, the history of offshore exploration and development in the Gulf of Mexico has been very different from the North Sea. By the late 1960s, when oil was first discovered off the Scottish coastline, the U.S. Continental Shelf in the Gulf of Mexico already constituted a mature oil and gas province with an extensive onshore network of supply bases, platform fabrication yards, pipe-coating facilities, refineries, and petrochemical plants. In these circumstances, the policy initiatives affecting offshore operations in the Gulf of Mexico have reflected somewhat different concerns and priorities on the part of impacted commmunities. In contrast to the North Sea experience, for example, environmental assessment and mitigation initiatives in the Gulf of Mexico have been directed primarily towards protection of the marine environment. Onshore, in contrast to the immediate difficulties experienced by local authorities in Scotland in siting support facilities and protecting amenity resources, the major concerns of U.S. Gulf coast communities over the long-term have proved to be fiscal, notably the timing of revenue flows relative to expenditure on services and capital improvements, the distribution of costs and benefits between federal, state, and local governments, and the magnitude and duration of local demand for new or expanded services (Williams and Zinn 1977).

Under the U.S. Outer Continental Shelf Lands Act of 1953, primary responsiblity for managing the mineral resources of those submerged lands that lie seaward of a state's territorial waters is assigned to the Department of the Interior (DOI). Protection of the marine environment has always been identified by DOI as an important consideration in the development of Outer Continental Shelf (OCS) oil and gas resources. However, prior to enactment of the National Environmental Policy Act (NEPA) in 1969, environmental quality per se received little formal attention in the leasing process, priority being given to economic criteria and to achieving other leasing objectives--notably the timely development of oil and gas resources to meet national energy needs and the receipt of a fair return to the nation from the leasing of federal lands. Passage of NEPA was intended to correct this situation by assigning protection of the environment the same priority as economic growth and by requiring that all federal agencies formally assess the environmental consequences of development decisions.

How successful has NEPA been in encouraging more environmentally sensitive decision-making? Some have suggested that the introduction of impact asessment has done little to create "conditions under which man and nature can exist in productive harmony for present and future generations." Fairfax (1978, 743) has argued that preoccupation with NEPA has "truncated preexisting and potentially significant developments in the definition of agency responsibility for environmental protection and in citizen involvement in agency deliberative processes. It turned environmentalists' efforts away from questioning and redefining agencies' powers and responsibilities and focused them instead on analyzing documents." Clearly, implementation of NEPA has not been without problems. Close examination of the impact of NEPA on offshore leasing and development in the Gulf of Mexico suggests, however, that the act has had a far greater and a far more positive impact on federal agency decision-making than Fairfax suggests.

An important change, and one which provides a sharp contrast with the Scottish experience, has been the expanded opportunities provided by NEPA for affected parties--state and local governments, as well as industry and environmental groups--to express their concerns prior to any final decision over the leasing and development of OCS lands. Such opportunities occur, for example, during the scoping meetings that precede prepartion of the impact statement for a lease sale, as well as during agency review of and public hearings on the draft EIS. Scoping has taken on added importance since the appearance in 1978 of revised guidelines intended to expedite the preparation of impact statements. The goal is to identify those significant issues that will require in-depth analysis prior to any leasing decision, and thereby avoid the introduction of last minute concerns such as delayed the preparation and evaluation of impact statements in the early days of NEPA. The scoping process may significantly alter the final proposal. In the course of scoping for the first area-wide lease offering in the Gulf of Mexico, for example, nine categories of special concern were initially identified for in-depth analysis, ranging from the potential impact of the lease offering on sensitive coastal habitats to possible repercussions for the Gulf Coast tourist industry (U.S. Department of the Interior 1983). Subsequently, additional concerns were expressed by Florida and local government agencies about the potential impact of leasing off South Florida. As a result, a new alternative was added to the proposal deleting all tracts in the eastern Gulf of Mexico located south of 26° N latitude.

This broadening of the OCS decision process to identify potentially adverse socioeconomic and environmental impacts has stimulated a search for innovative approaches to impact mitigation. Indeed, perhaps the best illustration of the way in which NEPA has led to substantive changes in Department of the Interior policies and programs is to be found in the use of special lease stipulations to protect valuable cultural and biological resources threatened by OCS development. Until the early 1970s, efforts to protect the marine environment relied almost exclusively upon OCS operating orders. However, OCS orders apply to all leases and deal primarily with technical issues of platform safety and pollution control. And, as Interior agencies began to implement the requirements of NEPA, it soon became apparent that a more flexible mechanism was needed to enable decision-makers to respond quickly and effectively to issues identified in the course of impact assessment.

The most extensive and innovative use of lease stipulations has been in the Gulf of Mexico. At first lease stipulations dealt primarily with issues of oil spill containment (specifying the equipment required of lessees drilling close to sensitive marine habitats) and platform placement (in an effort to minimize conflicts with commercial fishermen). The concept has been progressively expanded, however, to protect other resources and to accommodate new environmental concerns. In particular, military, archaeological, biological, and geological stipulations have been regularly invoked in recent Gulf of Mexico lease sales. The intent is (a) to mitigate the impact of oil and gas exploration and development on other prospective uses (military, fishing, recreation, research) by protecting the resources that sustain those activities; and (b) to reduce the risk of pollution by ensuring that oil-related structures are located away from hazardous areas.

The nature and effectiveness of stipulations is well illustrated by reference to the Flower Garden Banks (Figure 2). Located just over 100 miles off the Texas-Louisiana coastline on the edge of the continental shelf, the Flower Gardens are capped by the northernmost thriving tropical coral reef community in the Gulf of Mexico (Rezak et al. 1983). The prospect of offshore drilling in the vicinity of these coral reefs was viewed with apprehension by conservationists. In particular, it was noted that species at the very margin of their range are likely to prove particularly vulnerable to even a minor change in environmental conditions (U.S. Department of Commerce 1978). Thus, any change in turbidity or rate of sedimentation (as a result, for example, of the discharge of drilling effluents) would likely have significant implications for reef dynamics. In these circumstances, conservationists argued that the reefs required special protection.

The first lease sale to include tracts in the vicinity of the Flower Gardens was held in 1974, and it was during the preparation of the impact assessment for that sale that the basic format of the protective stipulation was determined in the course of discussions between the Interior Department, offshore operators, and independent scientists. Since that time the stipulation has imposed two major operational restricitions on lessees. First, no oil and gas activities of any form are allowed on the banks themselves in order to protect the reef from physical damage. Second, the "no-activity zone" is immediately surrounded by a buffer zone, within which operators are required to discharge all drilling effluents through shunting to the sea floor, rather than the more normal practice of surface disposal. This practice, which was proposed by offshore operators, disposes of effluents in such a way as to protect the biota from smothering, as well as from any toxic effects associated with the drilling fluids.

It is not suggested that the use of biological stipulations to protect the Flower Gardens has been without controversy. Indeed, the Environmental Protection Agency (EPA) continued to oppose leasing near the Flower Gardens and sought (unsuccessfully) to include a "not discharge" provision in National Pollutant Discharge elimination System (NPDES) permits for oil and gas facilities operating within nine miles of the banks on the ground that existing protection measures were inadequate (U.S. Environmental Protection Agency 1978). However, the use of stipulations in the Gulf of Mexico is symptomatic of the way in which NEPA has forced decision-makers to be more sensitive to the issue of environmental quality. As a result, there is now a greater effort

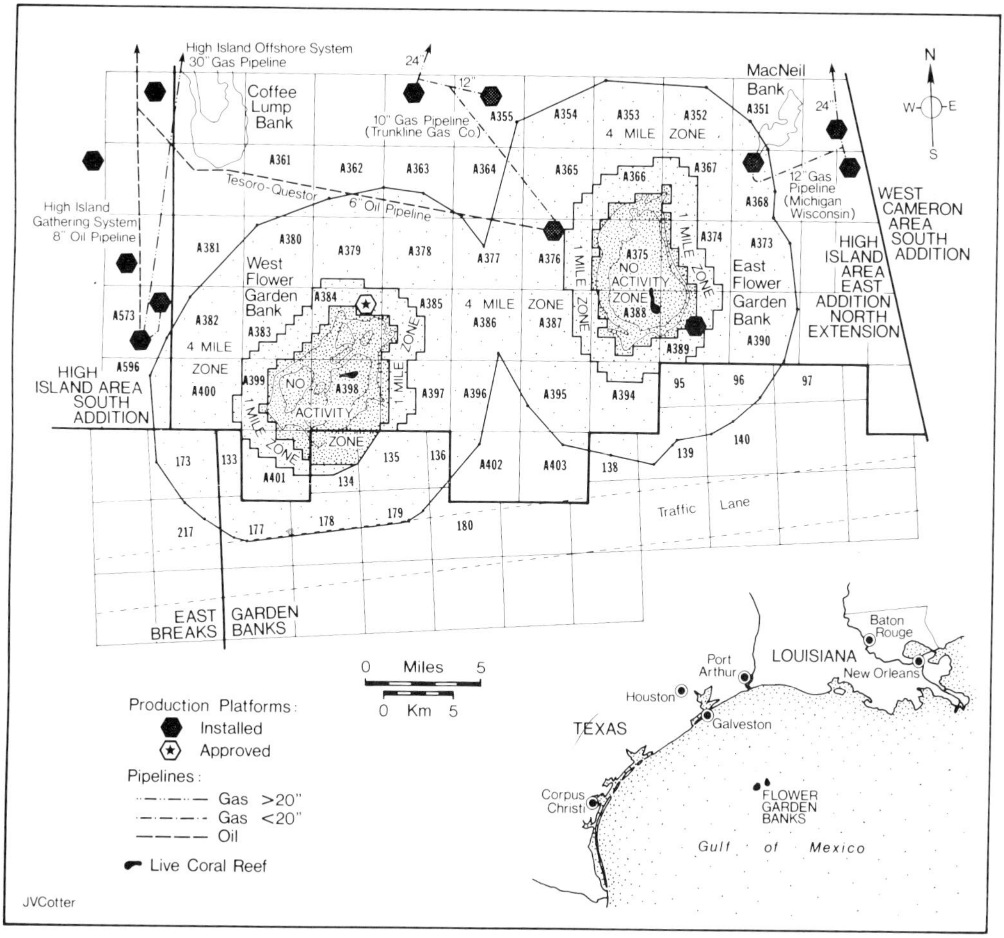

Figure 2. Oil and Gas Development and Lease Stipulations in the Vicinity of the Flower Garden Banks, Gulf of Mexico.

to identify and mitigate the potentially adverse effects of energy development on the marine environment. Moreover, efforts to protect the coral reefs were initiated by the Department of the Interior at a time when there existed considerable uncertainty about the nature and extent of the banks, let alone the most appropriate protective strategy. As more was learned about the banks, the stipulation has been modified to reflect the best available information. In the process, the prohibition on exploration and development has been extended to all or portions of 15 blocks around the Flower Gardens, while the buffer zone has been extended out to four miles from the 100-meter isobath surrounding the banks.

NEPA has clearly altered OCS decision-making, although not perhaps as much as some critics would like. Mitchell (1978), for example, has observed that impact statements for OCS lease sales tend to be preoccupied wtih offshore rather than onshore impacts and to focus on short-term effects to the exclusion of more significant long-term consequences. Moreover, as Anderson (1973, 268) has observed, agencies still retain considerable discretionary power over what to consider in an EIS and how to proceed. The Interior Department practice of adding tracts to a lease sale at the last moment, for example, was sharply criticized by the President's Council on Environmental Quality as a circumvention of NEPA, and undoubtedly contributed to suspicions that the agency was more interested in leasing tracts than in protecting them. This practice was particularly disturbing when the tracts added to the lease sale included such ecologically sensitive areas as the Flower Gardens.

The Scottish experience clearly illustrates the problems facing onshore planners in the absence of accurate and timely information about prospective developments, and the conflicts that are likely to occur over the siting of essential support facilities during the development phase of offshore operations. Although the Gulf of Mexico has a much longer history of offshore activity, such issues have taken on added significance for coastal communites following the decision of the Reagan administration to streamline OCS permitting procedures and to accelerate the pace of offshore leasing. And, as in Scotland, the most difficult questions have arisen in the context of assessing the cumulative impacts of phased development and reconciling competing national, regional, and local priorties in coastal zone management.

At least in theory, NEPA and the Coastal Zone Management Act (CZMA) should provide a framework within which such issues can be addressed. Thus assessment of potential impacts occurs not only prior to a lease sale, but at each stage in the offshore decision process, as a further environmental assessment is undertaken by DOI prior to approving any exploration or development and production plans. A notice that such plans have been filed and are available for public inspection and comment is published in the Federal Register; however, the sheer number of plans and revisions filed for the Gulf of Mexico, together with the difficulty of accessing such information, effectively means that postlease monitoring of OCS activity is conducted by Interior in human activities affecting the coast are conducted." Despite such reservations, it is clear that NEPA and CZMA provide coastal communities with opportunities to participate in and to influence OCS decision-making that are not available in Scotland.

Of greatest concern to coastal communities in Texas and Louisiana have been the fiscal impacts of offshore activity. The federal government's accelerated leasing program, for example, has drawn attention to the problem of front-end financing for those public services and infrastructure improvements necessary to sustain a higher level of OCS exploration and development. In communities such as Morgan City, Louisiana, the fiscal deficits experienced in the early years of offshore activity have been recouped over the long-term as a result of the increased revenue generated by onshore development (Louisiana State Planning Office 1977). Elsewhere, however, there has been apprehension that fiscal deficits might prove to be ongoing rather than short-lived. A study undertaken for Texas, for example, concluded that significant revenue shortfalls could be expected in communities along the Texas Gulf Coast over the long term because "there is not going to be sufficient onshore construction associated with this unique activity to cover the government's service costs" (Research and Planning Consultants 1977, 39).

Addressing these concerns was a major objective of the Coastal Energy Impact Program (CEIP) established by the U.S. Congress in 1976. In effect this program acknowledged that while rapid energy development was considered to be in the national interest, the costs of such development were likely to be experienced regionally and locally. CEIP was therefore basically a compensation program designed to redress some of the perceived inequities in OCS development, and thereby avoid the sorts of conflict that might delay offshore leasing and exploration. The program provided states and local governments with front-end loans for improvement of public services, as well as direct grants for impact assessment and mitigation. Both Texas and Louisiana were major beneficiaries under the program, receiving over $87 million in CEIP grants and loans, equivalent to 57 percent of the total funds distributed between 1977 and 1980 (Derheimer and Salmon 1982). Although CEIP was generally judged to be helpful in mitigating energy impacts in the coastal zone (see U.S. Department of Commerce 1980), the program has been a major casualty of efforts to reduce the size of the federal goverment's budget deficit. For similar budgetary reasons, attempts to replace CEIP with some form of revenue-sharing arrangement, whereby impacted states and communities would receive a percentage of OCS royalties, have been strongly opposed by the Reagan administration. In this situation, local communities have been seeking alternative mechanisms for redressing the fiscal deficits incurred in the course of servicing offshore enrgy development. Much of the dismay of offshore operators, one local authority in Port Arthur, Texas resorted to annexing and taxing oil and gas facilities in offshore waters. While the legality of offshore annexation is still being tested in the courts, the experience in the Gulf of Mexico suggests that the way in which the costs and benefits of offshore oil and gas development are distributed is likely to prove as great a source of controversy in well-developed regions as they are in frontier areas.

5. CONCLUSION

Coping with the diverse impacts that arise in the course of offshore oil and gas development clearly presents a considerable challenge to both government agencies and coastal residents. This chapter has sought to analyze and compare major policy initiatives intended to deal with those social and environmental concerns that have arisen in the course of exploring for and devloping energy resources in the North Sea and the Gulf of Mexico. Despite the contrasted history of these two offshore provinces, an underlying set of issues can be identified. These include (a) the adequacy, timing, and quality of information available to planners and other affected groups; (b) the nature of planning initiatives intended to minimize the adverse consequences of offshore drilling; and (c) the amount of fiscal support available to assist local communities prepare for and defray the costs of infrastructure development.

In Scotland and along the Gulf Coast these issues have been articulated and addressed in markedly different ways. In the United States, NEPA has provided decision-makers with an opportunity to analyze the complex, interdependent sets of social and environmental impacts resulting from OCS developments while there is still time to formulate appropriate management strategies. Yet in Texas and Louisiana there has been little effort to date to integrate impact assessment with comprehensive forward planning comparable to that now being undertaken in Scotland. In part, the reactive approach to OCS impact assessment and mitigation in these states reflects the long history and interdependent structure of the oil and gas industry along the Gulf Coast, as a result of which it is frequently difficult to link individual facilities and their effects with offshore activity. Moreover, fiscal concerns aside, the focus of attention in the Gulf of Mexico has been on providing greater protection for those marine resources threatened by OCS development. Somewhat paradoxically, while the Santa Barbara blowout undoubtedly encouraged support for NEPA and for a reassessment of development priorities in the United States, in the United Kingdom the perceived national interest in expedition oil and gas development reinforced resistance among policy makers at the national level to any formal requirement for impact assessment. The current approach to impact assessment in the United Kingdom offers a considerable degree of flexibility, yet the narrower focus on project appraisal has been at the expense of the policy context of development proposals. Thus, in contrast to the United States, there is no opportunity to formally assess the social and environmental implications of major policy decisions--as for example, a decision to undertake an accelerated leasing program or even to undertake a new round of leasing. In expressing concern about the limited scope of impact assesment in the United Kingdom, Grove-White (1984) notes the NEPA itself was a product of much soul searching about the future direction of industrial society. His conclusions, though directed to British policy makers, are worth reiterating in view of the growing global dependency on offshore energy resources. "What have been called 'the politics of development' are putting increasingly severe strains on our national--and global--physical and social resources. Yet it is still not public policy in this country to address systematically the implications of this fact for future development policies. The truly central importance of environment is still given minimal recognition by our political system, where future investment is at issue" (Grove-White 1984, 150).

6. REFERENCES

Anderson, F.R. 1973. NEPA in the courts: A legal analysis of the National Environmental Policy Act. Baltimore: John Hopkins University Press.

Baldwin, P.L., and Baldwin, M.F. 1975. Onshore planning for offshore oil: Lessons from Scotland. Washington, D.C.: The Conservation Foundation.

Chapman, K. 1976. North Sea oil and gas: A geographical perspective. London: David and Charles.

Chasis, S. 1980. An assessment of the Coastal Zone Mangement Program. In Coastal management: "Options for the 80's". Final Report of a Conference Sponsored by the coastal States Organization, Fredericksburg, Virginia, January 29-31, 1979, pp. 17-25. Washington, D.C.

Clark, B.D., Chapman, K., Bisset, R., Wathern, P., and Barrett, M. 1981. A manual for the assessment of major industrial proposals. London: Her Majesty's Stationery Office.

Clark, B.D., and Turnbull, R.G.H. 1984. Proposals for environmental impact assessment in the UK. In Planning and ecology, eds. R.D. Roberts and T.M. Roberts, pp. 135-44. London: Chapman and Hall.

Derheimer, D., and Salmon, J.D. 1982. Coastal energy impact mitiga tion in the Gulf: The past reviewed and alternatives for the future. Coastal Zone Mangement Journal 10: 161-81.

Editorial. 1978. What the offshore technology conferences is all about. Offshore 76: 113.

Fairfax, S.K. 1978. A disater in the enviromental movement. Science 199: 743-48.

Francis, J., and Swan, N. 1973. Scotland in turmoil: A social and environmental assessment of the impact of North Sea oil and gas on communities in the north of Scotland. Edinburgh: St. Andrew Press.

Grove-White, R. 1984. The role of environmental impact assessment in development control and policy decision-making. In Planning and ecology, eds. R.D. Roberts and T.M. Roberts, pp. 145-53. London: Chapman and Hall.

Husing, O. 1984. A matter of consistency: A Congressional perspective on oil and gas development on the Outer Continental Shelf. Coastal Zone Mangement Journal 12: 301-19.

Louisiana State Planning Office. 1977. Outer continental shelf impacts, Morgan City, Louisiana. Baton Rouge, Louisiana.

MacKay, D.I., and MacKay, G.A. 1975. The political economy of North Sea oil. Boulder: Westview Press.

Manners, I.R. 1978. Offshore oil: An overview. Policy Study No. 11. Austin, Texas: Center for Energy Studies, The University of Texas.

______. 1982. North Sea oil and environmental planning: The United Kingdom experience. Austin, Texas: University of Texas Press.

______. 1983 Offshore energy development and the Texas Gulf Coast. Proceedings of the 2nd International Conference on oil and the environment, pp. 3-11. Halifax, Nova Scotia: Technical University of Nova Scotia.

Mitchell, J.K. 1976. Onshore impacts of Scottish offshore oil: Planning implications for the Middle Atlantic State. Journal of the American Institute of Planners 42:4: 386-98.

______. 1978. Impact of offshore oil and gas developments on the coastal zone: Reforming the impact process. Coastal Zone Mangement Journal 4: 299-331.

National Research Council. 1983. Drilling discharges in the marine environment. Washington, D.C.: National Academy Press.

O'Riordan, T. 1981. Environmentalism. London: Pion.

Research and Planning Consultants. 1977. Offshore oil: Its impact on Texas communities. Austin, Texas.

Rezak, R., Bright, T.J., and McGrail, D.W. 1983. Reefs and banks of the northwestern Gulf of Mexico: Their geological, biological, and physical dynamics. Final Report submitted to the Minerals Management Service, U.S. Department of the Interior. College Station, Texas: Texas A&M.

Roberts, R.D., and Roberts, T.M. 1984. Planning and ecology. London: Chapman and Hall.

Scottish Development Department. 1974a. North Sea oil: Oil related development proposals. Circular No. 23/1974. Edinburgh: Her Majesty's Stationery Office.

______. 1974b. Appraisal of the impact of oil-related development. Technical Advice Note. Edinburgh: Her Majesty's Stationery Office.

______. 1974c. North Sea oil and gas: Coastal planning guidelines. Edinburgh: Her Majesty's Stationery Office.

______. 1975. North Sea oil and gas developments in Scotland: A physical planning resume. Edinburgh: Her Majesty's Stationery Office.

______. 1976. Environmental impact analysis: Scottish experience, 1973-1975. Edinburgh: Her Majesty's Stationery Office.

______. 1977. National planning guidelines for petrochemical developments. Edinburgh: Her majesty's Stationery Office.

Shetland Islands Council. 1976. Shetland structure plan. Lerwick.

Strathclyde Regional Council. 1977. Environmental impact appraisal: preliminary assessment of sites for a petrochemical complex. Department of Physical Planning.

Turnbull, R.G.H. 1979. Environmental appraisal for future petrochemical developments at national and regional levels in Scotland. Villach, Austria: UNECE.

U.K. Department of the Environment. 1976. Assessment of major industrial applications: A manual. Research Report No. 13. London: Her Majesty's Stationery Office.

U.S. Department of Commerce. 1978. Draft environmental impact statement prepared on the proposed East and West Flower Gardens marine sanctuary. Washington, D.C.: National Oceanic and Atmospheric Administration.

______. 1980. Coastal energy impact program: An evaluation.

prepared by the Office of Budget and Program Evaluation. Washington, D.C.: U.S. Government Printing Office.

U.S. Department of the Interior. 1983. Final regional environmental impact analysis: Gulf of Mexico. Metairie, Louisiana: Minerals Management Service.

______. 1984. Draft environmental impact statement: Proposed oil and gas lease sales 94, 98, 102. Metairie, Louisiana: Minerals Management Service.

U.S. Environmental Protection Agency. 1978. Recommendations and supporting rationale for NPDES permit conditions for oil and gas exploratory and production activities at the Flower Garden Banks. Washington, D.C.

Williams, D.C., and Hom, K.B. 1979. Onshore impacts of offshore oil: A user's guide to assessment methods. Washington, D.C.: U.S. Government Printing Office.

Williams, D.C., and Zinn, J.A. 1977. Onshore impacts of outer continental shelf oil and gas developments: Source Book. Washington, D.C.: The Conservation Foundation.

CHAPTER 7

LOCATION AND THE DEVELOPMENT OF ENERGY SUPPLIES FROM BIOMASS SOURCES*

Donald W. Jones
Energy Divison
Oak Ridge National Laboratory
Oak Ridge, TN 37831

John R. Krummel
Environmental Sciences Division
Oak Ridge National Laboratory
Oak Ridge, TN 37831

ABSTRACT. A Thunen land use model exploits well-known characteristics of biomass energy sources and development. A model is developed which determines the competitive location of a large-scale production facility devoted to biomass feedstock reduction and conversion to an alcohol fuel. It takes account of on-plantation feedstock assembly costs, delivery of processed fuel to sales point and competition from other land uses, while permitting land costs to be determined endogenously. Higher densities of feedstock cultivation are more profitable farther from final markets for the fuel. Larger plantations have higher feedstock costs; higher transport rates on converted fuel lower feedstock costs.

1. ISSUES IN THE ANALYSIS OF BIOMASS ENERGY DEVELOPMENT

Analysis of biomass sources for energy involve both resource availability questions and economic considerations of market prices for the final product. Past studies generally have focused on either resources (e.g., wood production) or market conditions that influence the feasibility of using biomass for energy (Solomon 1980; Spiewak <u>et al</u>. 1982). However, these studies have failed to resolve a number of critical issues related to biomass energy development. These issues include land use competition (e.g., biomass feedstock production versus crop production); conversion facility type, location or size; characteristics of the resource and their influence on effective supply prices and conversion facility size; and how market prices influence the structure of biomass use. These issues are critical to an evaluation of the future place of biomass energy in the mix of U.S and developing country energy markets (OTA 1980).

F. J. Calzonetti and B. D. Solomon (eds.), Geographical Dimensions of Energy, 79–99.

With few exceptions, previous studies of biomass energy have either been aggregated studies based on the mix and demand for fuels in the U.S. or disaggregated financial analyses of particular biomass systems. For example, Meekhof et al. (1980) studied the policy implications of gasohol production for U.S. agriculture. Their simulation evaluated the changes in corn and soybean prices, carryout stocks, and government price support programs. Hertzmark et al. (1980) also looked at the feasibility of alcohol production within the current mix of biomass systems in the U.S. Studies of this type generally look at potential changes in the aggregate agricultural or silvicultural sector of the U.S. economy. As an example of a disaggregated study, Felker (1984) analyzed intensively managed, short rotation mesquite biomass energy farms. He looked at the biological and economic constraints and opportunities, land and water needs, and environmental and industrial considerations of developing a commercial sized plantation in support of a wood-fired electric utility. The work of Medema et al. (1981), Weisz and Marshall (1980), and Wood et al. (1976) are other examples of studies that look at the financial feasibility of specific biomass systems. The primary characteristic of these analyses is that they explicitly consider the farm gate costs of producing biomass for use at a conversion facility. While both the aggregated and disaggregated approaches provide useful insights for a subset of the above issues, they fail to link important phenomena because of either the site-specific or more highly aggregated nature of their analytical framework. Locational and land use issues link both national and regional market prices with the specific plantation characteristics of biomass production systems.

The production of energy from biomass feedstocks shares two important geographical characteristics with agriculture. The feedstock production uses substantial amounts of rural land, and the product is generally transported from the production site to a distant market for consumption (OTA 1980; Overend 1982; McCollum and Hughes 1983; Felker 1984). The first characteristic suggests that useful insights about biomass energy production may be gained by studying it as a land use problem (Ranney and Cushman 1980). The second results in systematic spatial components in the production-site price of the converted fuel and, implicitly, of the feedstock, which is a feature of economic activity exploited by the Thünen land use model.

Dobie et al. (1977) reported costs for bailing, roadsiding, and transporting 2600 tons of rice straw per year. Costs for these three operations ranged from $16 to $25 (1976 dollars) per ton for a 10-mile transport distance. The Purdue study of agricultural biomass (Tyner and Bottum 1979) estimated harvest and transport costs of $15 to $23 per ton for crop residue in Iowa or Illinois. Also, their analysis of a similar system in Virginia showed harvest and transport costs increasing to $62 per ton. Most importantly, they found that transport costs accounted for about 70% of total costs. For woody biomass production, truck transport of 50 miles contributes 15-20% to total biomass costs (Howlett and Gamache 1977). Wood and charcoal transportation costs in

Liberia account for roughly 60% of the total energy costs in a system that converts small plots of naturally grown trees to charcoal for transport to the market (Perlack et al. 1984). Hitzhusen and Abdallah (1980) connected the farm gate price of corn stover, off-farm transport costs, and power plant retrofitting and operating costs to evaluate the economics of crop residue combustion in electrical energy production. They found that the economics of the power plant (eg., retrofitting costs and economies of scale in operating costs) directly influenced the percentage of costs associated with transportation.

In the typical biomass energy production facility, feedstock must be assembled from a large area and collected at a central site for processing into a fuel which must be transported subsequently to a final market. The transportation to the sale site is costly, and with a given sale price at its final market (or markets) the price net of transport cost received at the processing site from which the fuel is shipped declines with greater distances from market. The area from which cultivated or naturally occurring feedstock may be collected for a given processing facility is limited similarly by the haulage cost from where it is grown to the processing site. Thus there are dual cost constraints on the development of large scale, biomass energy production facilities. Barnard (1983) noted these links, appealing to a Löschian framework, and used them to study the feasibility of a spatially dispersed cropping and conversion strategy for biomass energy production. However, he did not formalize the model and, particularly, left the treatment of transport costs largely intuitive.

In this paper we develop a Thünen model of a biomass energy production facility composed of a conversion facility surrounded by a cultivated feedstock area. The harvested feedstock must be carried to the processing facility and the converted fuel must be hauled subsequently to a market site. Land costs involved in feedstock production are widely recognized as an important factor in biomass fuel costs (OTA 1980) and the present approach illustrates the endogeneity of the land cost component. Land costs are determined jointly by the location of the conversion facility relative to the sale site and the scale of operation -- or alternatively, by a combination of transport rates on the feedstock and on the converted fuel. The necessity to bid land away from other uses, particularly agricultural land, is another noted concern in the biomass energy field, and the model incorporates this phenomenon.

We develop the model algebraically and diagrammatically, but we limit the present analysis of its behavior to a diagrammatic one. The interested reader is referred to Jones and Krummel (1984) for an algebraic treatment of the behavior of a general plantation model.

Section 2 develops the locational theory of a biomass energy plantation intuitively, and Section 3 presents the theory algebraically. In Section 4, the consequences of several parameter differences are explored diagramatically. Section 5 considers some extensions of the present form of the model as well as its practical utility.

2. THE SPATIAL STRUCTURE OF A BIOMASS ENERGY PRODUCTION FACILITY

We begin with the usual assumptions used in an exposition of the Thünen model (eg., Jones 1984, pp. 124-25). A single, large city is located on an unbounded, homogeneous plain and is the sole market for all production in the region. Three options exist for employment of nonurban land in the region: food production, the biomass energy operation, and disuse. With an inconsequential loss in generality and substantial gain in expositional simplicity, assume that the production alternative to biomass feedstock cultivation captures land closer to the city. With some loss in generality, but with a corresponding gain in simplification, assume that the central conversion facility that transforms the raw feedstock into an alcohol fuel is located at what would be the center of a circular plantation.

Figure 1 shows the situation diagrammatically. Line R_A is rent to land in the alternative use, say, "prime" agriculture. Line R_pR_p is the market (city) value of the biomass fuel, minus processing costs, per unit of land; and R_pt_p is that market value minus transportation costs to market. The plantation is large relative to production units in the alternative land use, and there are substantial intraplantation assembly costs involved in delivering the feedstock to the conversion facility. Knowledge of the size (capacity) of the conversion plant and its production characteristics (ability to transform the feedstock into a transportable fuel), for any given market price of the fuels, determines R_pR_p. Processing plant capacity and its conversion technology determine the quantity of feedstock which the plantation must generate, and density of feedstock cultivation converts these requirements into a plantation size.

Diagrammatically, for a given plant size and set of processing characteristics, point P, representing the plant site, moves along line R_pt_p. From each potential plant site, assembly costs of the feedstock reduce its field-site value and, consequently, plantation land values. These land values decline away from the conversion plant site in the direction of the market until the location at which the alternative land use can outbid the plantation. This marks the inner edge of the plantation, k_i. Assuming no other land uses, the outer edge of plantation activity occurs where assembly costs completely erode the stump site value of the feedstock in the direction away from the market. The conversion plant, (almost) at the center of the plantation moves along R_pt_p until the quantity of feedstock which each correspondingly sized plantation can deliver equals the quantity which can be processed by the plant. As P moves inward, the altitude of the triangle Pk_or increases, increasing its base with a given transportation rate represented by the slopes of lines Pk_o and Pr from the vertical. Plantation areas would expand except for losses at the interior edge of the plantation to the alternative land use, the area of loss represented by the distance rk_i in Figure 1. As P moves outward, the altitude and base of plant shrink. As P moves inward, the altitude of the triangle Pk_or increases, increasing its base

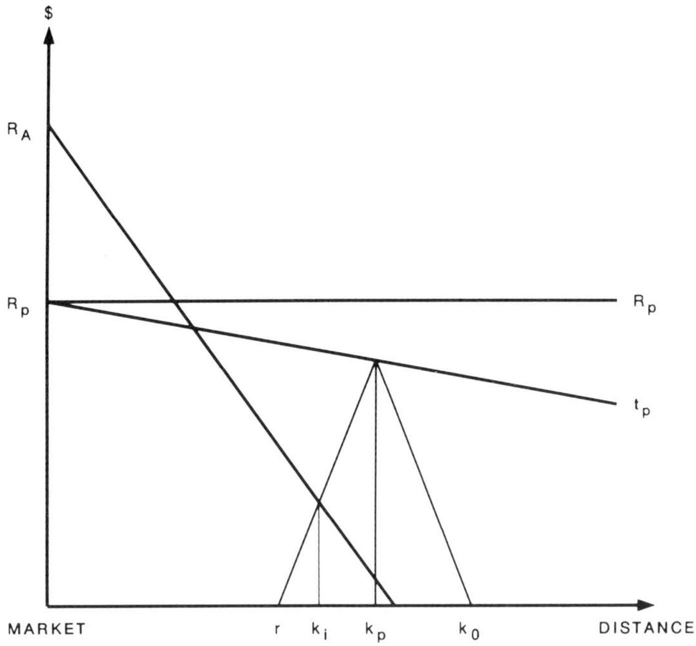

Figure 1. Location and size of a biomass plantation. Line R_A is agricultural land rents, and triangle Prk_o is land rents on the plantation. The conversion facility is located at k_p; the inner edge of the plantation is at k_i and the outer edge at k_o.

with a given transportation rate represented by the slopes of lines Pk_o and Pr from the vertical. Plantation areas would expand except for losses at the interior edge of the plantation to the alternative land use, the area of loss represented by the distance rk_i in Figure 1. As P moves outward, the altitude and base of the triangle shrink although some mitigation occurs through shrinkage of the territory between r and k_i. At some location k_p, the projection of P onto the land surface, the viable plantation area is maximized. With fixed plant size, conversion ratios and crop densities, this area may or may not be sufficient to keep the conversion facility fully employed. If sufficient land does exist to sustain a given sized, fixed coefficients conversion facility, the optimal size of operation may be less than the maximum attainable size. If plantation output is flexible, the concept of a simple maximum plantation size disappears. Higher crop input ratios raise line R_pt_p, raising the triangle altitude and plantation size, but diminishing marginal returns and a given market price of the processed plantation crop place a limit on such increases, making plantation area determinate.

The effect of variability in the size of conversion plant is similar, with the additional constraint of a given capital rental price.

We turn now to an algebraic development of the model presented in Figure 1.

3. A MODEL OF PLANTATION AGRICULTURE

Let food production be the production alternative to biomass energy development. The food production function is:

$$H = H(L,X) \ , \tag{1}$$

where L is land and X is farm capital equipment. Assuming homogeneous degree-one production in farming,

$$h \equiv \frac{H}{L} = h(\ell,1), \ \ell \equiv \frac{X}{L} \ , \tag{2}$$

and rent to land in farming at locations k is

$$r_H(k) = p_H(k)(h-\ell h'), \quad h' \equiv \partial h/\partial \ell > 0 \quad h'' \equiv \partial^2 h/\partial^2 \ell < 0 \tag{3}$$

and

$$p_H(k) = p_H(0) - \tau_H k \ , \tag{4}$$

where $p_H(0)$ is the price of food at the market, τ_H is the transport rate on food and k is distance from market.

The conversion of the biomass feedstock is done at a plant located at k_p. The conversion plant uses the harvested crop, W, and the capital embodied in the plant, Z, to produce a more transportable fuel which is subsequently carried to the market. Assume h•dl production at the plant:

$$Q = F(W,Z) \tag{5}$$

and

$$q \equiv \frac{Q}{Z} = f(z,1), \ z \equiv \frac{W}{Z}, \ f' > 0, \ f'' < 0 \tag{6}$$

The price of a unit of the feedstock at the plant site is

$$p_W(k_P) = p_Q(k_P)f' , \tag{7}$$

where

$$f' \equiv \partial f/\partial z > 0 \text{ and } p_Q(k_P) = p_Q(0) - \tau_Q k_P . \tag{8}$$

The field-site value of the feedstock at any location on the plantation is

$$p_W(k) = p_Q(k_P)f' - \tau_W (k_P-k)^2 , \tag{9}$$

where τ_W is the transportation rate involved in hauling it to the conversion plant, or what we call, for simplicity, the harvesting cost. At the edge of the plantation most distant from the market, $k \equiv k_o$, and

$$p_W(k_o) = p_Q(k_P)f' - t_W(k_o - k_P) = 0 . \tag{10}$$

The feedstock itself is produced according to

$$W = G(L,T) , \tag{11}$$

in which L is land and T represents the individual plants or trees. Then with an h•dl production function

$$w \equiv \frac{W}{L} = g(t,1), \quad t \equiv \frac{T}{L}, \; g' > 0, \; g'' < 0 . \tag{12}$$

Rent to land on the biomass facility is then

$$r_W(k) = p_W(k)(g-tg'). \tag{13}$$

The edge of the plantation closest to market is $k = k_i$ and is the location at which the feedstock can produce a land rent equal to that produced by food:

$$(p_H(0)-\tau_H k_i)(h-\ell h') = (p_Q(0)-\tau_Q k_p)(g-tg')f' - \tau_w(k_p-k_i)\ (g-tg'). \tag{14}$$

In any sized plantation, the supply of harvested feedstock to the conversion plant is

$$W^S = 2\pi \int_{k_p}^{k_o} g(t)kdk - \delta(\tau_F, \tau_Q, \tau_w, p_F, p_Q, f, g, h)g(t) ,$$

$$\frac{\partial\delta}{\partial\tau_j} < 0, \quad \frac{\partial\delta}{\partial p_H} > 0, \quad \frac{\partial\delta}{\partial p_Q} > 0, \quad \frac{\partial\delta}{\partial f} > 0, \quad \frac{\partial\delta}{\partial g} < 0, \quad \frac{\partial\delta}{\partial h} > 0,$$

in which W^S is feedstock supplied and δ is the area of overlap with the rent cone of food production. From equation (6), the demand for the feedstock, at any given z, is a function of plant size:

$$W^D = Zf^{-1}(q), \ \partial f^{-1}/\partial Z < 0 . \tag{16}$$

In equilibrium,

$$W^S = W^D = W , \tag{17}$$

and, altering the f^{-1} notation slightly,

$$[\pi(k_o-k_p)^2-\delta]g(t) = Zf^{-1}(Z) . \tag{18}$$

Knowledge of k_i and the harvesting cost, τ_w, will permit computation of k_p and δ, and hence of the feedstock required to feed a conversion plant at specified capacity, $f^{-1}(Z)$. Knowledge of k_i also identifies $p_w(k_p)$, which can be used to eliminate k_o from equation (18). Because $p_w(k_o) = 0$, equation (18) can be rewritten as

$$k_p = \frac{p_Q}{\tau_Q} - \frac{\tau_w}{\tau_Q f'} \sqrt{\frac{Zf^{-1}(Z)}{\pi g} + \frac{\delta}{\pi}} \ . \tag{19}$$

Then equations (7), (14) and (19) form a linear system which determines k_i, k_p and $p_W(k_p)$. The parameters are the market price of the fuel, $p_Q(0)$; transportation rates on the converted fuel, τ_Q, and the raw feedstock, τ_W; conversion facility size, Z; density of cultivation, g; market price, transportation rate and crop density in agriculture, $p_H(0)$, τ_H and h; and the input coefficient in the biomass fuel conversion, f'.

The three-equation model describes a constant-coefficients production activity, which, with h•dl production and constant factor prices, does not have a determinate output. In the present case, however, one cost curve slopes upward because the feedstock must be collected at an increasing marginal cost, and output is determinate. With both the input coefficients and conversion facility size fixed, most parameter changes shift the location of a constant sized plantation which produces a constant quantity of biomass fuel[1].

Modeling the variable input ratio in the conversion plant entails specifying the demand for the feedstock to be converted. The input ratio z is adjusted to equate the value of marginal product of the capital in the conversion plant and its rental price:

$$p_Z = (p_Q - \tau_Q k_p)(f - zf') \ . \tag{20}$$

The equilibrium value of z is governed by the relative prices of the feedstock and capital such that

$$z = \omega(p_W/p_Z), \tag{21}$$

where $\omega' < 0$. The elasticity of substitution between W and Z in production of Q derives from expression (21) as

$$\sigma \, d(p_W/p_Z) = \frac{-(p_W/p_Z)}{z} \, dz. \tag{22}$$

To make the conversion plant size a choice, formulate an implicit supply curve for the capital when the size of the conversion facility is a choice variable:

$$Zf = \Theta(p_Q), \ \Theta' > 0. \tag{23}$$

The appropriate argument in the θ function is p_Q^* rather than p_Q because k_p is endogenous; p_Q^* is the price that faces the plantation developer/operator, and p_Q^* is one of the optimizing choices.

The five-equation model is composed of equations (7), (14) and (19) of the fixed-coefficients model, appropriately generalized to variable coefficients, plus equations (21) and (23), which govern the factor substitution in the processing and the choice of plant size. In addition to determining k_i, k_p and $p_W(k_p)$, the five-equation model also determines profit maximizing values of Z and z. Conversion facility size, Z, is now an endogenous variable rather than a parameter, as is the variable input rate, z — equivalent to f' of the 3-equation model. Parameters are now $p_Q(0)$, τ_Q, τ_W, q, $p_H(0)$, τ_H, and h as in the 3-equation model, and p_Z, the rental price of capital in the conversion facility. We have examined the case of an endogenous g, density of feedstock cultivation, but the model results appear fundamentally unchanged and are more difficult to interpret. The model determines the location of the conversion facility, which is a costly investment. The processing plant location on any plantation would not be altered in response to a small parameter change, and possibly not even in the event of large changes. The utility of studying differences in k_p in response to such changes resides then in permitting a hypothetical comparison of entire plantations in identical circumstances save one. The questions which can be so addressed include: How would differences among regions in, say, potential crop yields affect the development of biomass energy resources in those regions? Would the availability of larger scale processing technologies facilitate development of plantations closer to or farther from market? What would be the consequences for competition with agricultural land?

4. ANALYSIS OF THE MODEL

We diagrammatically analyze the consequences of differences in three parameter values. The diagrams are restricted to the case of the fixed-size conversion facility operating a fixed rate of biomass feedstock input, although we refer to the consequences when facility size and input rate are endogenous. The parameters investigated are the intensity of cultivation, feedstock harvesting/assembly costs, and final market output price.

4.1 Differences in Intensity of Cultivation, g

The graphical analysis of differences in intensity of cultivation — or density of growth of a naturally occurring resource — is contained in Figure 2 for the case of a fixed-size conversion facility with fixed feedstock input proportions. In Figure 2, the base case is represented

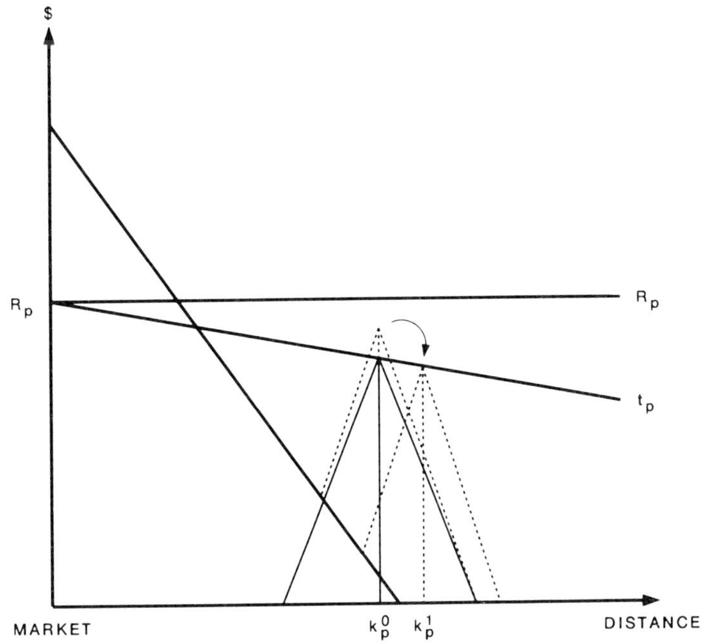

Figure 2. Consequences of differences in intensity of cultivation. With fixed production coefficients and capacity in the conversion facility, a plantation with higher cultivation intensity locates farther from the final market.

by the solid line depicting a plantation of given size and location. Suppose that density of feedstock growth is higher than in the base case. The land rent which the plantation can generate, as a result of its cultivation (if any), harvesting and conversion activities, rises to the broken-line triangle, initially keeping the conversion facility site, k_p, fixed. The angles of the base of the triangle representing land rent in the plantation become steeper, pivoting around their intersections on the abscissa of the diagram. Fixed conversion facility size and feedstock input rate keep line R_pR_p fixed, so at location k_p^o the plantation area would generate more feedstock than could be processed. Note that there would be a modest encroachment of the plantation into agricultural land if k_p remained unchanged. The plantation "moves" outward from the market until the apex of the new, more steeply angled rent triangle touches the output transportation rate line, R_pt_p. The conversion facility is located at k_p^1, and the entire plantation is located farther from the market, i.e., k_i also moves outward. The maximum land rent generated by the plantation is unequivocally lower, which is attributable to two causes: (1) the greater distance from market erodes the net price of the converted fuel at the conversiton facility, and (2) the maximum haul distance on feedstock is reduced, lowering the conversion-site value which the feedstock must yield in order to sustain trasportation costs on marginal units.

In the case of endogenous conversion facility size and feedstock input rate, line R_pR_p is variable, although its vertical movements are constrained by the fixed market price of biomass fuel and the degree of substitutability between capital and feedstock in the conversion process. When density of feedstock growth (or intensity of cultivation) increases, the feedstock input rate, z, increases and the conversion facility size may either increase or decrease, but the net effect generally will raise the intercept of R_pR_p. The plantation "moves" outward although with a diagrammatic analysis, the exact distance of the "move," or its distance relative to that of the fixed capacity case, is impossible to assess because of interactions between z and Z and their net effect on the intercept of R_pR_p. The value of the feedstock at the conversion plant can either increase or decrease as a result.

In general, an algebraic analysis is required to determine in which direction R_pR_p shifts in response to parameter changes, although the response to a change in cultivation density is particularly intuitive.

We have not distinquished between harvesting costs in general and assembly costs which involve moving the feedstock from its field sites to the conversion plant and is assumed here to be a pure transportation cost. In practice the two are distinguishable, and we have merged them in the present analysis only to sharpen the focus of the analysis on the transportation of the raw feedstock to the coversion facility. Under current assumptions therefore there are no interactions between crop density and harvesting/assembly costs - both are exogenous, in fact - but in actual operations, the harvesting component of the harvesting/assembly operation, for woody biomass in particular, is more costly at higher densities of given-sized trees. That is, for trees of a given diameter at harvest time, closer spacing increases harvesting cost because it is more cumbersome to operate the harvesting machinery at closer quarters. The present model abstracts from this effect.

4.2. DIFFERENCES IN FEEDSTOCK ASSEMBLY COSTS

A higher transportation rate in moving feedstock from field sites to the conversion plant is represented by pivoting the legs of the plantation rent triangle inward from the apex, as in Figure 3A. Holding K_p constant and leaving Rptp unchanged for the fixed capacity case, the plantation would lose area. An inward move of the plantation is required to acquire enough land to keep the fixed-coefficients conversion plant operating at full capacity. This involves a greater sacrifice of agricultural land to the biomass plantation. The net price of converted fuel received at the conversion site increases, and the shadow price of the feedstock increases on two counts: (1) the rise in the net output price is transmitted back to variable-price factors, and (2) the conversion site price of feedstock must rise to support the higher transport cost on a fixed quantity of required inputs cultivated at an unchanged intensity. If we determined intensity of cultivation endogenously, that density would increase somewhat but the plantation would still move inward and the shadow price of feedstock would still rise.

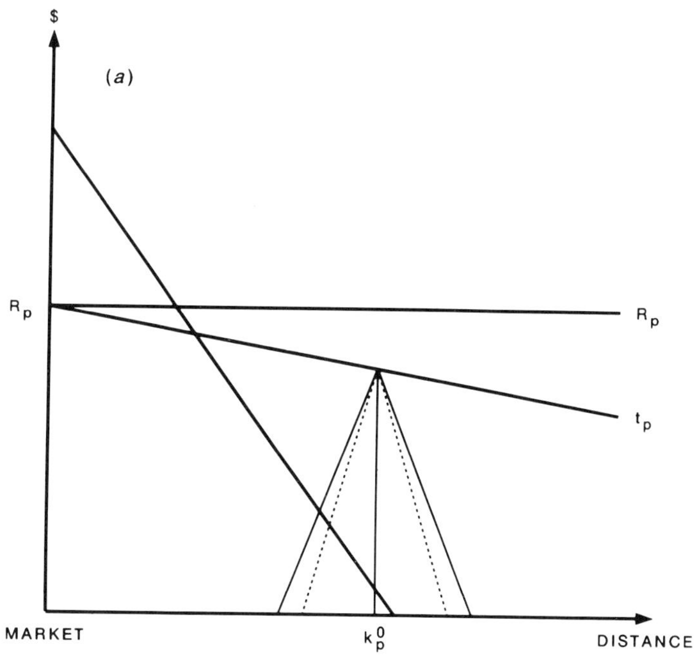

Figure 3A. Differences in feedstock assembly costs. The initial consequence of higher on-plantation transport (assembly) costs is a steeper plantation land rent gradient and a smaller plantation.

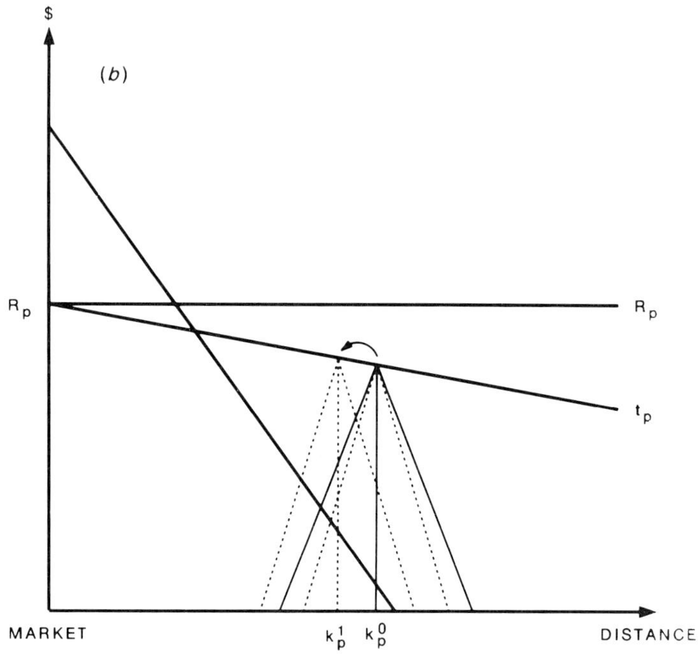

Figure 3B. Differences in feedstock assembly costs. Plantation location adjusts to compensate for higher assembly costs.

When conversion facility size and feedstock input rate are endogenous, the results are less clear cut. The feedstock input rate is lower when its assembly cost is higher, but facility size may increase or decrease according to the algebraic results. Ordinarily a higher input price will, given a fixed output price, reduce the quantity of output supplied. However in the present case, although the market price of biomass fuel, $p_Q(0)$, is fixed, the price upon which supply decisions are made, $p_Q(k_p)$, is variable and could conceivably increase if k_p shifts toward the market. Consequently, the diagrammatic analysis of the endogenous capacity case is inconclusive, and even algebraic results leave net direction of effect in doubt regarding differences in assembly costs.

A spatially based cost closely associated with on-plantation assembly costs of the harvested feedstock is the cost of delivering variable inputs to dispersed sites on the plantation. These costs are not well documented directly, but appear to be substantial enough to enter into plantation managers' decision calculi (see, eg., Bauer 1948, pp. 273-74; and Stryker 1976, for various citations). The incorporation of these variable input transport costs into land values has been treated by

Jones (1984b). Consideration of these costs, as well as the direct assembly cost, strengthens the motivation behind τ_W.

4.3 Differences in Market Prices of Biomass Fuel, $p_Q(0)$

The market price which biomass fuel commands is exogenous to the current model but is determined by prices of other, more widely used fuels such as oil, gasoline, natural gas and coal, and the substitutability of biomass fuels for these non-"alternative" fuels in various uses. In different regions of the United States, biomass energy could expect to compete predominantly with different fuels. For instance, in the Midwest, grain-based alcohol fuels would find gasoline and possibly some fuel oils as the principal competing fuels, while in Wisconsin there may be opportunities for wood to compete with coal in small utility plants. Felker (1984) has explored the economic viability of a range of sizes (5 to 500 megawatts) of mesquite-fired utility plants in south Texas. Changes in the prices of the major competitor fuels or differences in those prices among regions will affect the market price of biomass fuels facing actual or potential producers in a region.

With a fixed capacity, fixed coefficients conversion plant, a higher market price for the processed fuel brings forth a higher shadow price of the feedstock, although the necessity of this increase is not apparent from the diagrammatic analysis. In Figure 4A, line R_pR_p rises by the amount of the increase in $p_Q(0)$. If the plantation location remains unchanged, an excess supply of feedstock will be produced on a larger plantation. An outward "move" of the plantation will damp the increase in p_Q^* and hence in $p_W(k_p)$, keeping the feedstock produced on the plantation constant — because of fixed capacity in conversion. The broken rent triangle in Figure 4A shows the hypothetical increase in the plantation area when the plantation location is unchanged. In Figure 4B, an outward "move" of the plantation occurs, but stops short of driving the conversion-site rent on land back to its level at the lower $p_Q(0)$. The semi-circular "cut-off" area at the boundary between agriculture and the plantation will be slightly larger and must be compensated with a slightly higher central land rent and shadow price of feedstock to capture the same plantation area.

When conversion plant size and the feedstock input rate are endogenous, the plantation will locate closer to market with a higher market output price.

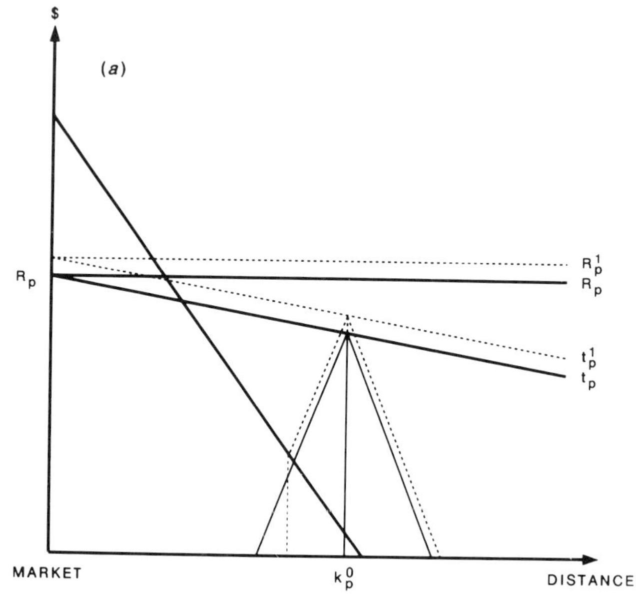

Figure 4A. Differences in market price of biomass fuel. With location unchanged, plantation size increases.

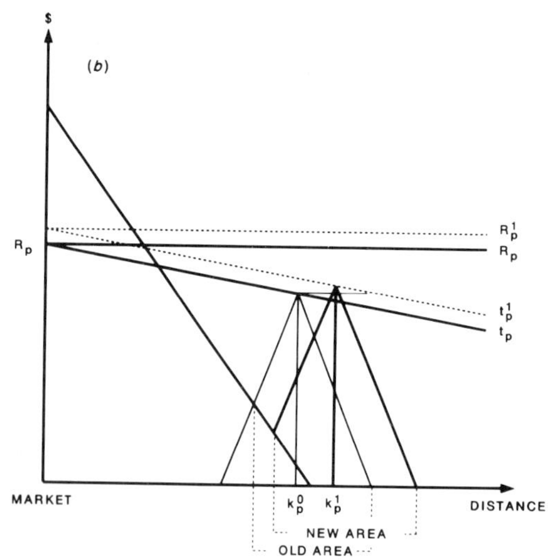

Figure 4B. Differences in market price of biomass fuel. Plantation moves outward to reduce size when conversion facility is of fixed capacity and has constant coefficients technology.

5. CONCLUSION

The model presented here has analyzed the case in which the converted biomass fuel is sent to a market some distance from the conversion facility. A major opportunity for using biomass energy sources is in cogeneration of thermal and electric power — use of the fuel in the facility from which it is derived, e.g., use of wood chips to provide power for sawmills. The problem is structurally very similar to the case analyzed above. Haulage costs restrict the area from which the material may be assembled, although if the fuel feedstock is simply a residual from primary-use material, the ability to use a larger percentage of the biomass hauled will have the effect of lowering its assembly cost. The value of the residue used for feedstock would increase with increased transportation costs on alternative fuels such as coal or oil.

Most analyses of the viability of biomass energy development have been aware of what we have called on-plantation transport costs or feedstock assembly costs (eg., Tyner and Bottum 1979; Medema et al. 1981; Barnard 1983). Studies of fuel wood viability have noticed primarily, although not exclusively, the costs of moving either unprocessed or processed wood to retail sale sites (Medema et al. 1981, pp. 25-28; Perlack et al. 1984). Barnard's recommendation for achieving economic viability for corn-based alcohol fuel in the American midwest involved exploiting a trade-off between on- and off-plantation transport costs, much as the present model specifies very generally.

The generality of the present model's structure may be underestimated by focusing on apparently missing steps in the assembly-conversion-market process. For instance, crude, on-site conversion of fuel wood to charcoal, as cited by Perlack et al. (1984), for Liberia, may appear to shortcut the assembly step, but may be thought of more usefully as involving exceptionally high on-plantation assembly costs relative to fixed costs in conversion. More efficient, but more costly, conversion facilities could be made economical by either reducing assembly costs or reducing transport costs on the trip to market. As Perlack has noted in personal communication, another alternative is the employment of mobile conversion facilities. The apparent absence of one of the three production stages formalized in the present model is likely to be an indicator of high relative cost of that stage and should focus attention on ways to reduce those particular costs. The particular contribution of this model is to distinguish these three phases of production so that the opportunities for trade-offs among them can be analyzed more effectively.

The present model suggests several conclusions for the development of biomass for energy use off the production facility. (1) Higher densities of feedstock cultivation are more profitable at locations more distant from final sales sites. (2) Increases in the size of the plantation area raise the cost of feedstock inputs. (3) Higher transport rates on feedstock assembly also raise feedstock input costs. Counterintuitively, higher transport rates on the converted fuel (not explicitly analyzed in this chapter) lower feedstock input costs — by reducing plantation size. (4) Land costs in biomass fuel production cannot be assessed without reference to the location of the proposed development site relative to the final sale site for the fuel. (5) As an extension of (4), total production cost of a biomass fuel cannot be assessed without such a locational reference.

The model is admittedly simplified, and it is reasonable to ask how useful it would be to policymakers in its present state or in various modified forms. The major element of abstraction appears to be in the specification of agricultural production costs on the plantation, with the collapsing of all labor, equipment, fertilizer, etc., costs into what have been called "land" and "trees." In a practical application of the model, say a simulation exercise to give a policymaker an idea of where, what kind of biomass energy developments (size, technology, type of biomass) can be expected to obtain what degree of economic viability, the two-input, on-plantation agricultural production function easily could be replaced with a more disaggregated one. In some developing countries, governments may have an employment objective as well as an energy substitution goal, in which case explicit incorporation of labor into the on-plantation agricultural production function, as well as the specification of an objective function including the two (or more) goals, would be important. In developed as well as developing countries, risk is prevalent in biomass energy production. First, agriculture and silviculture have unavoidable yield risk, which may require some compensatory structuring of a plantation. Second, world energy markets have been quite volatile in recent years and there appears little reason to expect that characteristic to change substantially in the discountable future. Consequently, price risk for the converted fuels may need to be accounted for. Jones (1983) has studied the spatial incidence of both price and yield risks in agriculture, finding the former increasing with remoteness from markets and a reasonable suspicion that the latter would as well. It could prove highly useful to incorporate risk modeling along the lines of Newbery and Stiglitz (1981) into the current production and location model.

In sum, the model developed here should prove useful to policymakers. It in no way replaces prior engineering-economic analyses of biomass feasibility. Rather, it assembles the results of that work regarding cultivation and conversion technology and costs into a unified analytical framework to facilitate the development of more complete answers to policymakers' questions about the practicality of developing biomass energy resources.

6. ACKNOWLEDGEMENTS

We have benefited from discussion with George Tolley and Peter Rogerson and from the comments of the Resources Workshop of the Department of Economics, University of Chicago. Anthony Turhollow and Robert Perlack, Energy Division, Oak Ridge National Laboratory, also provided useful comments on an earlier paper. Elizabeth Erickson, of USDA and the University of Akron, discussed a version of this chapter at the 31st North American Meetings of the Regional Science Association, Denver, November 11, 1984, and offered a number of useful insights. Barry Solomon and an anonymous reviewer also made very useful suggestions. We thank these people, but responsibility for any errors or other short-comings remaining in the paper is ours, of course.

7. FOOTNOTES

*Research sponsored by the Biomass Energy Technology Division, U.S. Department of Energy, under Contract DE-AC05-84OR21400 with Martin Marietta Energy Systems, Inc.

[1]Although f' is not defined for fixed-coefficients technology, we retain the notation of f' and f-zf' rather than switch to notations more common to Leontief technology descriptions. In the three-equation model, f' is properly interpreted as an input-output coefficient.

8. REFERENCES

Barnard, J.R. 1983. Gasohol/ethanol: A review of national and regional policy and feasibility issues. Regional Science Perspectives 13: 3-14.

Bauer, P.T. 1948. The rubber industry; a study in competition and monopoly. London: London School of Economics.

Dobie, J.B., Miller, Jr., G.E. and Parsons, P.S. 1977. Management of rice straw for utilization. American Society of Agricultural Engineers Technical Paper 77-1047.

Felker, P. 1984. Economic, environmental, and social advantages of intensively managed short rotation mesquite (*Prosopsis spp*) biomass. Biomass 5: 65-77.

Hertzmark, D., Flaim, S., Ray, D. and Parvin, G. 1980. Economic feasibility of agricultural alcohol production within a biomass system. American Journal of Agricultural Economics 62: 965-71.

Hitzhusen, F.J. and Abdallah, M. 1980. Economics of electrical energy from crop residue combustion with high sulfur coal. American Journal of Agricultural Economics 62: 416-25.

Howlett, K. and Gamache, A. 1977. Silvicultural biomass farms: forest and mill residues as potential sources of biomass. Mitre Corporation Report TR-7347, Vol. III.

Jones, D.W. 1983. Location, agricultural risk, and farm income diversification. Geographical Analysis 15: 231-46.

______. 1984a. A land use model with a constant-utility spatially variant wage. Geographical Analysis 16: 121-33.

______. 1984b. Nonland factor markets in the Thunen model. Papers of the Regional Science Association 55: forthcoming.

______, and Krummel, J.R. 1984. The locational theory of the plan tation. Energy Division, Oak Ridge National Laboratory.

McCollum, M.P. and Hughes, C.M. 1983. An equation for predicting harvesting costs on second-growth southern yellow pine sites in the midsouth. Southern Journal of Applied Forestry 7: 89-92.

Medema, E.L., Hatch, C.R. and Christophersen, K.A. 1981. Investment analyses of fuelwood plantations in Sri Lanka. Forest, Wildlife and Range Experiment Station, University of Idaho, Contribution No. 221.

Meekhof, R.L., Tyner, W.E. and Holland, F.D. 1980. U.S. agricultural policy and gasohol: a policy simulation. American Journal of Agricultural Economics 62: 408-15.

Newbery, D.M.G. and Stiglitz, J.E. 1981. The theory of commodity price stabilization; a study in the economics of risk. Oxford: Oxford University Press.

Office of Technology Assessment, Congress of the United States 1980. Energy from Biological Processes. Washington, D.C.

Overend, R.P. 1982. The average haul distance and transportation work factors for biomass delivered to a central plant. Biomass 2: 75-79.

Perlack, R., Barron, W., Samuels, G. and Rhinelander, E. 1984. A study of the costs of fuel supply for wood-fired electric power plants in rural Liberia. (draft mimeo) Oak Ridge, TN: Oak Ridge National Laboratory and National Energy Committee, Government of Liberia.

Ranney, J.W. and Cushman, J.H. 1980. Regional evaluation of woody biomass production for fuels in the southeast. In Biotechnology and Bioengineering Symposium No. 10, pp. 109-20. New York: Wiley.

Solomon, B.D. 1980. Gasohol, economics, and passenger transportation policy. Transportation Journal 20: 67-64.

Spiewak, I., Nichols, J.P., Alvic, D. Delene, J.G., Fitzgerald, B.H., Hightower, J.R., Klepper, O.H., Krummel, J.R. and Mills, J.B. 1982. Technical Analysis of the use of Biomass for Energy Production. ORNL/TM-7919. Engineering Technology Division, Oak Ridge National Laboratory.

Stryker, J.D. 1976. Population density, agricultural technique, and land utilization in a village economy. American Economic Review 66: 347-58.

Tyner, W.E. and Bottum, J.C. 1979. Agricultural energy production: economic and policy issues. Purdue University Agricultural Experiment Station Bulletin, No. 240.

Wood, B.W., Carpenter, S.B. and Witter, R.F. 1976. Intensive culture of American sycamore in the Ohio River Valley. Forest Science 22: 338-42.

Wrisz, P.B. and Marshall, J.F. 1980. Fuels from Biomass: A Critical Analysis of Technology and Economics. Mobil Research and Development Corporation - Central Research Division.

CHAPTER 8

THE DEVELOPMENT POTENTIAL OF NEW ENERGY RESOURCES: A SPATIAL-ECONOMIC ANALYSIS OF COALBED METHANE

Jeffrey P. Richetto
Department of Geography
The University of Alabama
University, Alabama 35486

ABSTRACT. The issues enveloping the development of new and experimental energy resources are interdisciplinary in scope. Such considerations as supply, demand, price, production cost, substitutability, investment recovery, availability of technology and public-environmental acceptability are exemplary of the myriad of concerns influencing the timing and development potential of novel resources. This chapter examines the potential for coalbed methane gas development in the Warrior Coal Field for the state of Alabama. A data base which identifies geographical areas containing the highest densities of logical coal gas users will help evaluate the desirability of coalbed methane as a developable energy resource. In a similar way, the development potential of other new and experimental energy resources may be evaluated.

1. INTRODUCTION

Since the 1973 oil crisis, there has been an on-going emphasis in the development of new, unconventional energy resources (e.g., solar, geothermal, tidal and wind power, and unconventional natural gas). The marginal economics of these unconventional energy resources will restrict their use to regional and local markets. This is due to the highly localized availability of these resources, and the limits to transmitting usable energy over long distances. Most geographers' energy research efforts have focused on siting analysis, energy transmission, and impact assessment of various energy-producing technologies (Osleeb and Sheskin 1977; Hanham and Calzonetti 1983; Pijawka and Chalmers 1983; Pasqualetti 1983; Richetto 1984). However, few geographic studies address served energy markets. Since the size and nature of served markets are closely linked with such basic spatial concepts as distance, intervening opportunities, competitive locations, and spatial interactions, this lack of consideration represents a void in energy geography, which may be explained in part by the type of

F. J. Calzonetti and B. D. Solomon (eds.), Geographical Dimensions of Energy, 101–121.
© 1985 by D. Reidel Publishing Company.

served market. For example, electric utilities are required to provide electrical service within markets over which they have sole responsibility (Richetto 1983). Suppliers of both oil and natural gas have established long-term contracts with customers, thus locking in a large proportion of their available supplies. In effect, the conventional energy producing resources have been and will continue to serve markets that are characteristically well established, slightly competitive, and reinforced by the public/private structure of the energy industry as a whole.

As new energy resources are developed, associated markets will need to be identified either on a competitive or cooperative basis; these markets will be smaller in size and based initally on private venture capital. Such considerations as supply, demand, investment recovery, substitutability, availability technology, and public-environmental acceptability exemplify the concerns influencing the timing and development potential of new energy resources. By identifying the regional and site-specific nature of the above considerations, the interrelationships between timing, level, and location of new energy resources may be determined. In turn, public policy and the extent of private sponsorship may be evaluated and implemented.

Energy geographers can contribute to the development potential of new energy resources by identifying served markets. The geographic perspective is illustrated by the following questions: Who are the potential users? What is the expected level of demand? Which geographic areas contain the highest densities of potential users? Do compatible energy distribution systems exist? If so, what is their spatial configuration? These and other economic geography-based questions may serve to identify the spatial extent and nature of associated markets and also those supply sites exhibiting the greatest potential for resource production.

This chapter examines the market potential for integrated coalbed methane recovery systems in the Warrior Coal Basin in Alabama. Within the past four years over 175 well permits were let to corporations in the Basin, making it the most active coalbed methane region in the country. This study: (1) discusses coalbed methane as an example of unconventional natural gas resources, (2) identifies potential coalbed methane users, (3) assesses potential demand for coalbed methane, (4) illustrates the locational association between production sites and demand areas by type of user and level of demand, and (5) analyzes both direct and indirect economic effects accompanying the development of the coalbed methane energy resource.

2. NATURAL GAS RESOURCES

At current rates of consumption, proven gas reserves in the United States, estimated at 200 Tcf, will be consumed within the next decade. Although natural gas usage increased annually at a rate of 7 percent between 1947 and 1971, consumption growth is expected to continue at one-tenth that rate through 1990. This lower rate of consumption, along with the recent surplus of deliverable natural gas, indicates that only one-half of the United States' demand for gas in 1990 will be supplied by this nation's conventional natural gas resources (Osleeb and Sheskin 1977; American Gas Association 1984).

Annual additions to U.S. gas reserves are steadily declining and a long-term reversal of this trend is not expected. Eventually the decrease in reserve additions will be reflected in production decreases. The major options for maintaining domestic gas production past 1985 and continuing the pipeline network and infrastructure of the gas industry are limited to additional conventional gas sources, coal gasification, and new unconventional gas sources. Since providing natural gas to the residential sector is the highest priority, industrial and commercial users must increasingly turn toward alternative sources of natural gas. Four sources of unconventional natural gas are identified: (1) tight gas sands; (2) Devonian shale; (3) geopressurized aquifers; and (4) coal seams (Figure 1). Conventional and recently developed technology sources, including the incentive of price decontrol for coal seams exploitation, could augment significantly the supply of natural gas (Satriana 1980).

Twenty major tight sands basins with large quantitites of natural gas stretch from New Mexico northward into Canada, and eastward into Arkansas and Louisiana. Estimates of in-place gas are as high as 1,200 Tcf, of which nearly 50 percent are recoverable. However, many sandstones throughout these basins have a high resistance to gas flow and pose a technical challenge for stimulating gas production. The eastern Devonian gas shales, located in New York southward to Alabama and westward into the Ohio River Valley, also contain sizeable volumes of natural gas. In-place resource estimates involve a high degree of uncertainty, and range between 50 and 520 Tcf. In many areas the concentration of free gas within the shales is too low to be commercially produced at competitive prices. Shale-derived gas has been produced in limited quantities since 1820, and present annual production is about 0.1 Tcf. Geopressurized aquifers are found in various regions of the world, including China, the Soviet Union, and the North Sea. The largest known aquifer lies under the Texas and Louisiana Gulf Coast both on- and off-shore. These water-bearing reservoirs are found at depths ranging from 5,000 to 18,000 feet, with temperatures of 200 to 400 degrees Fahrenheit, and are characterized by abnormally high pressures of 4,000 to 15,000 psi. Due to different assumptions about the supply of the resource (3,000-115,000 Tcf), the amount of dissolved natural gas per barrel of water, the extent and thickness of the deposits, and the porosity and permeability of the reservoirs, recent studies have been quite pessimistic about the recoverability of this resource. Finally, perhaps the most promising near-term unconventional gas resource is methane from coal seams.

Coalbeds, especially in the eastern United States, contain significant quantitites of natural gas and are currently producing small but locally important amounts of gas. Until recently, the coal industry

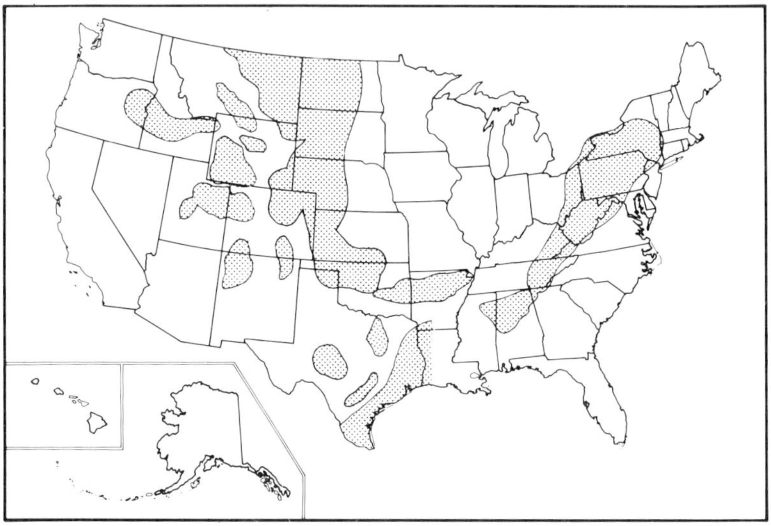

Figure 1a. The Geographical Dimension of Tight Gas Sands Deposits in the United States.

Figure 1b. The Geographical Dimension of Devonian Gas Shales in the United States.

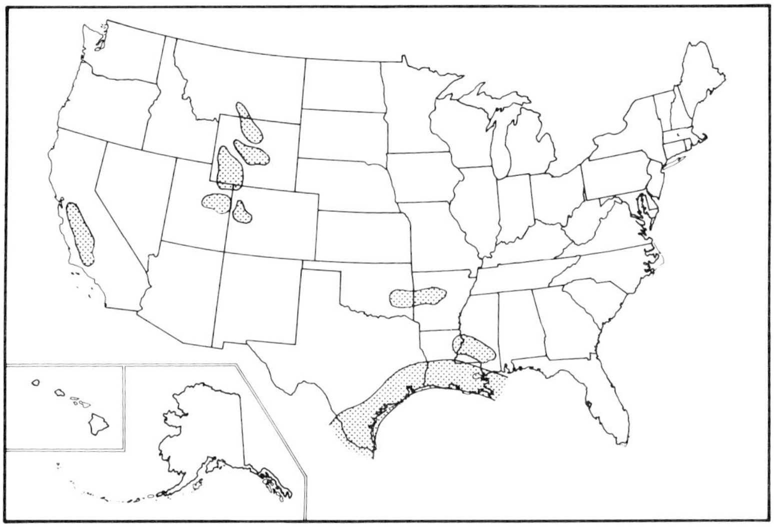

Figure 1c. The Geographical Dimension of Geopressurized Aquifers in the United States.

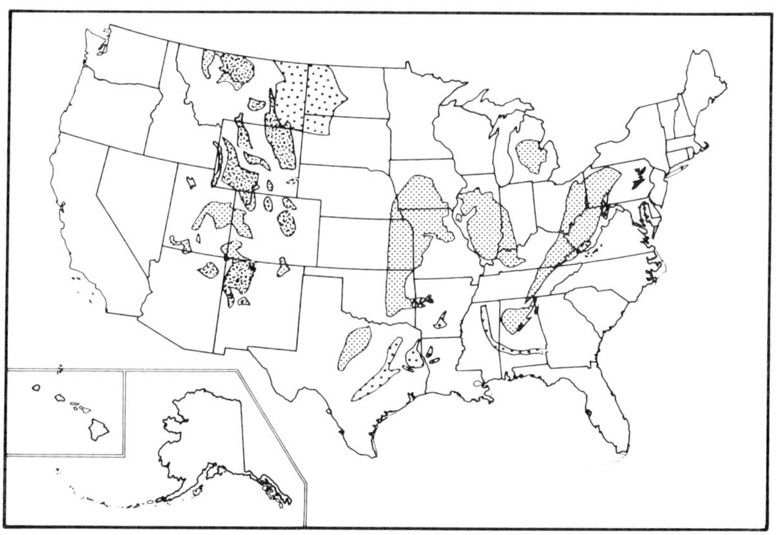

Figure 1d. The Geographical Dimension of Coal Deposits in the United States.

was interested in the removal of methane (the major component of natural gas) from coal mines as a safety precaution. Because of the low permeability of strata associated with coalbeds, methane is retained either as gas absorbed on the surface of the coal, free gas in vertical joints and fractures in the coalbed, or both. Due to the natural association of methane with coalbeds, the geographical dimensions of coalbed gas are defined by the extent of sub-surface coalbeds (Figure 1d). The amount of methane available within any coalbed varies considerably and is a function of the rank of the coal and the depth of its burial (Figure 2),

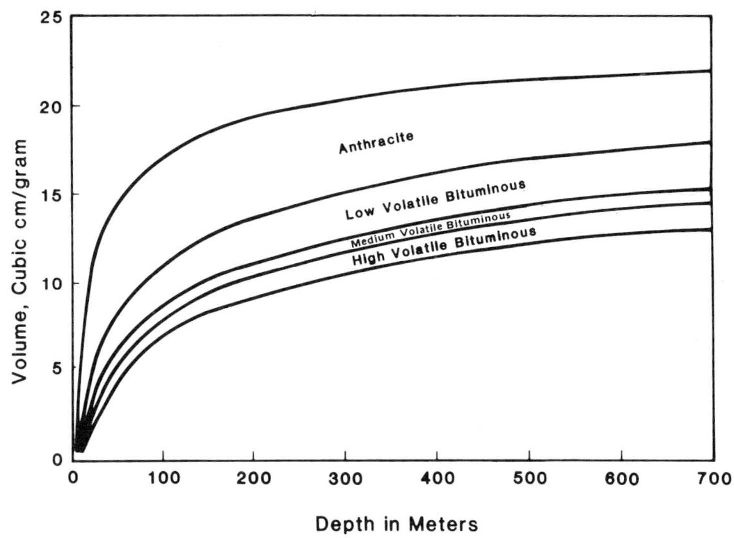

Figure 2. Estimated Methane Content of Coal With Depth and Rank

among other things. The amount of methane is inversely proportional to the moisture content of the coal, i.e., moisture decreases as the rank of the coal increases. The rank of a coalbed depends upon its degree of coalification, which is influenced by time, temperature, pressure, and tectonic and burial history of the organic matter present. Finally, the depth of burial determines the pressure of a coal seam, i.e., gas content increases with increasing pressure (Table 1). Recent estimates for coalbed methane in the United States range from 650-850 Tcf, and approximately 300 Tcf appear to be technologically and economically recoverable.

Once coal gas is recovered, it can be used in at least three ways: (1) when the quality and methane content of the recovered gas is sufficiently high, coalbed gas can be gathered, purified, pressurized, and injected into a commercial natural gas pipeline (Table 2); (2) coalbed gas can be converted into liquefied natural gas; and (3) coalbed gas can be used for local power generation and industrial and commercial

Table 1. Gas Content Of U.S. Coals by Rank and Depth

Coal Rank	Depth (feet) Under 1000 cf/t	1000-3000 cf/t	Over 3000 cf/t
Anthracite and Bituminous	150	250	500
Sub-Bituminous	60	100	200
Lignite	30	50	100

Source: Kuurskraa, V., and Meyer, R. 1980. "Review of world resources of unconventional gas," IIASA Conference on Conventional and Unconventional World Natural Gas Resources, Proceedings, Laxenburg, Austria, pp. 27-43.

usage. In spite of these and other possible uses, along with the probable magnitude of this resource, there are numerous technological, economic, environmental, and legal obstacles inhibiting the commercialization of coalbed gas (Deul 1976; Harbough 1977; Gillies and Snygg 1980). Coal companies, public utilities, and industry have limited experience in the commercialization and economic evaluation of coalbed methane. Criteria for deliverability, quality of supply, and terms of purchase for coalbed gas have not been established.

3. MARKET POTENTIAL OF COALBED GAS IN ALABAMA

The Warrior Coal Basin comprises over 9,420 square miles and includes all or parts of twenty counties (Figure 3). With over 40 coalbeds, this basin is the most commercially mineable and structurally least complex coal-bearing area in Alabama. Based on the geological studies of the Basin, nearly 10 Tcf of in-place coalbed methane exists, 50 percent of which is technologically and economically recoverable. Recent drilling in the Warrior Coal Basin has produced useful data for interpreting the potential methane resource. Samples of gas have been collected from numerous coreholes and active wells (especially in Jefferson and Tuscaloosa counties) and analyzed, using gas chromatography (Table 2).

Table 2. Comparison of Coalbed Methane With Conventional Natural Gas

	Percentage of Coalbed Gas					
	Pocahontas No. 3	Pittsburgh	Kittanning	Hartshorne	Mary Lee	% of Natural Gas
CH_4	96.87	90.75	97.32	99.22	96.05	94.40
C_2H_6	1.39	0.29	0.01	0.01	0.01	3.80
C_3H_8	0.0147	-	-	-	-	0.6
C_4H_{10}	0.0008	-	-	-	-	0.3
C_5H_{12}	-	-	-	-	-	0.2
O_2	0.17	0.20	0.24	0.10	0.15	
N_2	1.7	0.59	2.3	0.6	3.5	0.4
CO_2	0.36	8.25	0.14	0.06	0.10	
H_2	0.01	-	-	-	-	
He	0.03	-	-	-	0.27	
Btu	1,059	973	1,039	1,058	1,024	1,068

Source: Duel, M., and Kim, A. 1978. Methane in coal: From liability to asset. Mining Congress Journal; Moore, B. 1976. Analysis of natural gas in the United States, U.S. Bureau of Mines, IC8302.

In general, the methane content is of a high quality, averaging 95 percent for all sites sampled. The heat of combustion for the samples ranges between 900-1,100 Btu per cubic foot. It is anticipated that with moderate development[1] drilling could increase to nearly 350 wells per year by 1990. Therefore, a careful delineation of the market potential (type, size, and location) and economic impact for developing coalbed gas is imperative.

In an effort to identify and assess geographical areas containing the highest densities of logical coalbed gas users, a questionnaire was developed to elicit responses from potential users to questions about location, level of consumption, size, plans for expansion, type of activity, and expressed interest in utilizing coalbed methane. Since coalbed gas can be used most efficiently on-site, information was solicited from industries, school systems, medical facilities, municipal buildings, and gas supply systems. A total of 341 completed surveys by

Figure 3. The Warrior Coal Basin.

229 industrial firms, 56 major medical facilities, 22 municipal buildings, 17 municipal gas suppliers, and 17 school systems provided baseline data for this study's analysis of the market potential for coalbed gas. Discussion of the market potential for coalbed gas in this chapter is confined to industry (Richetto et al. 1984).

4. MARKET ANALYSIS OF INDUSTRY

The overall pattern in the consumption of natural gas by county in the Warrior Coal Basin is illustrated in Figure 4. A noticeable northern and southern concentration emphasizes the industrial character of the included counties. At the same time, other counties form a west-to-east centralized arc, are characteristically rural, and fall well below their urban-industrial counterparts in natural gas consumption. Reporting firms were divided into two categories, general and industrial park; the significance of industrial parks is reflected in the close spatial proximity among tenants, which offers the opportunity for on-site coalbed gas development. That is, an industrial park may be fitted with an array of wells, allowing tenants to use the gas directly without an intervening distribution system. Figure 5 illustrates the location of industrial parks and the corresponding number of tenants in the study area.

Figure 4. Natural Gas Consumption in the Warrior Coal Basin, 1983.

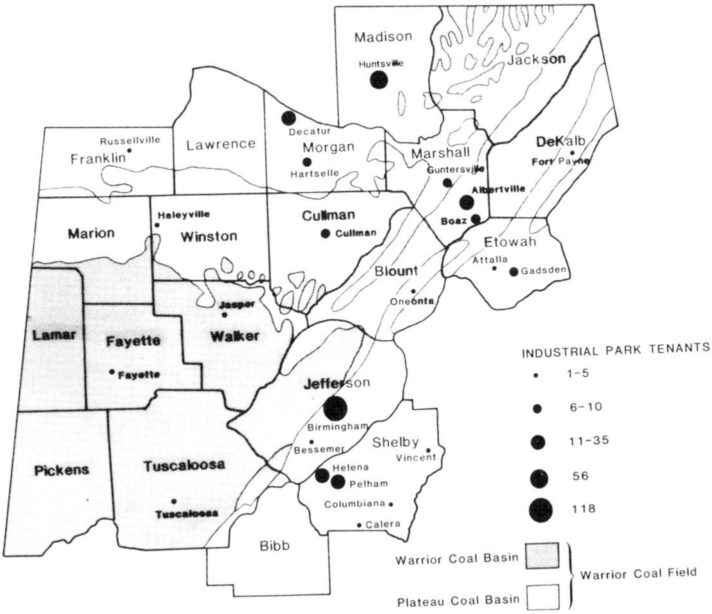

Figure 5. The Number and Location of Industrial Park Tenants.

The resulting pattern is a southeastern to northeastern ribbon-like concentration, while the central and western portions of the study area exhibit little, if any, industrial park activity. There are three useful indicators for examining the potential utilization of coalbed gas by industrial parks: (1) existing level of gas consumption; (2) planned expansion of existing operations; and (3) level of expressed interest in using coalbed gas. Industrial park tenants in Jefferson, Madison, and Marshall counties accounted for nearly 95 percent of the natural gas consumed in 1983 by all reporting tenants. Nearly 45 percent of existing tenants planned to expand their operations within the next five years, and 62 percent of industrial parks anticipated the location of new tenants over this same five-year period. Finally, almost three-fourths of all industrial park tenants expressed a moderate or high level of interest in adopting coalbed gas in their production process. In fact, 87 percent of all reporting firms indicated a moderate or high interest in coalbed gas development (Table 3).

Firms reporting a high interest were concentrated primarily in the counties of Jefferson and Tuscaloosa and, to a lesser extent, in Madison and Morgan. Firms expressing a moderate interest were located in DeKalb, Etowah, Madison, Marshall, and Morgan counties. This pattern not only reflects the location of active coalbed methane projects in northern Tuscaloosa and southwestern Jefferson counties, but also underscores the market potential of coalbed gas throughout the eastern portion of the Warrior Coal Basin. Potential high-use industries are involved in manufacturing textiles, chemicals, primary

Table 3. Number of Firms Expressing Interest in Utilizing Methane, By County

	County																			
Level of Interest	Bibb	Blount	Cullman	DeKalb	Etowah	Fayette	Franklin	Jackson	Jefferson	Lamar	Laurence	Madison	Marshall	Morgan	Pickens	Shelby	Tuscaloosa	Walker	Winston	
None			1				1		5			3		1	2		1			
Low					3				16	1	1	1		1	7		2	1		
Moderate	3	1	2	6	4		1	3	37	2	2	16	7	10	10	2		8	4	4
High		1	6		1	1	6		23			5	3	2	7	1	1	8	1	2

and fabricated metals, and food and kindred products. These types of industries accounted for nearly two-thirds of total industrial consumption of natural gas. This finding agrees with other national studies in determining which industries consume inordinate amounts of natural gas (York 1979; Adams and Hewitt 1982; Van Schonfeldt et al. 1982).

In sum, Alabama industries in this study were favorably impressed with the possibility of using coalbed methane in the operation of their facilities. Moreover, if the extraction of methane gas from coal seams becomes economically competitive with conventional natural gas, development efforts in the Warrior Coal Basin should focus on supplying the large and enthusiastic market along its eastern perimeter, extending from Madison through Tuscaloosa counties. However, the identification of geographic areas containing the highest densities of logical coalbed gas users by type and level of demand is only one of the critical issues bearing on the timing and development of the gas resource. The associated direct and indirect economic impacts are of equal importance.

5. ECONOMIC IMPACTS

Research efforts on the size and marketability of coalbed methane in the United States since the mid-1970's have produced a small industry which can make potential contributions to both state and local economies. Under a moderate development pace, there could be substantial supply-demand impact in those regions which have a large volume of coal. While methane gas could provide a stable, clean energy source conducive to industrial development and expansion, there would be supply constraints in the form of equipment and trained personnel, construction, employment, and industrial development potential.

Major operations bearing on the supply of coalbed methane are drilling and well completion. Equipment and trained personnel to perform these operations are not normally subsumed in the general construction activity and, at present, there are no more than two or three active rigs in the Warrior Coal Basin. Three scenarios in drilling rates were developed to examine the potential impacts on Alabama's economy and gas supply (Figure 6). All three cases begin at the 1980 rate of 30 wells per year, extending through a rate of 100 wells per year (4-5 rigs) by 1985 and reach a variable rate of 200, 350, and 600 wells per year (10, 17, and 30 rigs, respectively) by 1990. High potential coalbed methane subareas within the Warrior Coal Basin were identified for analysis (Figure 7) to project the associated economic impacts for the three drilling rates. These subareas were limited to localities where the Mary Lee coals are at least 1.83 meters thick and 213.36 meters deep. The development and operations costs for the moderate development pace are provided in Table 4 for intangible, tangible, and operations and maintenance categories.[2] Expansion pipeline transmission was included. Estimates of the associated costs for these lines total $26,000,000 (1982 $), which would be incurred in the early years of development. No expansion of the current high-pressure transmission systems was included since it was assumed that any growth in demand for gas is not directly related to the availability of this new unconventional gas source.

Major economic impacts associated with the demand for coalbed methane, in the form of direct and indirect benefits, have already been accuring to the State. Foreign and local investments in the drilling and completion of wells, installation of gas processing equipment, new and expanded local support services, taxes, and savings in coal production costs from reduced levels of methane in pre-drained mines represent a partial list of benefits related to coalbed methane development. Estimates of these and other impacts related to coalbed methane development were calculated by using the Alabama Econometric Model (AEM)[3] (Figure 8). The AEM is a simultaneous equation model consisting of nearly 120 behavioral equations and identifies. The basic driving sector for the model is the output sector, as measured by value added in production. Construction expenditures and the value of output

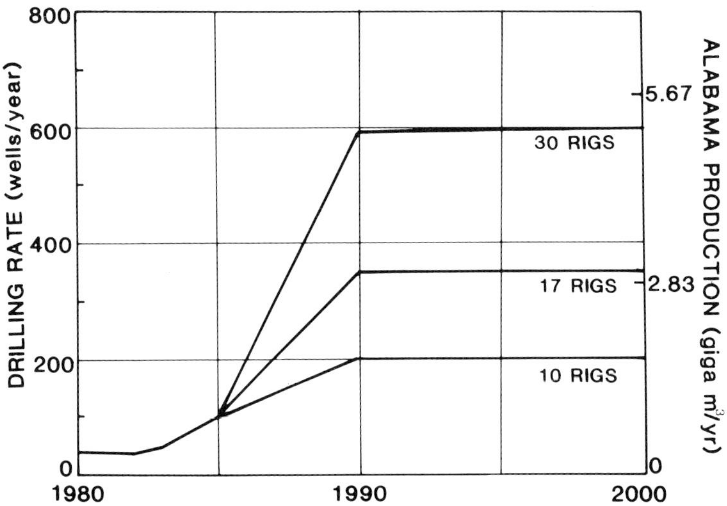

Figure 6. Variable Drilling Rates and Corresponding Production.

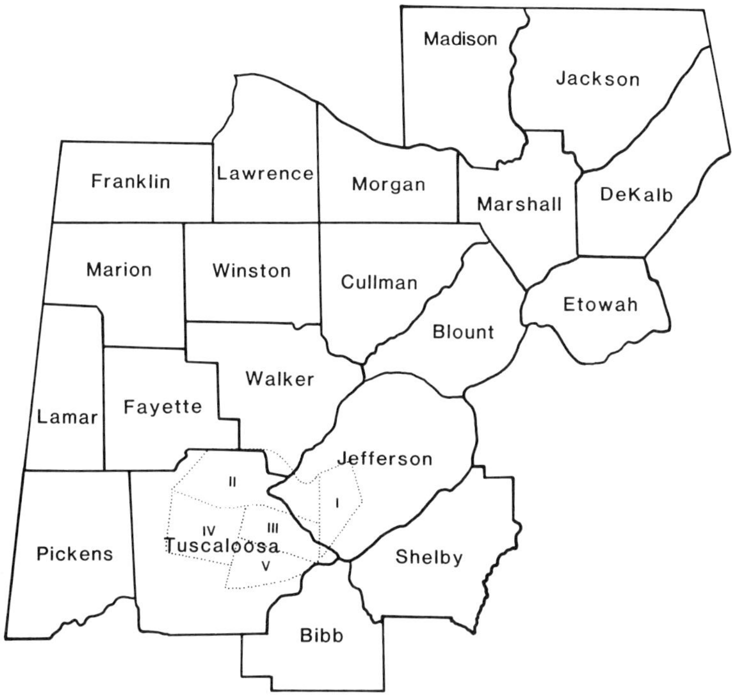

Figure 7. High-Potential Coalbed Methane Sub-Areas.

Table 4. Intangible, Tangible, and Operations and Maintenance (O&M) Expenditures for the Moderate Development Pace

($ million 1982)

	Sub-Areas					
Year	I	II	III	IV	V	Total
1980						
Intangible	1.2	3.5	1.7	2.2	1.5	10.1
Tangible	0.4	1.1	0.7	0.8	0.7	3.7
O&M	-	-	-	-	-	-
1985						
Intangible	5.1	14.8	7.5	9.9	6.6	43.9
Tangible	1.8	5.0	2.9	3.5	2.8	16.0
O&M	0.8	2.3	1.3	1.5	1.3	7.2
1990						
Intangible	23.2	69.0	35.9	46.3	32.0	206.4
Tangible	8.1	23.4	13.7	16.6	13.7	75.5
O&M	5.3	16.1	9.2	10.8	9.4	50.8
1995						
Intangible	31.1	92.4	48.0	61.9	42.8	276.2
Tangible	10.8	31.3	18.3	22.2	18.4	101.0
O&M	13.4	40.9	23.7	27.6	24.2	129.8
2000						
Intangible	41.6	123.6	64.3	82.9	57.3	369.7
Tangible	14.4	41.9	24.5	29.7	24.6	135.1
O&M	18.4	56.0	32.5	37.8	33.2	177.9

from well production for coalbed methane development were converted to a value added equivalent. The AEM utilizes projections of several exogeneous variables from Wharton Econometric Forecasting Associates. production.

The exact level and nature of economic impact from coalbed gas development in Alabama is largely a function of severance tax structure, and whether or not there will be induced industrial growth. The remainder of this study identifies the economic impact of developing coalbed methane and assumes: (1) no moratorium (moratorium) or severance tax for ten years and no induced industrial development; scenarios I and II; and (2) no moratorium (moratorium) on severance tax for ten years and induced industrial development scenarios III and IV. Baseline projections without coalbed gas development were calculated and compared with new projections for developing this gas resource for each scenario. The difference represents a measure of the associated economic benefits.

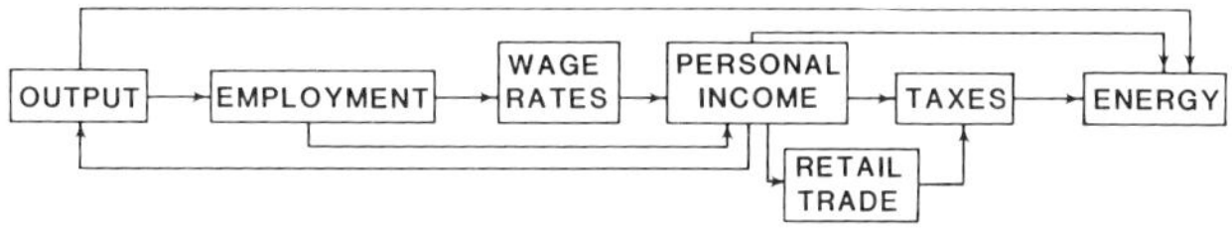

Figure 8. A Schematic Diagram of the Alabama Econometric Model.

Scenarios I and II highlight the impact of the severance tax without induced industrial growth on the relative levels of economic activity in Alabama. These impacts for gross state product (GSP), state tax revenue (STR), and employment (E) are summarized in Table 5. Assuming coalbed gas development with existing severance taxes (scenario I), the impact on Alabama's gross state product ranges from $12,000,000 in 1983 to over $500,000,000 by 1991. Throughout this same period, state tax revenue is projected to increase from $3,000,000 to nearly $85,000,000 and the number of new jobs will be almost 7,000 by 1991. Alternatively, if coalbed gas development proceeds under a ten-year moratorium on severance tax (scenario II), there will be a considerable impact on all three economic indicators. For example, by 1991, over $2,000,000,000 will be added in gross state product and 7,300 new jobs, at a cost of $36,000,000 in state tax revenue. A severance tax equal to one-tenth of the value of gas produced would be reinstated and a slower rate of change in state tax revenue would occur.

Scenarios III and IV highlight the impact of induced industrial growth as a result of coalbed gas development. These impacts are also summarized in Table 5. Assuming coalbed gas development with both induced industrial growth and existing severance taxes (scenario III), the impact on gross state product, state tax revenue, and employment are substantially higher than the estimates calculated for the previous two scenarios. By 1988, for example, the change in gross state product amounts to $803,000,000, over $500,000,000 more than the same case without induced industrial growth. On the other hand, if coalbed gas development is accompanied by industrial growth under a ten-year severance tax moratorium (scenario IV), both state tax revenue and employment would be lower and gross state product would be slightly higher in 1991.

Table 5. Estimated Economic Impact of Coalbed Methane Development

($ million 1982)

Year	Change in GSP ($)				Change in STR ($)				Change in E ('000)			
	I	II	III	IV	I	II	III	IV	I	II	III	IV
1983	12	32	31	33	3	1	4	1	.3	.3	.4	.3
1984	29	78	75	78	6	1	6	1	.7	.7	.8	.7
1985	54	150	168	173	9	4	12	4	1.2	1.3	1.7	1.6
1986	94	264	308	318	16	6	20	8	1.8	2.0	3.0	2.6
1987	152	437	512	528	23	11	31	14	2.6	2.9	4.6	3.9
1988	229	694	803	829	34	17	47	22	3.7	3.9	6.0	5.4
1989	323	1050	1195	1235	49	25	69	32	4.7	5.0	8.3	7.0
1990	428	1523	1674	1732	66	36	94	43	5.8	6.2	10.1	8.3
1991	538	2125	2275	2357	85	49	123	58	6.8	7.3	11.9	9.4

6. SUMMARY AND CONCLUSIONS

This study illustrates the nature and role of the market concept in understanding the timing and utilization of newly developing energy resources. In particular, questions addressing the pattern and level of demand, type of user, geographical concentration of expected users, and institutional environment were investigated within the context of coal-bed methane development in Alabama. Technological advancements for this unconventional natural gas resource can be brought quickly to the marketplace because of this nation's unique gas transportation infrastructure. With over 1,000,000 miles of gas collecting, transmission, and distribution pipeline and an extensive gas storage network, the uninterruptible purchase of gas year-round enables production and transportation facilities to operate at near capacity. The American Gas Association (1982) anticipates that methane from coal seams may represent as much as 10 percent of the total U.S. gas supply by the end of this century. While the timing and development potential of this resource are tied to supply/demand economics, mining and gas technology, and environmental and political considerations, federal and state policies will control the pace of coalbed gas commercial utilization. At the national level, research and development funding may accelerate recoverability, and recent price decontrol will help determine the share of total energy supplied by coalbed methane. State government can strongly influence the commercialization process in a variety of ways. Some examples include: (1) supplemental R&D funding--by targeting research support to specific technical/geographic areas; (2) environmental policy--the enactment of policies specific to coalbed methane development (e.g., produced water and air pollution standards); and (3) economic development policy--the establishment of policies emphasizing the comparative economic advantages through the co-ordinated use of natural resources. For example, within the Warrior Coal Basin, the combination of uninterruptible gas supplies, coal, extensive water-borne transportation, and a central location within the Southeastern United States will be more attractive to some industries than others. If exploited, this locational advantage would justify public sector investment in resource recovery technology by expanding the industrial, tax, and employment bases and by providing an exportable technology.

7. ACKNOWLEDGEMENTS

Research for this paper was supported by a grant from the U.S. Department of Energy (DE-AC 21-82MC19334) and administered through the School of Mines and Energy Development. The University of Alabama. Appreciation is extended to James Barnard and Richard Glover for their assistance in data collection. All figures were prepared by the Cartographic Lab, Department of Geography, The University of Alabama.

8. FOOTNOTES

[1]Moderate development is defined as following: drilling will increase from 40 to 100 wells per year by 1985 and stabilize at about 350 wells per year by 1990.

[2]Intangible costs include: support fuel and water, equipment rental, cementing, chemical treatments, stimulation and transportation. Tangible costs include: production casing, pumps, wellhead equipment, powerlines, compressors, separators, water collection, and major gas collection lines. Operation and maintenance costs include: workovers, inspections, power and fuel, sampling and testing, and equipment maintenance.

[3]Specific assumptions used in determining the economic impact associated with coalbed methane development include: (1) moderate pace;

(2) the price paid for coal gas reflects the ratio of the price of such gas to the average price for all gas in 1982 (representing a 5 percent growth rate in real price from a base price of $5.50/Mcf); (3) the ratio of value added to total revenue is .66 (in the absence of a severance tax) and .56 (with a 10 percent severance tax); (4) severance tax revenues are spent in the same year as collected at a rate of .502; (5) initially 75 percent of the tangible construction expenditures will occur out-of-state and thereafter decline uniformly to zero by 1993; (6) initially 25 percent of the intangible construction expenditures will occur out- of-state and thereafter decline uniformly to zero by 1988; (7) the ratio of value added to total value of new construction is .4242; (8) four scenarios were examined relative to impact on the state's industrial development; and (9) baseline economic data reflect coalbed gas activity for the period 1980-82.

9. REFERENCES

Adams, M., and Hewitt, J. 1982. Coalbed methane potential of the Appalachians. Unconventional Gas Recovery Symposium, Proceeding, pp. 125-29.

American Gas Association. 1982. The gas energy supply outlook: 1980-2000. Arlington, Virginia.

American Gas Association. 1984. The role of natural gas in achieving national energy priorities. Paper presented at International Gas Research Conference, Washington, D.C., September.

Bernstein, H., Hinkle, B., and Basquez, E. 1975. Projections of energy availability, cost, and aggregate demand for 1975, 1980, 1985, and 1990.

Comptroller General of the United States. 1980. Help for declining natural gas production seen in the unconventional sources of natural gas. NTIS PB 80-128929.

Duel, M. 1976. Natural gas from coalbeds. Natural gas from unconventional geologic sources NAS: 193-245.

Ford Foundation. 1974. A time to choose: America's energy future. Cambridge, Massachusetts: Ballinger.

Gillies, A., and Snygg, A. 1980. Development of technology for coalbed methane recovery program planning. TRW Energy, Quarterly Report.

Hanham, R., and Calzonetti, F. 1983. Regional and temporal trends in power plant unit siting, 1912-1978. The Professional Geographer 35: 416-26.

Harbough, M. 1977. Methane in Coal. Supply-Technical Advisory Task Force.

Osleeb, J., and Sheskin, I. 1977. Natural gas: A geographical perspective. The Geographical Review 67: 71-85.

Pasqualetti, M. 1983. Nuclear power impacts: A convergence/divergence schema. The Professional Geographer 35: 427-36.

Pijawka, D., and Chalmers, J. 1983. Impacts of nuclear generating plants on local areas. Economic Geography 59: 66-80.

Richetto, J. 1983. The location of electric energy-producing facilities: An examination of structuural and environmental components. Research Series in Geography and Regional and Urban Planning, no. 7, University of Alabama.

Richetto, J. 1984. Locating nuclear electric energy facilities: Structural relationships and the environment. In Nuclear power: Assessing and managing hazardous technology, eds. M. Pasqualetti and D. Pijawka, pp. 103-120, Boulder, Colorado: Westview Press.

Richetto, J., Barnard, J., and Glover, R. 1984. Utilization potential of coalbed methane gas in Alabama. Research Series in Geography and Regional and Urban Planning, no. 9, University of Alabama.

Satriana, N. 1980. Unconventional natural gas: Resources, potential and technology. New Jersey: Noyes Data Corporation.

Van Schonfeldt, H., et al. 1982. Production and utilization of coalbedmethane gas in Island Creek Coal Company Mines. Unconventional Gas Recovery Symposium, Proceeding, pp. 235-38.

York, T. 1979. Commercialization of coalbed methane. Methane Recovery From Coalbeds Symposium, Proceedings, pp. 17-22.

CHAPTER 9

PLANNING FOR NUCLEAR POWER PLANT ACCIDENTS: SOME NEGLECTED SPATIAL AND BEHAVIORAL CONSIDERATIONS[1]

James H. Johnson, Jr.,
Department of Geography,
University of California - Los Angeles,
Los Angeles, CA 90024

ABSTRACT. Federal regulations governing the operation of commercial nuclear power plants contain unrealistic assumptions about human spatial behavior in crisis situations which, in the event of a major reactor accident, are likely to result in inadequate and ineffective emergency preparedness and response. In this paper the extensive and substantive literature in geography, sociology, and psychology which deals with crisis decision-making and behavior, including pertinent studies conducted in the aftermath of the accident at Three Mile Island in 1979, are drawn on in an effort to (1) provide a realistic picture of the actual range of emergency spatial behaviors a major nuclear reactor accident is likely to evoke among both the general population and designated emergency personnel; (2) demonstrate the potential negative consequences of failing to plan for these emergency spatial behaviors; and (3) propose revisions to the existing federal regulations in order that radiological emergency preparedness and response plans might more accurately reflect the spatial and behavioral factors which appear to distinguish nuclear emergencies from other types of crises.

1. INTRODUCTION

Federal regulations governing the operation of commercial nuclear power plants contain unrealistic assumptions about human spatial behavior in crisis situations which, in the event of a major reactor accident, are likely to result in inadequate and ineffective emergency preparedness and response. In this paper the extensive and substantive literature in geography, sociology, and psychology which deals with crisis decision-making and behavior, including pertinent studies conducted in the aftermath of the near catastrophic accident at Three Mile Island (TMI) near Harrisburg, PA., in 1979, are drawn on in an effort to (1) provide a realistic picture of the actual range of emergency spatial behaviors a major nuclear reactor accident is likely to evoke

F.J. Calzonetti and B.D. Solomon (eds.), Geographical Dimensions of Energy, 123–154.
© 1985 by D. Reidel Publishing Company.

among both the general population and designated emergency personnel; (2) demonstrate the potential negative consequences of failing to plan for these emergency spatial behaviors and related factors, such as jurisdictional conflicts, which are likely to arise in a nuclear emergency; and (3) propose revisions to the existing federal regulations in order that radiological emergency preparedness and response plans might more accurately reflect the spatial and behavioral factors that appear to distinguish nuclear emergencies from other types of crises.

2. AN OVERVIEW OF PLANNING FOR NUCLEAR POWER PLANT ACCIDENTS

Until recently the federal government assigned a low priority to offsite emergency planning and preparedness for nuclear power plant accidents. Nuclear facilities were assumed to be designed and constructed with such a high level of integrity, including "multiple barriers ... to separate radioactive materials from man and the environment" (FEMA 1980, I-2), that the chances of a major reactor accident with offsite radiological consequences were extremely remote (NRC 1975). As a result of this defense-in-depth design and construction philosophy, in order to obtain an operating license, facility owners or managers were required to develop only onsite emergency response plans covering the 2-to-3 mile low population zone around their plants.

Despite the widely held belief that a major nuclear accident was a highly unlikely event, federal guidelines for the development of offsite emergency response plans in support of nuclear power plants were issued in 1973 (Office of Emergency Preparedness 1973; AEC 1973). A number of federal agencies, including the Atomic Energy Commission [predecessor of the Nuclear Regulatory Commission (NRC)], were instructed to assist state and local governments in their planning efforts. In 1974 the guidelines were revised and updated (AEC 1974). In particular,

> Leading agency responsibility for "reviewing and concurring" in state and local government radiological emergency response plans was assigned to the NRC and the planning assistance function was expanded to include accidents involving the transportation of radioactive materials (FEMA 1980, I-5).

In 1977 the guidelines were reprinted, and a supplement was added by the NRC which listed 70 essential elements offsite plans must have to earn NRC concurrence, i.e., federal approval (NRC 1977). State and local government emergency response plans were not required by law, however, and primarily for this reason only 11 of 34 states with operating nuclear power plants had NRC-approved offsite plans when the accident at TMI took place (FEMA 1980).

The state of Pennsylvania was not one of them.

The TMI accident and earlier incidents at nuclear facilities (e.g., the fire at Brown's Ferry Nuclear Power Station 2 in 1975) prompted the federal government to seriously reconsider the priority and status it had assigned to emergency preparedness and response planning. In its final report to the President on the accident at TMI, the Kemeny Commission recommended major changes in the emergency planning regulations governing the nuclear industry (Kemeny 1979). And in August, 1980, pursuant to a Presidential directive, revised guidelines were issued that promulgated, among other regulatory changes, federally approved state and local emergency response plans as a requisite for granting and maintaining a nuclear power plant operating license (NRC 1980). Under the revised and upgraded regulations, FEMA (the Federal Emergency Management Agency) was delegated main responsibility for offsite radiological emergency planning. The NRC retained primary responsibility for licensing nuclear power plants and for onsite emergency planning and was directed to assist FEMA, in conjunction with other federal agencies, with the coordination of state and local emergency planning and preparedness activities (FEMA 1982). FEMA and the NRC demonstrated their mutual agreement on the respective duties they were assigned under the revised regulations by jointly issuing a report that outlines the criteria to be used in both the preparation and evaluation of state and local governments' as well as the licensee's emergency preparedness response plans in support of nuclear power plants (NRC/FEMA 1980).

The revised emergency planning regulations purportedly build on the lessons learned from the TMI accident to ensure that "adequate protective measures can and will be undertaken in the event of [another] radiological emergency" (NRC 1980, 55402). They call for offsite plans covering two emergency planning zones (EPZs): a ten mile plume exposure pathway zone and a 50 mile ingestion exposure pathway zone (NRC/FEMA 1980). In devising the generic boundaries of these two EPZs, however, FEMA and the NRC appear to have relied solely on technical information related to the timing, release characteristics, and potential consequences of a range of possible reactor accidents that existed prior to the TMI crisis. The revised regulations stipulate, for example, that detailed planning encompassing a full range of protective actions, including evacuation, is only required for the resident, transient, and institutional population within the plume exposure pathway zone, "because the probability of large doses of radiation drops off substantially at about 10 miles from the reactor" (Collins, Grimes, and Galpin 1978, I-37). Planning for the 10-50 mile EPZ is limited to "the identification of major exposure pathways from contaminated food and water and the associated control and interdiction points and methods", in large part because reactor safety studies indicate that "much of any particulate material in a radioactive plume would have been deposited on the ground within about 50 miles of a facility" (NRC/FEMA 1980, 13). In revising and upgrading the regulations, FEMA and the NRC failed to consider the findings and recommendations of studies by geographers (Barnes, et al. 1979; Brunn, Johnson, and Zeigler 1979; Zeigler, Brunn, and Johnson 1981; Cutter and Barnes 1982) and other social scientists (Chenault, Hilbert and Reichlin 1979; Flynn 1979, 1981; Hu and Slaysman 1981; Kasl, et al. 1981 a, b; Maxwell 1982; Macleod 1981; Smith and Fisher 1981; Fischer 1981) that investigated the behaviors of both the general population and organizations that were expected to perform emergency roles during the TMI crisis.

3. SOME NEGLECTED SPATIAL AND BEHAVIORAL CONSIDERATIONS

In part as a consequence of FEMA and the NRC's failure to fully take into account the behavioral lessons learned from the TMI accident, the revised emergency preparedness and response regulations contain at least three unrealistic planning assumptions. They are

1. that the population within the vicinity of nuclear power plants will adhere to official prote action advisories in a nuclear reactor emergency;

2. that people who have been assigned emergency work roles will report promptly to perform these tasks;

   and

3. that pre-emergency education information and emergency notification messages will minimize, if not all together eliminate, the propensity for people to adopt extreme coping behaviors in the event of a radiological emergency.

In this section these assumptions are addressed within the broader context of the existing literature on crisis decision-making and behavior.

3.1 Public Behavior

Implicit in the revised emergency planning regulations is the notion that people's behavior in a nuclear emergency will not be unlike reactions to natural disasters (e.g. floods and hurricanes) and non-radiological technological emergencies (e.g., accidental explosions and transport accidents involving toxic materials) (Jaske 1983). The evidence amassed over the last three decades in studies of these types of crises indicate that, although people initially "react in a stunned and bewildered fashion for varying periods on time" (Cohen and Ahearn 1980, 32), most adhere to official protective action advisories, especially in crisis situations where there is ample intervening time before disaster strikes (Chapman 1954; Baker and Chapman 1962; Grosser, Weschsler, and Greenblatt 1964; Burton, Kates, and White 1978; White 1974; White and Haas 1975; Foster 1980; Perry, Lindell, and Green 1981; Leik, Carter, and Clark 1981; Cornell 1982). In other words, in non-radiological disasters a majority, if not all, of the potentially impacted population generally look to emergency response officials for guidance and assistance and only a minimum, if any, of extreme behavior (i.e., over-reaction and under-reaction) occurs (Cornell 1982). Studies by Perry (1979) and others (e.g., Christensen and Ruch 1980) have shown that compliance with protective action recommendations in natural disasters is related to, among other factors, the type, content, and frequency of the warnings and to the potentially impacted population's ability to confirm this emergency information through direct sensory evidence (e.g., the winds associated with a hurricane) and/or social and family ties. Research by disaster psychologists (Wolfenstein 1957; Cohen and Ahearn 1980; Smith 1983) indicate that the minority who adopt extreme coping behaviors in natural disasters tend to be either anxiety prone under normal conditions or have recently experienced some type of personal crisis.

Owing largely to the high degree of fear (Nowotny 1976; Otway 1977; Cook 1982) and distrust (Hohenemser, Kasperson, and Kates 1977; Kasperson, et al. 1980) of nuclear power in the U.S., however, it appears unreasonable for emergency management officials not to anticipate significantly more extreme public behavior in the event of a radiological emergency than has been observed in other types of crises. Previous research indicates that nuclear power is viewed with more dread than any other activity except terrorism and warfare (Slovic and Fischhoff 1983). Factors contributing to this state of affairs include the following: (1) peaceful uses of nuclear power, such as the generation of electricity, are associated with the ability to develop nuclear weapons (Cook 1982; Kasperson, et al. 1980; Office of Technology Assessment 1984); (2) major problems associated with the siting, design, construction and operation of commercial nuclear power plants, which often are cited as evidence of the federal government's inability to protect the public from the hazards of nuclear power (Cook 1982; National Academy of Sciences 1979; Nelkin 1981; Otway 1977; Udell 1982; Sheridan 1980; Mayo, et al. 1980; Otway and Thomas 1978; Mackenzie 1984; Office of Technology Assessment 1984); and (3) the unique characteristics of the disaster agent - radiation - which the public associates with, among other things, birth defects and cancer (Slovic, Fischhoff, and Lichtenstein 1979; Slovic and Fischhoff 1983; Lifton 1968; Perry 1983). Unlike natural and other technological disaster agents, such as flood waters and noxious gases from accidental explosions, radiation is an ambiguous hazard (Erikson 1982); for the most part it is imperceptible to the senses (except, of course, in doses sufficient to induce radiation sickness). In large part for these reasons, warnings of an impending nuclear emergency are likely to generate significantly more stress in humans than emergency notifications of other types of disasters. Support for this view include studies conducted by psychiatric epidemiologists and psychologists which reveal a high degree of prolonged stress among residents within the vicinity of the crippled TMI reactor (Baum, et al. 1981; Cunningham 1982; Dohrenwend, et al. 1981; Kasl, et al. 1981a; Bromet, Schulberg, and Dunn 1982; Houts and Goldhaber 1982; Goldstein and Schorr 1982; Schwebel and Schwebel 1981; Fabrikant 1983).

In planning for nuclear reactor accidents, therefore, emergency response planners should anticipate a high level of "hypervigilant" behavior (Janis 1962; Janis and Mann 1977); that is, for large numbers of people to "take protective and more drastic actions sooner than the situation may warrant" (Johnson and Zeigler 1983, 387). They should also expect a small but nonetheless significant amount of population underreaction. It is highly likely that, as in other kinds of crises, some people will purposefully ignore the warnings of a nuclear emergency and continue normal activities, rationalizing that the threat will not materialize (denial reaction) or that they will be safe even if it does (defense avoidance) (Janis 1962). It is also very likely that some people, not unlike in natural disasters, will freeze, go numb, or become immobile and, although they may wish to comply with the official protective action advisory, these individuals will be incapable of doing so because of the "psychic numbing" effect of the warnings (Lifton 1968, 1982).

Surveys of actual behavior during the TMI crisis and intended behavior in the event of an accident at the Shoreham Nuclear Power Station (SNPS) in Suffolk County, New York, offer strong empirical support to the notion that a nuclear emergency is likely to generate a substantial amount of extreme public behavior. During the TMI crisis, the governor of Pennsylvania advised pregnant women and preschool children within 5 miles of the crippled reactor to evacuate, and everyone else within 10 miles to shelter or stay indoors. This protective action advisory should have precipitated the evacuation of only about 2500 people. However, it has been estimated that 144,000 people, or 39% of the population within 15 miles of the plant (Flynn 1979, 1981), and 9% of the population in three communities approximately 25 miles from the site (Brunn, Johnson, and Zeigler 1979), actually evacuated. Chenault, Hilbert, and Reichlin (1979) characterized the process whereby individuals and families evacuated the TMI area even though they were not advised to take such protective actions as spontaneous evacuation. Analyzing human responses to the accident from a geographic perspective, Zeigler, Brunn, and Johnson (1981) termed the tendency for people beyond the designated danger zone to over-react by spontaneously evacuating as the evacuation shadow phenomenon. As indicated earlier, this type of behavior has occurred in other emergency situations, but not to the order of magnitude observed during the TMI crisis. In fact, the gap between expected and actual behavior was so large at TMI that Chenault, Hilbert, and Reichlin recommended that

> The problems of spontaneous evacuation and anticipating public behavior and response should be further analyzed ... state and local agencies need methods for estimating the flow and extent of spontaneous evacuation movements from the time a crisis begins (Chenault, Hilbert, and Reichlin 1979, x).

Not only was the gap between actual and expected behavior at TMI much greater than one would have anticipated given the protective action advisory, but the evacuees tended to travel distances much greater than had been observed in previous studies of evacuation behavior. According to Hans and Sell's (1974) study of events requiring evacuation from 1960-1973, the longest evacuation on record was in response to Hurricane Carla in 1961 where the evacuees travelled a median distance of 80 miles. At TMI, the median distance travelled by evacuees was determined to be 85 miles in one study (Zeigler, Brunn, and Johnson 1981) and 100 miles in another (Barnes, et al. 1979). A fairly strong directional bias was also evident in the pattern of evacuation. Most of the evacuees displayed a preference for destinations in the mountains and in sparsely settled communities north and west of the plant, in all probability reflecting a desire to place a physical barrier between themselves and the crippled reactor and to avoid the possibility of having to subsequently evacuate from population centers to the east which are downwind of the plant and more densely settled (Zeigler, Brunn, and Johnson 1981).

At the same time this high level of public over-reaction in the form of spontaneous evacuation was taking place, some under-reaction also occurred at TMI. Hu and Slaysman (1981) estimate, for example, that about 1% of the families within five miles of the plant in which a pregnant woman was present did not evacuate.

In light of observed public behaviors during the TMI crisis, Suffolk County, New York, sponsored a social survey to ascertain how the residents of Long Island are likely to behave in the event of an accident at the SNPS (Social Data Analysts 1982a), which is located approximately 60 miles east of New York City on the north shore of the island (Figure 1). Designed by an interdisciplinary team of social and behavioral scientists, the survey elicited responses to, among other questions, three reactor accident scenarios which are summarized at the bottom of Table 1. In devising the scenarios, the objective was to discover how the Long Island population is likely to respond to nuclear reactor accidents of different degrees of severity. Toward this end, the TMI protective action advisory was used as the benchmark scenario and then scenarios postulating accidents less severe (scenario 1) and more severe (scenario 3) than that accident were developed.

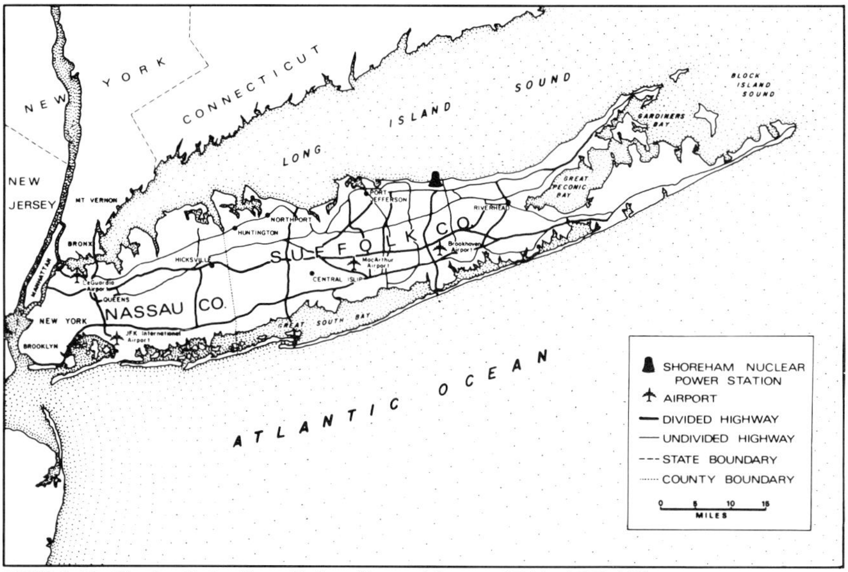

FIGURE 1. The Shoreham Nuclear Power Station. (Source: Johnson and Zeigler (1983, 388). Reprinted by permission of Economic Geography.)

Table 1 summarizes the responses to the three scenarios. The results strongly support the proposition that people will not adhere to official protective action advisories during a radiological emergency. The stated behavioral intentions of the 2595 survey respondents, averaged across the three scenarios, indicate that only 27% of the households on Long Island (approximately 217,000 families) are likely to follow official protective action recommendations if an accident occurs at the SNPS. Some would under-react, but most would over-react, primarily by evacuating voluntarily or spontaneously. Disaggregated by distance from the plant, the survey results indicated that, depending on the scenario, one fourth (215,000 families) to nearly one half (430,000 families) of the households on Long Island are likely to evacuate, casting an evacuation shadow extending in excess of 50 miles from the plant (Johnson and Zeigler 1983; Zeigler and Johnson 1984; Johnson and Zeigler 1984; Johnson 1984a).

TABLE 1. Propensity of Long Island Residents to Follow Instructions, Underreact, and Overreact in the Event of a Radiological Emergency at the Shoreham Plant

Behavioral Intention	I	II	III	Average Response
Follow Orders	31.0	31.6	19.0	27.2
Under-React	.05	1.4	.68	.7
Over-React	66.0	63.8	78.0	69.2
Don't Know	3.0	3.2	3.0	3.1
Number of Families Not Following Orders	569,907	593,337	683,801	615,023
% of all Families	65.7	68.4	78.7	70.9

Source: Johnson and Zeigler (1983, 392). Reprinted by permission of Economic Geography.

Scenario I: all people within 5 miles advised to stay indoors;

Scenario II: all pregnant women and pre-school children within 5 miles advised to evacuate and everyone else within 10 miles to stay indoors;

Scenario III: everyone within 10 miles advised to evacuate.

The results of the Shoreham survey are judged to be reasonably accurate estimates of actual behavior in the event of a real radiological emergency at the SNPS for the following reasons. First, multivariate statistical analyses of the survey data revealed that the residents' stated behavioral intentions are intricately related to the locational characteristics (i.e., distance and direction) of their homes in relation to the plant, their stage in life cycle, and their perceptions of the risks associated with nuclear power plants (Johnson and Zeigler 1983). The fact that these variables satisfactorily discriminated among those who are most likely to follow orders, those most likely to underreact, and those most likely to overreact in the event of a radiological emergency at the SNPS suggests strongly that the residents of Suffolk and Nassau counties were not responding casually or thoughtlessly to the survey (Johnson 1984a). Second, the behavioral intentions of Long Island residents closely parallel the actual behaviors of TMI area residents approximately five years earlier under analogous conditions (Johnson and Zeigler 1984; Zeigler and Johnson 1984). Third, the theory of "reasoned action" (Ajzen and Fishbein 1980) posits that a person's intended behavior is the best predictor of his or her actual behavior as long as the intention does not change before the individual has a chance to act out the behavior. Given the controversial and violent social history of nuclear power (Hohenesmer, Kasperson, and Kates 1977; Kasperson, et al. 1980; Cook 1982), it is highly unlikely that pre-existing perceptions, beliefs, and attitudes toward nuclear power (i.e., the factors which previous research (Johnson and Zeigler 1984) has shown to be the major determinants of people's behavioral intentions in the event of a nuclear emergency), can be easily changed (Slovic, Fischoff and Lichtenstein 1980b).

In sum, the data from TMI and Shoreham suggest that, if implementable emergency preparedness and response plans for commercial nuclear power plants are to be developed, strategies or methods must be devised to either accommodate or mitigate extreme public behavior, especially the propensity of households to overreact by spontaneously evacuating. If extreme public behavior is <u>not</u> taken into account, then the traffic conditions that are likely to exist during a radiological emergency cannot be adequately assessed and the evacuation time estimates which the regulations require licensees to devise for the 10 mile plume exposure pathway zone will most likely be grossly inaccurate.

In a recent study prepared for the Long Island Lighting Company, KLD Associates (1983) attempted to assess the effects of, among other factors, the evacuation shadow phenomenon (i.e., the propensity for people to evacuate voluntarily or spontaneously) on an official order to evacuate the 10 mile EPZ of the Shoreham plant, in the event of a major accident. Specifically, evacuation time estimates were derived for nine hypothetical cases: planned and uncontrolled evacuation of the 10 mile EPZ under normal and inclement weather conditions, assuming no evacuation shadow; planned and uncontrolled evacuation under normal weather conditions, assuming 25% and 50% spontaneous evacuation; and uncontrolled evacuation under inclement weather conditions, assuming 50% spontaneous evacuation. Table 2 lists KLD's evacuation time estimates for six of the hypothetical cases; it shows that, in the event of a major reactor accident at the SNPS, the evacuation shadow phenomenon will increase the evacuation time for the 10 mile EPZ (approximately 31,000 families) by only one to two hours, depending on weather conditions and the type of evacuation. KLD concludes on the basis of these estimates that the additional traffic generated by spontaneous evacuees will <u>not</u> significantly encumber efforts on the part of emergency management officials to quickly evacuate the plume exposure pathway zone population.

TABLE 2. Evacuation Time Estimates for the Shoreham 10 Mile EPZ

Scenario	Type of Evacuation	Weather Conditions	Evacuation Time (Hours-Minutes)
No Evacuation Shadow	Uncontrolled	Normal	6 - 10
50% Evacuation Shadow			7 - 15
No Evacuation Shadow	Uncontrolled	Inclement	7 - 35
50% Evacuation Shadow			9 - 45
No Evacuation Shadow	Planned	Normal	4 - 35
50% Evacuation Shadow			6 - 15

Source: KLD Associates, 1983.

KLD's evacuation time estimates are believed to be grossly inaccurate, however, because the analyses rest on unrealistic assumptions about human behavior in crisis situations. First of all, KLD assumed erroneously that all people within the 10 mile EPZ will respond promptly to an official evacuation advisory. Not unlike human behavior in other crisis situations, including the TMI accident, the Shoreham evacuation survey indicates that, in the event of an accident requiring a full scale evacuation of the plume exposure pathway zone, approximately 20% of the households within 10 miles of the SNPS (6000 families) would ignore the evacuation advisory (Zeigler and Johnson 1984; Johnson and Zeigler 1984). Getting these underreactors to evacuate the danger zone will increase the evacuation time significantly.

Secondly, KLD assumed, unrealistically, that the effect of the evacuation shadow will be limited to within 20 miles of the plant. Results of the Shoreham survey also indicate that, in the event of a major accident at the plant, 45% of the households beyond 20 miles are likely to evacuate. A major accident at the SNPS, as indicated above, will most likely precipitate the development of an evacuation shadow extending in excess of 50 miles from the site (Johnson 1984a). Voluntary evacuation, in short, will be geographically more extensive than KLD assumed and this, too, will increase the plume exposure pathway evacuation time.

Thirdly, implicit in KLD's analyses is the incorrect assumption that the evacuation shadow is composed of randomly dispersed households of voluntary evacuees. Analyses of the Shoreham evacuation survey data indicate, to the contrary, that the percent of potential spontaneous evacuees declines with increasing distance from the SNPS, especially beyond the ten mile radius, which is not unlike the spatial behavior of TMI area residents (Zeigler, Brunn, and Johnson 1981). This means that the majority of the spontaneous evacuees will originate closer to rather than farther from the 10 mile EPZ. High levels of traffic congestion immediately outside of the 10 mile plume exposure pathway zone would severely encumber evacuation of the people closest to the plant who are in the greatest danger of radiation exposure.

Finally, in attempting to assess the impact of the evacuation shadow, KLD failed to incorporate in its traffic simulation model estimates of the number of vehicle accidents, mechanical failures, and fatalities which can be expected to occur <u>outside</u> of the 10 mile EPZ, and to determine the impact of these on evacuation time. Given the potential magnitude and geographic extent of spontaneous evacuation in case of an accident at the SNPS (Johnson 1984a), it is unrealistic not to expect a large number of vehicle mechanical breakdowns and other traffic related incidents in the evacuation shadow which, in turn, would significantly impede the flow of traffic out of the plume exposure pathway zone.

In sum, the behavioral input needed to devise realistic evacuation time estimates for the 10 mile EPZ of commercial nuclear power plants includes estimates of (a) the proportion of the population within the plume exposure pathway zone who may be inclined to ignore official evacuation orders (underreactors), (b) the magnitude and geographic extent of spontaneous evacuation (overreactors), and (c) the number of traffic related incidents likely to occur not only <u>inside</u> but also <u>outside</u> of the 10 mile EPZ. To derive realistic evacuation time estimates, it is especially important that a distance decay function, that is, a measure of the rate by which the proportion of voluntary evacuees is likely to decline as distance from the site increases, be incorporated in the evacuation simulation model. Otherwise, as in the analyses by KLD (1983) and other researchers (Sheffi, Mahmassani, and Powell 1981; Sheffi 1982; Walsh, et al. 1983), the evacuation time estimates will be grossly understated and thus an inadequate basis for arguing that evacuation is a feasible protective action in the event of a major reactor accident.

Preliminary analyses for the SNPS by PRC Voorhees (Polk 1983), which take into account spontaneous evacuation and the other factors mentioned above, yielded evacuation time estimates for the 10 mile EPZ in the 12 to 18 hour range (depending on weather conditions). Ignoring the evacuation shadow phenomenon, Vorhees' evacuation time estimates were substantially lower, between four and six hours. Clearly, the evacuation shadow phenomenon must be taken into account in planning for a radiological emergency.

3.2 Emergency Personnel Behavior

In revising the radiological emergency preparedness and response regulations, FEMA and the NRC also assumed that, if a major reactor accident were to occur, people who have been assigned emergency work roles will report promptly to perform those duties. No consideration was given to the potential impact of multiple group membership on the behavior of designated emergency personnel in crisis situations. Previous studies have shown that when disaster - natural or human induced - strikes, some emergency workers may experience <u>role conflict</u>, a situation in which they are torn between multiple group membership loyalties. As Killian noted more than three decades ago

> When catastrophe strikes a community, many individuals find that latent conflict between ordinarily nonconflicting group loyalties suddenly becomes apparent and that they are faced with the dilemma of making an immediate choice between various roles (1952, 310-11).

His research in four southwest U.S. communities strickened by disasters revealed that dilemmas of loyalty may arise in crisis situations among the following roles: family <u>vs</u> community, rescue worker <u>vs</u> occupation, workplace <u>vs</u> fellow employees, and community <u>vs</u> extra community groups. Among these potentially conflicting group loyalties, Killian found that the most common conflict reported was loyalty to family versus other obligations and that this dilemma, in most instances, was resolved in favor of the health, safety, and welfare of the family. Other disaster studies corroborate this finding. It is well established, for example, that families generally evacuate as a unit in emergencies (Drabek 1969; Drabek and Boggs 1968; Perry, Greene, and Lindell 1980; Perry, Lindell and Greene 1981; Smith 1983).

Depending on the type and timing of the event, then, role conflict may result in either of two responses on the part of emergency workers: <u>delayed response</u>, where the individual reports for duty only after ascertaining either through direct or indirect contact that family members are safe, or <u>non-response</u>, where the individual relocates family members from the danger zone and stays with them for the duration of the crisis. However, previous studies of organization behavior in both natural and non-nuclear, technological disasters suggest that emergency workers rarely, if ever, abandon their roles in crisis situations (Quarantelli and Dynes 1978; Dunning and Silva 1980).

But, contrary to the findings of studies of organization behavior in non-nuclear disasters, accounts of emergency personnel behavior in the TMI crisis suggest that role conflict may seriously encumber emergency response efforts if a nuclear reactor accident occurs in the future (DeMuth and Miller 1982; DeMuth and Trautlein 1979; Breo 1979; Haglund 1979; Strohl 1979; Chenault, Hilbert, and Reichlin 1979; Smith and Fisher 1981; Maxwell 1982; Weidner, et al. 1980; Macleod 1981; Kuntz 1979; Kasl, Chisholm, and Eskenazi 1981b). Describing the situation at area hospitals during the TMI accident, for example, Smith and Fisher noted that "during the [hospital emergency response] planning process, a new problem arose - the exodus of people included physicians, nurses, and technicians required to staff both the short term and long term medical facilities" (1981, 1656). Maxwell, in another analysis and evaluation of hospital emergency planning and response during the TMI accident, stated that "... the conflicting responsibility to family and work resulted in escalating staffing problems as the crisis continued" (1982, 276). Some of the local hospitals' personnel moved their families outside of the danger zone and returned to work for extended periods of time (delayed response). Others left the area and stayed away until the immediate crisis was over (non-response). At one local hospital, for example, only six of the 70 physicians who were scheduled for weekend emergency duty reportedly showed up for work (Maxwell 1982, 278). On the basis of these observed behaviors, Maxwell concluded that, in case of another radiological emergency, "administrators can expect significant absences from staff members who have family responsibilities and should anticipate a shortage of physicians as well" (1982, 276).

Hospital personnel were not the only group who experienced role conflict during the TMI crisis. Problems of multiple group membership loyalty, especially between work and family obligations, reportedly also occurred among the Pennsylvania National Guard (1979) and TMI nuclear power plant workers (Kasl, Chisholm, and Eskenazi 1981b). In a comparative study of, among other issues, the behavioral responses of supervisory and nonsupervisory personnel at TMI and the nearby Peach Bottom Nuclear Power Plant (40 miles away), Kasl and colleagues noted

> ... The greater conflict experienced and recalled by TMI workers: their felt need to be in different places at the same time, the intrusion of work demands on their overall planning of how to react to the accident, and the specific conflict with spouse over arriving at an agreed upon decision regarding the family's response to the accident ... for nonsupervisory TMI workers, the overwhelming majority (about 90 percent) of instances of conflict involved the work role <u>vs</u> the family role; for TMI supervisors, some 25 percent of instances of conflict involved two or more aspects of

their work demands (Kasl, Chisholm, and Eskenazi, 1981b, 478).

Kasl and colleagues estimated that 4% and 11% of TMI supervisory and nonsupervisory personnel, respectively, evacuated during the crisis. Most of those who left lived within 5 miles of the plant, the designated zone of selective evacuation during the reactor crisis.

Additional evidence of the extent to which role conflict is likely to be a problem in a radiological emergency is derived from social surveys of school bus drivers and volunteer fire fighters within the vicinity of the Shoreham plant (Social Data Analysts 1982b), and of public school teachers in the San Luis Coastal Unified School District near the Diablo Canyon Nuclear Power Plant in California (Johnson 1983). The emergency response plans for the respective nuclear power plants assign these groups emergency work roles which they are expected to perform if a major reactor accident were to occur.

In each of the three surveys, the respondents were asked what do you think you would do first if an accident requiring a full scale evacuation of the population within 10 miles of the plant were to occur? Consistent with role conflict theory, the behavioral options were (a) perform your emergency work role, (b) make sure your family was safely out of the evacuation zone, (c) leave the evacuation zone immediately to make sure you were in a safe place, or (d) do something else. The results of the three surveys are summarized in Table 3.

Table 3. Initial Reactions of Selected Emergency Personnel to An Evacuation Advisory[a]

Behavioral Intentions	School Teachers (n=232)	Bus Drivers (n=246)	Volunteer Firemen (n=291)
Perform Emergency Work	67	24	21
Check on Family	24	73	68
Leave the Area	1	3	1
Do Something Else	8	-	11

Source: Zeigler, Johnson, Brunn (1983, 81). Reprinted by permission of the Association of American Geographers.
[a]Data in percentages.

Clearly evident in the results, as role conflict theory postulates (Killian 1952), is the importance of family ties or obligations in crisis situations. Over two-thirds of the volunteer firefighters (68%) and nearly three fourths (73%) of the school bus drivers in Suffolk County indicated that, if a major reactor accident were to occur at the Shoreham plant, family obligations would take precedence over their duties as emergency workers (Social Data Analysts 1982b). By contrast, less than one fourth of the school teachers in San Luis Obispo County, California, indicated that they would check on family first. Most said they would first help to evacuate school children from the designated danger zone. A significant proportion of this group qualified their responses, however, by stating that participation would be (a) contingent upon being able to contact family members by telephone, (b) restricted to the evacuation of their class only, or (c) limited to a specified length of time. These "intangible" or "hard to quantify" measures of intended behavior suggest that role conflict may be even more of a problem among public school teachers than the raw numbers in Table 3 indicate for the following reasons. First, it is unlikely that teachers will be able to contact family members by telephone during a nuclear power plant accident. Telephone exchanges are likely to be overloaded, as in the TMI crisis when more than two million calls were attempted on a system designed to handle only half as many (Chenault, Hilbert, and Reichlin 1979). Under such conditions, it is not inconceivable that those teachers whose assistance is predicated upon being able to contact family members by phone would leave immediately, since rarely, if ever, have emergency personnel reported for duty without first contacting family members (Erikson 1982). Second, an efficient and effective evacuation of schools would require the cooperation of all parties involved until the last child is safely relocated outside of the designated danger zone and released to his/her parents or guardian (Johnson 1983). Teachers who are willing to evacuate only their class, or to participate for only a limited period of time could severely hamper and complicate evacuation efforts, and thereby further endanger the health, safety, and welfare of the school children.

In sum, the data presented in this section lead to the conclusion that emergency planners can expect some emergency personnel, in particular those with strong family ties who live within the plume exposure pathway zone (Henderson 1981; Lathrop 1980), not to be available immediately, if at all, to provide assistance and guidance in the event of a radiological emergency. In their analysis and evaluation of organization behavior in the TMI crisis, Chenault, Hilbert, and Reichlin (1979) concluded that, if a radiological emergency occurs in the future, administrators of institutions (hospitals, nursing homes, schools, day care centers, etc.) may opt to evacuate early - perhaps at the initial warning stage, to advert the staffing problems that are most likely to develop if they wait until an anticipated official evacuation order is issued. They note further that "early or premature evacuation of institutions can serve to trigger large movements by the general public..." (Chenault, Hilbert, and Reichlin 1979, 139).

3.3 The Role of Emergency Information

Also implicit in the revised emergency planning regulations is the notion that, not unlike human behavior in other crisis situations, extreme public behavior will be minimal, if it occurs at all, in a nuclear reactor emergency primarily because the population within the vicinity of commercial nuclear power plants will have been exposed to two types of emergency information. The first of these is pre-emergency education information which the revised regulations require licensees, in conjunction with state and local governments, to distribute (at least annually) to the population within the plume exposure pathway zone of commercial nuclear power plants.

> This information shall include, but not necessarily be limited to : (a) educational information on radiation; (b) contact for additional information; (c) protective measures, e.g., evacuation routes and relocation centers, sheltering, respiratory protection, radio-protective drugs; and,(d) special needs of the handicapped (NRC/FEMA 1980, 49).

The second is emergency notification messages, that is, information which is to be issued at the onset and during a reactor crisis. The revised regulations require

> the licensee in conjunction with state and local organization [to] establish the contents of the initial [and follow-up] emergency messages to be sent from the plant. These messages shall contain information about the class of emergency, whether a release is taking place, potentially affected population and areas, and whetherprotective measures may be necessary (NRC/FEMA 1980, 43).

Additionally, the upgraded regulatory guidelines stipulate that

> State and local government organizations shall establish a system for disseminating to the public appropriate information contained in initial and follow-up messages received from the licensees including the appropriate notification to broadcast media, e.g., the Emergency Broadcast System (EBS) (NRC/FEMA 1980, 45).

It is unreasonable to assume that these two sources of emergency information will minimize or eliminate extreme public behavior, especially the propensity of households to spontaneously evacuate, in the event of a radiological emergency for the following reasons. First, the licensees are not required under the revised regulation to distribute pre-emergency education information in communities where spontaneous evacuees are likely to originate in large numbers. These communities lie outside of the ten mile emergency planning zone (see Figure 12.5 in Johnson and Zeigler 1984, 289). Second, even if such information were widely disseminated, previous research has shown that special programs designed to educate people about simultaneously low probability and high risks events have not been particularly successful in changing pre-existing perceptions, beliefs, attitudes and behavioral predispositions (Mileti, Drabek, and Haas 1975; Saarinen 1979; Sims and Baumann 1983; Sorensen 1983). With respect to nuclear power, moreover, the evidence suggests that where public opposition is widespread, emergency education information most likely will be rejected or dismissed, characterized as inaccurate, erroneous, or unrepresentative (Slovic, Fischhoff, and Lichtenstein 1980a). Third, recent surveys conducted within the vicinity of several nuclear power plants indicate that a substantial proportion of the households would have little or no faith in emergency information issued during a real radiological emergency, regardless of whether it would come from utility company representatives or public officials (Social Data Analysts 1982a; Yankelovich, Skelly, and White 1983; also see Gricar and Baratta 1983). As in the case of the TMI crisis (Kemeny et al. 1979; Brunn, Johnson, and Zeigler 1979; Flynn 1979, 1981; Cutter and Barnes 1982), it is therefore very likely that the decision with respect to the most appropriate protective action(s) to be taken will be strongly influenced by a range of other sources of information (e.g., friends, relatives, and the news media), whose analyses and interpretations of the events surrounding a reactor accident may be extremely unreliable, biased, and /or contradictory (The Media Institute 1979, 1980a), thus leading to what Dupont refers to as irrational "phobic reactions" (The Media Institute 1980b).[2]

4. A REVISED PLANNING APPROACH

The behavior of the public and members of selected organizations who were expected to perform emergency roles during the TMI crisis, combined with the evidence amassed in social surveys of the behavioral intentions of both the general population and designated emergency personnel around other nuclear power plants, suggest strongly that offsite emergency plans developed according to the guidelines established by FEMA and the NRC are likely to result in ineffective and inadequate preparedness and response if a nuclear reactor accident occurs in the future. More specifically, the evidence points clearly to the need for a revised planning approach which, at the very least, takes into account the following behavioral considerations : (1) that a proportion (probably very small) of the population within the designated danger zone (defined on the basis of technical criteria) will not adhere to official protective action advisories; (2) that a substantial proportion of the technically not-at-risk population, defined as such as a result of either their stage-in-life cycle or where they live in relation to the plant, are likely to evacuate voluntarily or spontaneously; and (3) that some designated emergency personnel may not report promptly, if at all, to perform assigned emergency duties.

These and related considerations, especially the distance decay evacuation intention curves for the SNPS (see Figure 12.2 in Johnson and Zeigler 1984, 285), suggest strongly that detailed emergency planning should be mandatory for the 50 mile ingestion exposure pathway zones of commercial nuclear power plants. As Figure 2 shows, however, requiring detailed planning within a 50 mile radius, instead of for the more limited 10 mile plume exposure pathway zone as the current regulations stipulate, would result in overlapping planning zones in areas with high concentrations of nuclear power plants (e.g., along the North Atlantic Seaboard, around the Great Lakes, and in the Carolina Piedmont) (Zeigler, Johnson, and Brunn 1983, 73-75). For these areas more generalized emergency planning regions might be established as in Figure 3, which also identifies major cities that could serve as regional headquarters from which command and control activities would be handled in the event of a radiological emergency.

Expanding the emergency planning zone to 50 miles would increase the chances of recruiting personnel who most likely would report promptly to carry out emergency work roles: primarily older individuals without close family ties who live outside of the plume exposure pathway zone (Johnson 1983). Such an expanded planning zone would also advert the types of jurisdictional conflicts which nearly emerged during the TMI nuclear emergency (Chenault, Hilbert, and Reichlin 1979). Cumberland County, located about 10 miles from TMI, established plans to barricade five roads leading into the county from neighboring Dauphin and York counties which are closest to the plant. Had the situation at the plant deteriorated to a level that required an extension of the evacuation zone out to 10 miles, this plan, by prohibiting the flow of traffic through Cumberland County, would have substantially increased the probability of residents closest to the plant being exposed to potentially dangerous radiation (Zeigler, Johnson, and Brunn 1983). In addition, extending the detailed planning region to 50 miles would allow emergency management officials to formally incorporate in offsite plans the host

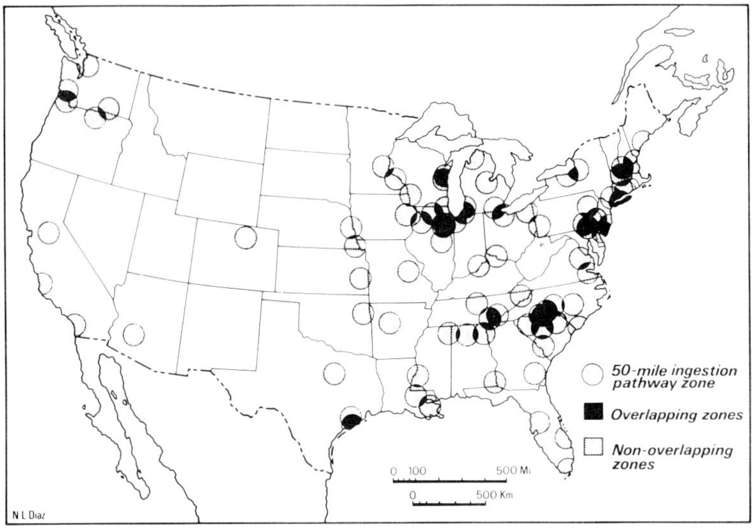

FIGURE 2. Ingestion exposure pathway zones of U.S. nuclear reactors. (Source: Zeigler, Johnson, and Brunn (1983, 75). Reprinted by permission of the Association of American Geographers.)

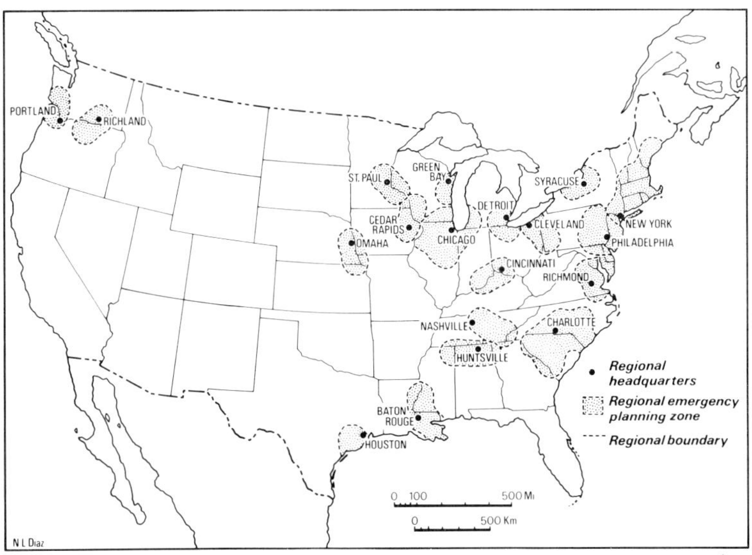

FIGURE 3. Proposed emergency planning regions. (Source: Johnson (1984b, 119). Reprinted by permission of John W. Frazier, editor, Papers and Proceedings of Applied Geography Conferences, Vol. 7.)

communities and institutions to which the population in institutions within the designated danger zone (hospitals, nursing homes, schools, etc.,) would be transferred in the event of a nuclear emergency.

Previous research indicates that the public and technical experts possess radically different perceptions of the risks associated with nuclear power plants (Thomas, et al. 1980; Otway 1977; Lichter and Rothman 1983; Cohen and Lichter 1983; Otway and Thomas 1978; Manning 1982; Firebaugh 1981). Within each of the EPRs proposed here, therfore, surveys that have been carefully designed by social and behavioral scientists to assess public perceptions of the risk associated with nuclear power plants, like the one administered to the Long Island population (Social Data Analysts 1982a), should be conducted and the results in turn evaluated vis a vis the technical emergency planning criteria established by the NRC and FEMA. As Lindell and Perry point out "These comparisons would ... provide an understanding of the distances from a reactor at which a given proportion of a population would believe themselves to be at risk and, thus, inclined to undertake protective actions" (1983, 50). More specifically, the survey data "would provide a basis for estimating the number of people that would take shelter or evacuate, even though such protective actions were not recommended for them" (Lindell and Perry 1983, 50). As indicated above, by incorporating data on population underreaction and overreaction in the evacuation time models, more realistic estimates of the time it would take to evacuate the plume exposure pathway zone could be developed. In addition the survey results could be used to identify optimal locations for emergency relocations centers. At TMI, the emergency relocation center established at Hershey, PA (approximately 11 miles from the site) was grossly under-utilized, partly, in all probability, because it was perceived to be too close to the threatening reactor (Zeigler, Brunn, and Johnson 1981).

In conclusion, in the absence of sufficient empirical evidence pertaining to actual human behavior in radiological emergencies, the alternative approach outlined briefly here, if adopted, would go a long way toward simultaneously dealing with the behavioral issues which heretofore have been ignored by the federal government and establishing more realistic criteria for evaluating radiological emergency preparedness and response plans. The proposed approach is also justifiable economically since utility companies within the generalized EPRs could share the costs of emergency planning. This concept should appeal to licensees given the financial problems that plague the nuclear industry at present (Staff Report 1984).

NOTES

[1] This is a substantially revised and expanded version of a paper which was presented at the 7th Annual Applied Geography Conference, held at Florida State University, Tallahassee, Florida, November 1-3, 1984 (Johnson 1984b). I am especially indebted to Ms. Althea Silvera for editorial assistance, and to Donald Zeigler and an anonymous reviewer for their comments on an earlier draft of this paper.

[2] In large part on the basis of evidence accumulated in studies of human behavior in natural disasters, Sorensen (1984) has recommended a number of steps which might be taken to improve emergency information dessimination in a nuclear reactor crisis; however, the extent to which these steps will minimize the propensity of individuals and families to adopt extreme coping behaviors is unknown.

REFERENCES

AEC, 1973. Guide and checklist for the development and evaluation of state and local government radiological emergency response plans in support of fixed nuclear facilities. WASH-1293, Nov. 1. Available at NRC Public Documents Room.

AEC, 1974. Guide and checklist for the development and evaluation of state and local government emergency response plans in support of fixed nuclear facilities WASH-1293-Rev. 1, Dec. 1. Available at NRC Public Documents Room.

Ajzen, I., and Fishbein, M. 1980. Understanding attitudes and predicting social behavior. Englewood Cliffs, NJ: Prentice Hall.

Baker, G.W., and Chapman, D.W. ed. 1962. Man and society in disaster. New York: Basic Books, Inc.

Barnes, K., et al. 1979. Responses of impacted population to the Three Mile Island nuclear reactor accident: an initial assessment. New Brunswick, NJ: Department of Geography, Discussion Paper No. 13, Rutgers University.

Baum, A. et al. 1981. Chronic stress and acute stress associated with the Three Mile Island accident and decontamination: preliminary findings of a longitudinal study. Technical report submitted to the U.S. Nuclear Regulatory Commission.

Breo, D.L. 1979. Nuclear scare tests hospital's disaster plan. Hospitals, J.A.H.A. May 1:33.

Bromet, E., Schulberg, H.C., and Dunn, L. 1982. Reactions of psychiatric patients to the Three Mile Island nuclear accident. Archives of General Psychology 39:24-40.

Brunn, S.D., Johnson, J.H., Jr, and Zeigler, D.J. 1979. Final report on a social survey of Three Mile Island area residents. East Lansing, MI: Department of Geography, Michigan State University.

Burton, I., Kates, R., White, G.F. 1978. Environment as hazard. London: Oxford University Press.

Chapman, D.W. ed. 1954. Human behavior in disaster: a new field of social research. Journal of Social Issues 10:1-73.

Chenault, W.W., Hilbert, G.D., and Reichlin, S.D. 1979. Evacuation planning in the TMI accident. Washington, DC: Federal Emergency Management Agency.

Christensen, L., and Ruch, C.E. 1980. The effect of social influence on response to hurricane warnings. Disasters 4:205-10.

Cohen, R., and Lichter, S.R. 1983. Nuclear power: the decision makers speak. Regulation March/April:32-37.

Cohen, R.E., and Ahearn, F.L. 1980. Handbook for mental health care of disaster victims. Baltimore, MD: The Johns Hopkins University Press.

Collins, H.E., Grimes, B.K., and Galpin. F. ed. 1978. Planning basis for the development of state and local government radiological emergency response plans in support of light-water nuclear power plants. Washington,DC: U.S. Nuclear Regulatory Commission (NUREG-0396).

Cook, E. 1982. The role of history in the acceptance of nuclear power. Social Science Quarterly 63:3-15.

Cornell, J. 1982. The great international disaster book. New York: Charles Scribner's and Sons.

Cunningham, A.M. 1982. Is there a seismograph for stress? Psychology Today 16(October):47-52.

Cutter, S., and Barnes, K. 1982. Evacuation behavior and Three Mile Island. Disasters 6:116-24.

DeMuth, W.E., and Trautlein, J.J. 1979. The luck of Three Mile Island. Journal of Trauma 19:792-94.

DeMuth, W.E., and Miller, K.L. 1982. A perspective on Three Mile Island. Continuing Education December:18-24.

Dohrenwend, B.P., et al. 1981. Stress in the community: a report to the President's commission on the accident at Three Mile and. In The Three Mile Island nuclear accident: lessons and implications, ed. T.H. Moss and D.L. Sills, pp. 159-74. New York: The New York Academy of Sciences.

Drabek, T.E., and Boggs, K.S. 1968. Families in disaster: reactions and relatives. Journal of Marriage and the Family 30:443-51.

Drabek, T.E. 1969 Social processes in disaster: family evacuation. Social Problems 16:336-49.

Dunning, C., and Silva, M. 1980. Disaster-induced trauma in rescue workers. Victimology: An International Journal 5:287-97.

Erikson, K.T. 1982. Human response in a radiological accident. In The Indian Point book: a briefing of the safety investigation of the Indian Point nuclear power plant, pp. 55-59. Cambridge, MA: Union of Concerned Scientists; and New York, New York Public Interest Research Group, Inc.

Fabrikant, J.I. 1983. The effects of the accident at Three Mile Island on the mental health and behavioral responses of the general population and nuclear workers. Health Physics 45:579-86.

FEMA 1980. Report to the president: state radiological emergency planning and preparedness in support of commercial nuclear power plants. Washington, DC: Federal Emergency Management Agency, June 30.

FEMA 1982. Radiological emergency planning and preparedness: final regulations. Federal Register 47 (March 11): 10758-62.

Firebaugh, M.W. 1981. Publicattitudesand information onthe nuclear option. Nuclear Safety 22:147-56.

Fischer, D.W. 1981. Planning for large scale accidents: learning from the Three Mile Island accident. Energy 6:93-108.

Flynn,C.B. 1979. Three Mile Island telephone survey: preliminary report of procedures and findings. Washington, DC: U.S. Nuclear Regulatory Commission (NUREG/CR-1093).

Flynn, C.B. 1981. Local public opinion. In The Three Mile Island nuclear accident: lessons and implications, ed. T.H. Moss and D.L. Sills, pp. 146-58. New York: The New York Academy of Sciences.

Foster, H.D. 1980. Disaster planning: the preservation of life and property. New York: Springer-Verlag.

Goldstein, R., and Schorr, J.K. 1982. The long-term impact of a man-made disaster: an examination of a small town in the aftermath of the Three Mile Island nuclear reactor accident. Disasters 6:50-59.

Gricar, B.G., and Baratta A.J. 1983. Bridging the information gap at Three Mile Island: radiation monitoring by citizens. The Journal of Applied Behavioral Science 19:35-49.

Grosser, G.H., Weschsler, H., and Greenblatt, G. ed. 1964. The threat of impending disaster: contributions to the psychology of stress. Cambridge, MA: The M.I.T. Press.

Haglund, K. 1979. At Hershey: medical system near "failure" during Three Mile Island. New Physician 28: 24-25.

Hans, J.M., Jr., and Sell, T.C. 1974. Evacuation risks -an evaluation. Las Vegas, NV: U.S. Environmental Protection Agency.

Henderson, O.K. 1981. Radiological emergency response planning in Pennsylvania. In Current nuclear power plant safety issues, pp. 331-41. Vienna: IEA (IAEA-CN-39/7.1).

Hohenesmer, C., Kasperson, R., and Kates, R. 1977. The distrust of nuclear power. Science 196 (April):25-34.

Houts, P.S., and Goldhaber, M.K. 1981. Psychological and social effects surrounding Three Mile Island after the nuclear accident on March 28, 1979. In Energy, environment and the economy, ed. S. Majumdar, pp. 151-64. PA:Pennsylvania Academy of Sciences.

Hu, T., and Slaysman, K.S. 1981. Health related economic costs of the Three Mile Island accident. Center for Research on Human Resources, Institute for Policy Research and Evaluation. University Park, PA: The Pennsylvania State University.

Janis, I.L. 1962. Psychological effects of warnings. In Man and Society in Disaster, ed. G.W. Baker and D.W. Chapman, pp. 55-92. New York: Basic Books.

Janis, I.L., and Mann, L. 1977. Emergency decision making: a theoretical analysis of responses to disaster warnings. Journal of Human Stress 3:35-48.

Jaske, R.T. 1983. Emergency preparedness: status and outlook. Nuclear Safety 24: 1-11.

Johnson, J.H., Jr. 1983. Reactions of public school teachers to a possible accident at the Diablo Canyon nuclear power plant, a report prepared for the California Teachers Association, San Luis Obispo County Chapter. Available upon request from the author.

Johnson, J.H., Jr., and Zeigler, D.J. 1984. A spatial analysis of evacuation intentions at the Shoreham Nuclear Power Station. In Nuclear power: assessing and managing hazardous technology, ed. M.J. Pasqualetti and K.D. Pijawka, pp. 279-301. Boulder, CO: Westview Press.

Johnson, J.H., Jr. 1984a. Planningfor spontaneous evacuation during a radiological emergency. Nuclear Safety 25: 186-94.

Johnson, J.H., Jr. 1984b. Towards a regional approach to radiological emergency response planning. Papers and Proceedings of Applied Geography Conferences 7:114-22.

Johnson, J.H., Jr., and Zeigler, D.J. 1983. Distinguishing human responses to radiological emergencies. Economic Geography 59:386-402.

Kasl, S., Chisholm, R.F., and Eskenazi, B. 1981a. The impact of the accident at the Three Mile Island on the behavior and well-being of nuclear workers - part I: perceptions and evaluation, behavioral responses and work-related attitudes and feelings. American Journal of Public Health 71:472-83.

Kasl, S., Chisholm, R.F., and Eskenazi, B. 1981b. The impact of the accident at the Three Mile Island on the behavior and well-being of nuclear workers - part II: job tensions, psychophysiological symptoms, and indices of distress. American Journal of Public Health 71:484-95.

Kasperson, R.E., et al. 1980. Public opposition to nuclear

energy: retrospect and prospect. Science, Technology and Human Values 5:11-23.

Kemeny, J.G., et al. 1979. Report of the president's commission on the accident at Three Mile Island Washington, DC: U.S. Government Printing Office.

Killian, L.M. 1952. The significance of multiple group membership in disaster. American Journal of Sociology 57:309-14.

KLD Associates 1983. Estimated evacuation times for the entire population within the emergency planning zone for the Shoreham Nuclear Power Station, considering the effects of uncontrolled evacuation, voluntary evacuation, inclement weather and accidents. Huntington Station, NY:KLD Associates.

Kuntz, E. 1979. Hospitals prepared radiation plans in wake of nuclear plant accident. Modern Healthcare 9:16.

Lathrop, J.W. 1980. An open discussion of problems in nuclear accident preparedness. In Planning for rare events: nuclear accident preparedness and management, ed. J.W. Lathrop, pp. 13-44. Oxford: Pergamon Press.

Leik, R.K., Carter, T.M., and Clark, J.P. 1981. Community response to natural hazards warnings. Summary final report for National Science Foundation, Federal Emergency Management Agency, and National Weather Service. Minneapolis: University of Minnesota.

Lichter, S.R., and Rothman, S. 1983. Scientists' attitudes toward nuclear energy. Nature 305:91-94

Lifton, R.J. 1968. Death in life: survivors of Hiroshima. New York: Random House.

Lifton, R.J. 1982. Beyond psychic numbing: a call to awareness. American Journal of Orthopsychiatry 52:619-29.

Lindell, M.K., and Perry, R.W. 1983. Nuclear power plant emergency warning: how would the public respond? Nuclear News 26:49-53.

MacKenzie, J.J. 1984. Finessingtherisks of nuclearpower. Technology Review 87(Feb/March):34-39.

MacLeod, G.K. 1981. Some public health lessons from Three Mile Island: a case study in chaos. Ambio 10:18-23.

Manning, D.T. 1982. Post-TMI perceived risk from nuclear power in three communities. Nuclear Safety 23:379-84.

Maxwell, C. 1982. Hospital organizational response to the nuclear accident at Three Mile Island: implications for future oriented disaster planning. American Journal of Public Health 72:275-79.

Mayo, L.H., et al. 1980. Public concerns and alternative nuclear power systems. Prepared for the Nonproliferation Alternative System Assessment Program, U.S. Department of Energy. Washington, DC: George Washington University.

Mileti, D.S., Drabek, T.E., and Haas, J.E. 1975. Human systems in extreme environments: a sociological perspective. Boulder, CO: Institute of Behavioral Science.

National Academy of Sciences. 1979. Risks associated with nuclear power. Nuclear Safety 20:664-70.

Nelkin, D. 1981. Some social and political dimensions of nuclear power: examples from Three Mile Island. American Political Science Review 75:132-42.

Nowotny, H. 1976. Social groups - fear of nuclear power and economic interest. Transactions of the American Nuclear Society 24:99-100.

NRC 1975. Reactor safety study: an assessment of accident risks in U.S.commercial nuclear power plants. NRC Report WASH-1400 (NUREG 75/104), NTIS, October.

NRC 1977. Guide and checklist for the development and evaluation of state and local government radiological emergency response plans in support of fixed nuclear facilities. NRC Report NUREG - 75/111, March.
Supplement No. 1, Essential guidance elements and associated checklist elements for NRC concurrence in state and local government emergency response plans. Available at the NRC Public Documents Room.

NRC 1980. Emergency planning: final regulations. Federal Register 45:55402-18.

NRC/FEMA 1980. Criteria for preparation and evaluation of radiological emergency response plans and preparedness in support of nuclear power plants. Washington DC: U.S. Nuclear Regulatory Commission and Federal Emergency Management Agency (NUREG 0654).

Office of Emergency Preparedness 1973. Nuclear incident planning

- fixed facilities. Federal Register 38(January 24): 2356.

Office of Technology Assessment 1984. Nuclear power in an age uncertainty. Washington, DC: U.S. Government Printing Office.

Otway, H.J. 1977. Risk assessment and the social response to nuclear power. Journal of the British Nuclear Energy Society 16:327-33.

Otway, H.J., and Thomas, K. 1978. The contribution of safety issues to public perceptions of energy systems. In Pro-abilistic Analysis of Nuclear Reactor Safety, pp. IV.1-1 - IV.1-10. Los Angeles, CA: American Nuclear Society, European Nuclear Society, and OCED Nuclear Energy Agency.

Perry, R.W. 1979. Evacuation decision-making in natural disasters. Mass Emergencies 4:25-38.

Perry, R.W., Greene, M.R., and Lindell, M.K. 1980. Enhancing evacuation warning compliance: suggestions for emergency planning. Disasters 4:443-49.

Perry, R.W., Lindell, M.K., and Greene, M.R. 1981. Evacuation planning in emergency management. Lexington, MA: Lexington Books.

Perry, R.W. 1983. Population evacuation in volcanic eruptions, floods, and nuclear power plant accidents: some elementary comparisons. Journal of Community Psychology 11:36-47.

Polk, P.A. 1983. Testimony of behalf of Suffolk County regarding contentions 23.D (evacuation shadow phenomenon) and 65 (evacuation time estimates). In the matter of Long Island Lighting Company (Shoreham Nuclear Power Plant, Unit 1).

Quarantelli, E.L., and Dynes, R.R. 1978. Structural factors in the minimization of role conflict: a re-examination of the significance of multiple group membership in disasters. Preliminary paper 49, Disaster Research Center, Ohio State University.

Saarinen, T.F. 1979. The relationship of hazard awareness to adoption of approved mitigation measures. Boulder: University of Colorado Natural Hazards Research and Applications Center IBS 6.

Schwebel, M., and Schwebel, B. 1981. Children's reaction to the threat of nuclear power plant accidents. American Journal of Orthopsychiatry 51:260-70.

Sheffi, Y., Mahmassani, H., and Powell, W. 1981. Evacuation studies for nuclear power plant sites, a new challenge for traffic engineers. Inst. Transportation Engineering Journal 51:25-29.

Sheffi, Y. 1982. A note on computer graphics in network evacuation studies. Computer Graphics and Applications 2:57-62.

Sheridan, T. 1980. Human error in nuclear power plants. Technology Review. 82:22-33.

Sims, J.H., and Baumann, D.D. 1983. Educational programs and human response to natural hazards. Environment and Behavior 15:165-89.

Slovic, P., Fischhoff, B., and Lichtenstein, S. 1979. Rating the risks. Environment 21:14-39.

Slovic, P., Fischhoff, B., and Lichtenstein, S. 1980a. Facts and fears: understanding perceived risk. In Societal risk assessment: how safe is safe enough? ed. R.C. Schwing and W.A. Albers, pp. 181-214. New York: Plenum Press.

Slovic, P., and Fischoff, B. 1983. How safe is safe enough? In Too hot to handle?, ed. C.A. Walker, et al., pp. 112-50. New Haven, CN: Yale University Press.

Slovic, P., Fischhoff, B., and Lichtenstein, S. 1980b. Informing people about risk. In Product labeling and health risks. Banbury Report 6, ed. A.L. Morris, M.B. Mazis, and I. Barofsky, pp. 165-81. New York: Cold Spring Harbor Laboratory.

Smith, J.S., and Fischer, J.H. 1981. Three Mile Island: the silent disaster. JAMA, The Journal of the American Medical Association. 245(April 24):1656-59.

Smith, S.M. 1983. Disaster: family disruption in the wake of natural disaster. In Stress and the Family, Volume II: Coping with Catastrophe, ed. C.F. Figley and H.I. McCubbin, pp. 120-47. New York: Brunner/Mazel.

Social Data Analysts, Inc. 1982a. Attitudes towards evacuation: reactions of Long Island residents to a possible accident at the Shoreham Nuclear Power Plant. New York: Social Data Analysts, Inc.

Social Data Analysts, Inc. 1982b. Responses of emergency personnel to a possible accident at the Shoreham Nuclear Power Plant. New York: Social Data Analysts, Inc.

Sorensen, J.H. 1983. Knowing how to behave under the threat of disaster: can it be explained? Environment and Behavior 15:438-57.

Sorensen, J.H. 1984. Evaluating the effectiveness of warning systems for nuclear power plant emergencies: criteria and application. In Nuclear power: assessing and managing hazardous technology, ed. M.J. Pasqualetti and K.D. Pijawka, pp. 259-77. Boulder, CO: Westview Press.

Staff Report 1984. Nuclear power: bombing out? Time February 13, 1984:34-45.

Strohl, G. 1979. Nuclear threat: hospitals need to know. Osteopathic Hospitals 23:6, 8-9.

The Media Institute 1979. Television evening news covers nuclear energy: a ten year perspective. Washington, DC: The Media Institute.

The Media Institute 1980a. The public's right to know: communicator's response to the Kemeny commission report. Washington, DC: The Media Institute.

The Media Institute 1980b. Nuclear phobia-phobic thinking about nuclear power: a discussion with Robert L, DuPont, M.D. Washington, DC: The Media Institute.

The Pennsylvania National Guard 1979. Three Mile Island nuclear accident 28 - March - 4 April 1979 after action report. Annville, PA: Department of Military Affairs, June 7.

Thomas, K., et al. 1980. Nuclear energy: the accuracy of policy maker's perception of public beliefs. Behavioral Science 25:332-44.

Udell, R. 1982. Public citizen's nuclear power plant safety report. Washington, DC: Critical Mass Energy Project.

Walsh, S.J., et al. 1983. Application of population projections and remote sensing for nuclear power plant licensing. The Social Science Journal 20:89-102.

Weidner, W.A., et al. 1980. The impact of nuclear crisis on a radiology department. Radiology 135:717-23.

White, G.F. ed. 1974. Natural hazards: local, national, global. New York: Oxford University Press.

White, G.F., and Haas, J.E. 1975. Assessment of research on

natural hazards. Cambridge, MA: The M.I.T. Press.

Wolfenstein, M. 1957. Disaster: a psychological essay. Glencoe, IL: Free Press, 1957.

Yankelovich, Skelly, and White, Inc. 1983. Status report on public response to emergency planning efforts. Prepared for the Long Island Lighting Company, Long Island, NY.

Zeigler, D.J., Brunn, S.D. and Johnson, J.H., Jr. Evacuation from a nuclear technological disaster. Geographical Review 71:1-16.

Zeigler, D.J., Johnson, J.H., Jr., and Brunn, S.D. 1983. Technological hazards. Resource Publications in Geography. Washington, DC: Association of American Geographers.

Zeigler, D.J., and Johnson, J.H., Jr. 1984. Evacuation behavior in response to nuclear power plant accidents. The Professional Geographer 36:207-15.

PART III

Power Plant Siting and Land Use

CHAPTER 10

EXISTING AND FUTURE SITING PATTERNS OF ELECTRIC UTILITY POWER PLANTS

J. Sherman Feher
Electric Power Research Institute
3412 Hillview Avenue
Palo Alto, California 94304

ABSTRACT. Many siting studies have been done at national, regional, and local scales for selected technologies. Yet, none have tried to project all of the sites that are needed for virtually all electricity generating technologies. In this chapter existing sites are inventoried and new sites are projected on national and regional bases. An analysis based on projections of approximately 3.0% per year electricity demand growth rate concludes that between 4630 and 5660 total sites may be required by the year 2000. This range results from two scenarios based on asumptions of plant size and expanded use of existing plant sites. Regional allocation of sites and various siting factors are presented for many generating technologies.

1. INTRODUCTION

Electric utility power plant siting is a complex spatial decision-making process that integrates a variety of considerations, including environmental and socioeconomic factors, utility system and resource requirements, and public opinion. The chapter addresses the number of projected sites that may be required and their associated spatial patterns as well as siting factors that are unique to each electricity generation technology.

Many siting studies have been done at national, regional, and local scales (selected examples include: Dobson 1977: Eagles et al. 1980; Hanham and Calzonetti 1983; Hillsman et al. 1983; Honea et al. 1979; Meier et al. 1979). Yet, none have looked at the total number of projected power plant sites that may be required nationally for most electricity generating technologies between now and the year 2000. Most national studies have focused on medium- to large-size coal and nuclear technologies while overlooking siting requirements and regional siting patterns for smaller technologies. Because renewable technologies are becoming more popular, their siting needs are also deserving of study. Also, the present financial condition of many

F. J. Calzonetti and B. D. Solomon (eds.), Geographical Dimensions of Energy, 157–177.
© 1985 by D. Reidel Publishing Company.

utilities coupled with lower electricity growth rates than before have resulted in utilities considering smaller-sized power plants. A primary purpose of this study is to compare the number of power plant sites required for two scenarios, one of larger and the other of smaller power plants.[1]

The remainder of this paper presents overall requirements for new power plant sites and their generalized siting patterns. Siting factors, along with projections of new sites, and their regional allocation are also presented for specific types of generation.

2. POWER PLANT SITING REQUIREMENTS

In 1980, the electric utility industry had a total installed generating capacity of 576 gigawatts (GWe) located on some 3,133 sites. Many of these existing sites are used for small hydroelectric, petroleum, and natural gas-fired plants with capacities less than 10 megawatts (MWe).

Analysis of future requirements for power plant siting is predicated in large part on electricity demand projections. However, projecting electricity demand is fraught with uncertainty. In recent years, annual load growth has slackened to 1-3% per year from previous values of 7-8% per year experienced in the 1960s and early 1970s. For strategic planning purposes, EPRI analyzes the ramifications of various electricity demand growth rate scenarios on its research and development (R&D) program (Electric Power Research Institute 1981). For this study, an annual electricity demand growth rate of approximately 3.0% is used (modified from projections in Electric Power Research Institute 1981). This growth rate is based on the following 1980-2000 average annual growth rate assumptions: a realized gross national product of 2.5% and total energy consumption of approximately 1.4%. In 1980 electricity consumed 33% of total primary energy consumption. This fraction is projected to increase to 45% or more in 2000. Electricity consumption is projected to be 25% lower by 2000 than it would have been because of the effect of price increases and more productive use of energy (conservation). Electricity imports from Canada and Mexico and industrial self-generation are projected to account for approximately 5% of the electricity generation.

---

[1]The Electric Power Research Institute (EPRI), as a part of its strategic planning process, examined future power plant siting requirements. EPRI conducts a research and development program for its member electric utilities. It encourages the development of new and improved technologies to help the utility industry meet present and future electric energy needs in environmentally and economically acceptable ways. EPRI conducts research on all aspects of electric power production and use, including fuels, generation, delivery, energy management and conservation, environmental effects, and energy analysis.

Based on this growth rate, U.S. utility installed capacity in the year 2000 is projected to be approximately 900 GWe requiring some 324 GWe of totally new capacity plus replacement capacity for retired units. Coal-fired technology will provide the majority of new capacity (213 GWe) with hydro (15 GWe), nuclear (56 GWe), and other renewable and alternative generating sources (15 GWe) providing the remaining new capacity (capacity mix modified slightly from scenarios in Electric Power Research Institute 1981).

The projections of siting requirements and patterns are not to be construed as an absolute prediction of the number and location of power plant sites, but as a "what if" scenario. In other words, given the growth rates, capacity requirements, and our present knowledge of power plant technologies and their siting needs, the number of required sites are postulated along with their spatial patterns.

The primary reason for this study was to compare future siting requirements for two different philosophies of plant size and technologies. The large-plant scenario is based on utility siting trends of the 1950s, 1960s, and early 1970s. During this time utilities enjoyed the benefits of economies of scale--where technology improvements permitted them to install power plants of ever increasing sizes at lower costs. The average new coal-fired power plant unit size increased from roughly 50 MWe in 1940 to 650 MWe in 1970. But in the 1970s requirements to add more extensive environmental control equipment, coupled with rampant inflation and perceived technology limits, brought economies of scale into question. During this time utilities also had problems obtaining sufficient funds to finance construction of these large plants. To counter these trends, utilities began to consider smaller-size generating units, typically 100 to 400 MWe, that could be easily installed without a significant financial burden on themselves. At the same time R&D was producing encouraging results for several new smaller-size technologies such as geothermal, wind turbines, and low-head hydroelectric. These trends are generally represented by the small-plant scenario. By making this comparison one can gain a general understanding of the siting implications of these two different plant size philosophies.

Based on the projected electricity requirement of approximately 900 GWe, between 4,630 and 5,660 sites may be needed; of which 1,500 to 2,530 would be new sites. Table 1 shows existing and projected sites, by fuel type, in the continental United States to the year 2000. The large-plant scenario, which projects approximately 1,500 new sites, reflects the use of larger plant capacities. The small-plant scenario, which projects some 2,530 new sites, reflects the use of smaller plant capacities.

The majority of new sites in either scenario will probably be hydroelectric and other renewable resource-based technologies, though the majority of capacity probably will be in larger conventional plants. Table 2 compares the percentage of existing and projected sites and generating capacity for various technologies. The actual number of sites in the year 2000 will probably fall between 4,630 and 5,660 sites.

Table 1. Existing and projected utility sites in the Continental United States of the year 2000

Technology	Existing Sites 1980	Projected New Sites	
		Large-Plant Scenario	Small-Plant Scenario
Nuclear	38	30	30
Coal	287	150	475
Coal-derived fuels, petroleum and natural gas	1268	0	0
Hydroelectric and pumped storage	1243	1005	1410
Other renewable resources	16	165	515
Combined fuel types	281	150	100
Total	3133	1500	2530
Total large-plant scenario		4633	
Total small-plant scenario			5663

Replacement of retired power plant capacity has not been fully taken into account in this study because of the complexity of the issue. For instance, the number of sites could increase depending on the reuse of retire sites. While reuse of existing sites is considered for some technologies in this study, further research is needed for a better understanding of this siting option.

Table 2. Comparison of existing and projected sites and generating capacity for various technologies

	Percent of Total Existing (1980)		Perception of Total New Projected (2000)		
	Sites	Capacity	Large-Plant Scenario Sites	Capacity	Small-Plant Scenario Sites
Nuclear	1	9	2	17	1
Coal	10	41	11	66	20
Coal-derived fuels, petroleum and natural gas	45	6	0	*	0
Hydroelectric and pumped storage	44	14	75	12	58
Other renewable resources	<1	<1	12	5	21
Total number of sites and amount of capacity (GW)	3133	576	1500	324	2530

Note: Combined fuel type sites and capacity are not included in this table because the capacity is included with the respective technology.
*No new capacity, replacement of old capacity is not included.

3. POWER PLANT SITING PATTERNS

From a geographical perspective, the number of required sites and unique siting factors for various technologies is interesting, but the most important question is <u>where</u> will these sites actually be located? As a <u>first approximation</u>, in order to provide a rough spatial perspective, projected siting requirements were allocated on a large regional basis. The regions used for this purpose are based on aggregated North American Electric Reliability Council (NERC) regions, referred to as EPRI regions (see Figure 1). NERC was formed in 1968 by the electric utility industry to promote the reliability and adequacy of bulk power supply in electric utility systems of North America. NERC consists of nine regional reliability councils which are also referred to as NERC regions. EPRI has consolidated the nine regions into six for regional R&D assessments and analysis.

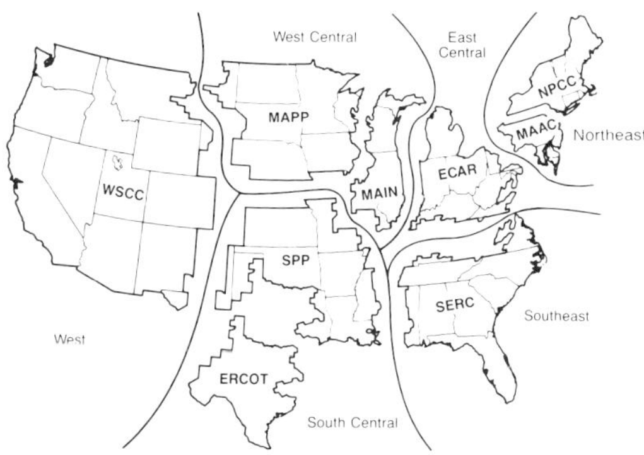

ECAR: East Central Area Reliability Coordination Agreement
ERCOT: Electric Reliability Council of Texas
MAAC: Mid-Atlantic Area Council
MAIN: Mid-America Interpool Network
MAPP: Mid-Continent Area Power Pool
NPCC: Northeast Power Coordinating Council
SERC: Southeastern Electric Reliability Council
SPP: Southwest Power Pool
WSCC: Western Systems Coordinating Council

Figure 1. EPRI regional systems.

The diversity of technologies necessitates various methods to allocate new sites regionally. A regional generation expansion model was used for coal-fired power plants to help determine siting patterns. Virtually no new petroleum or natural gas-fired plant sites are expected because of the reuse of existing sites and laws limiting the use of petroleum and natural gas. New nuclear plant sites are defined as plants already under construction and are readily identified. Hydroelectric, geothermal, solar, wind, and biomass technologies were located by an examination of resource potential. More details on the means of regional allocation of sites are provided with each technology.

Figure 2 shows existing (1980) and projected utility sites by region to the year 2000. The West Central region has the largest number of existing utility sites, followed by the West and Northeast regions. The West Central region also has the most existing sites for coal, petroleum and natural gas technologies and sites with combined fuel types. The West region has the largest number of existing hydroelectric and pumped storage sites. The West region has the largest projected siting requirements for both large- and small-plant scenarios followed by the Northeast region. Both of these regions have a large number of hydroelectric and other renewable and alternative resources.

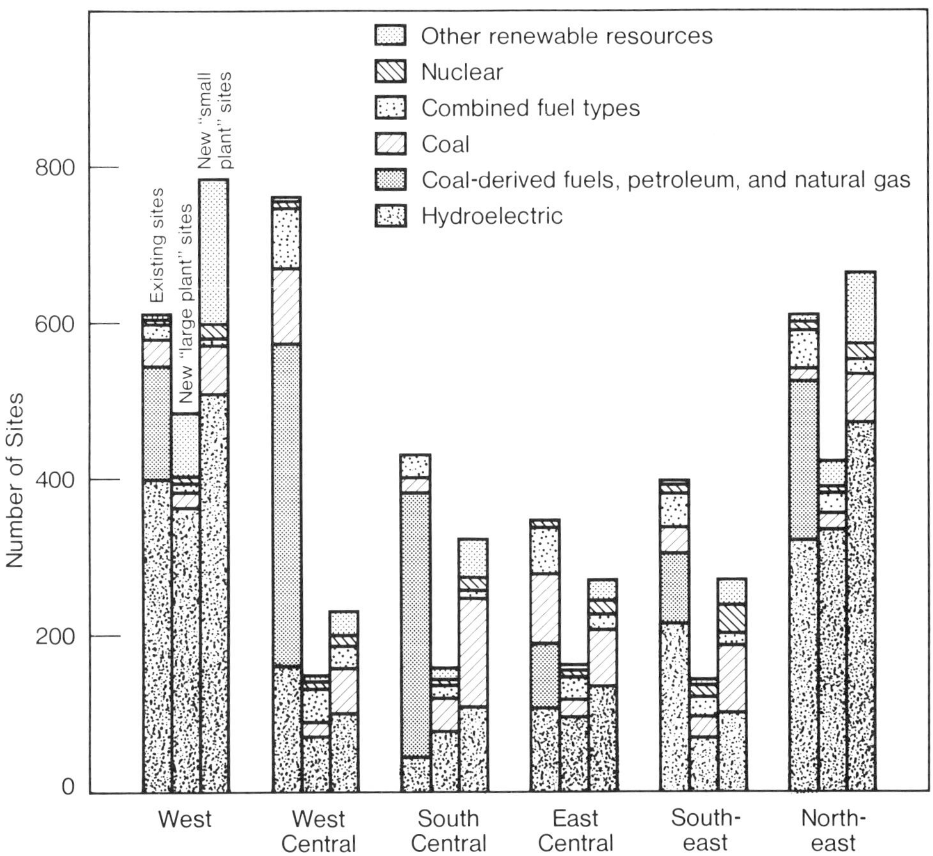

Figure 2. Regional existing and projected utility sites to the year 2000.

4. SITING CONSIDERATIONS AND REGIONAL ALLOCATION OF SITES

Siting factors, projections of new sites, and their regional allocation will be discussed for each generation type.

4.1. Coal

Coal-fired plant installed capacity was 237 GWe at some 287 sites in 1980. Figure 3 shows the 1980 coal steam electric capacity by county. In Figures 3, 5, and 6 the capacity shown by each circle is the total of all power plants within a county. The projected siting requirement for 213 GWe of new installed conventional (pulverized coal plant) or advanced (e.g., atmospheric fluidized-bed combustion and gasification-combined cycle power plants) coal-fired capacity is 150 new sites (large-plant scenario) and 475 new sites (small-plant scenario).

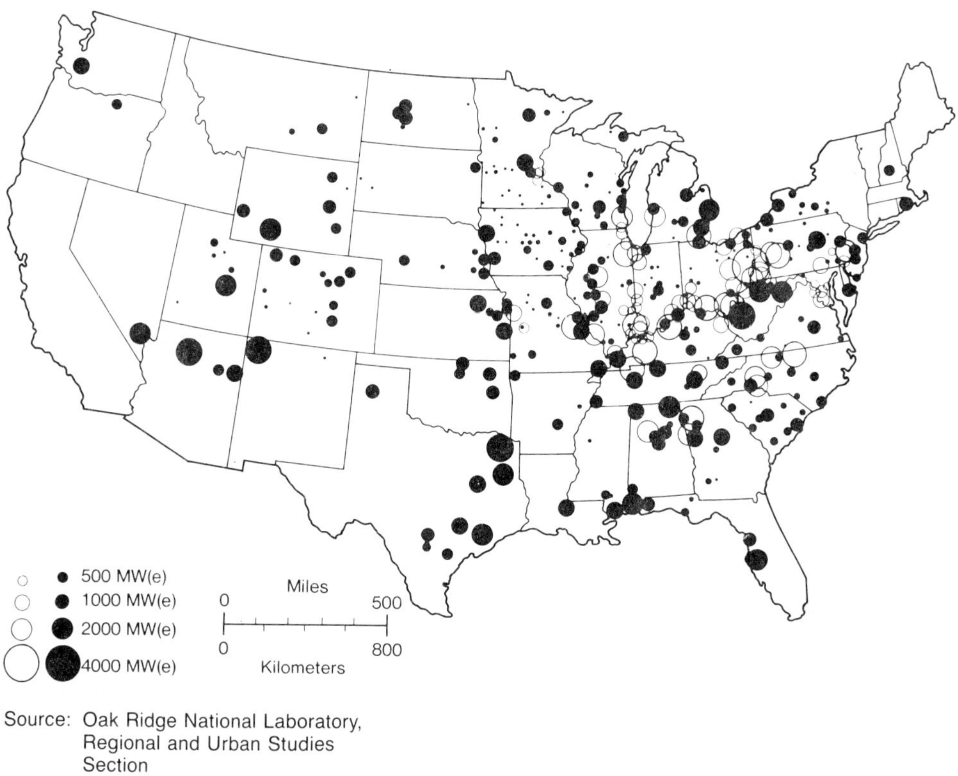

Figure 3. Coal steam electric capacity in 1980 at the county level.

The large-plant scenario assumes that each new site will have 1000 MWe of coal-fired capacity. Some 150 new sites would be constructed while the remaining capacity would be located on existing sites. This is based on analysis of plans for new power plant construction reported by the nation's utilities (U.S. Department of Energy 1982), which indicates that 30% of new coal-fired capacity will be located at existing sites between 1980-1989. Taking the 30% and extrapolating to the year 2000 results in 64 GWe of new coal-fired capacity being located on existing sites. Further study would be needed to determine which existing sites would be used for new coal-fired capacity.

The small-plant scenario assumes that an average size unit of approximately 450 MWe will be located on each site resulting in 475 new sites for new coal-fired capacity by 2000. The assumption of 450 MWe per site was based primarily on the average utility coal-fired unit size planned for between 1980-1989 (U.S. Department of Energy 1982). This average continues to the year 2000, with each unit located on each site, rather than the current utility practice of aggregating 2-4 units in clusters. This scenario explicitly assumes that no new capacity will be located on existing sites except replacement of retired units, which is roughly equivalent to 30% of new coal-fired units built between 1980-1989.

For the regional allocation of projected coal-fired plant sites, regional coal-capacity additions representing actual utility planned additions for the 1980-1989 time period similar to those from (U.S. Department of Energy, 1982) were added to hypothetical coal-capacity additions from 1990-2000 taken from EPRI Regional Systems, Reference Generation Expansion Plans 1990-2020. The EPRI Regional Systems are expansion plans, developed using a simulation model, that take into account the existing supply system, retirement schedule, and forecasted load growth. Given technology constraints and economic factors, the model provides the optimum technology mix of generating additions for specified planning horizons. The model's optimization performs a trade-off between capital investments and operating costs for various technologies to minimize overall electricity costs. The percent of coal-fired capacity for each region, calculated by this method, was used to determine the number of coal-fired power plant sites for each region. The South Central region had the highest percentage of total projected coal-fired capacity (29%) which resulted in 43 and 138 sites for the large- and small-plant scenarios, respectively. Higher economic growth in Texas coupled with the increased use of coal and lignite due to noneconomic use of oil and gas produced this result. The remaining regions had the following percentages of total projected coal-fired capacity: Southeast = 18%, East Central = 15%, Northeast and West = 13%, West Central = 12%. These percentages were applied to 150 sites for the large-plant scenario and 475 sites for the small-plant scenario.

Air quality and solid waste disposal are two factors that play a major role in coal-fired power plant siting. Air quality laws and regulations for coal-fired power plants are very complex, and a detailed discussion is beyond the scope of this paper. However, federal air quality laws and regulations may affect siting of new coal-fired power plants. Figure 4 shows nonattainment areas for selected pollutants (air quality standards have not been met) as well as Class 1 areas (very stringent air quality standards associated with national parks and wilderness areas). These regulatory designations may constrain siting of new coal-fired plants, particularly in the West. State and local air quality laws can also affect siting decisions. For example, California has air quality regulations that are more stringent than federal standards, making it more difficult and costly to site coal-fired power plants. Future acid rain legislation, if enacted, could affect the siting of coal-fired power plants depending on the final form of the legislation.

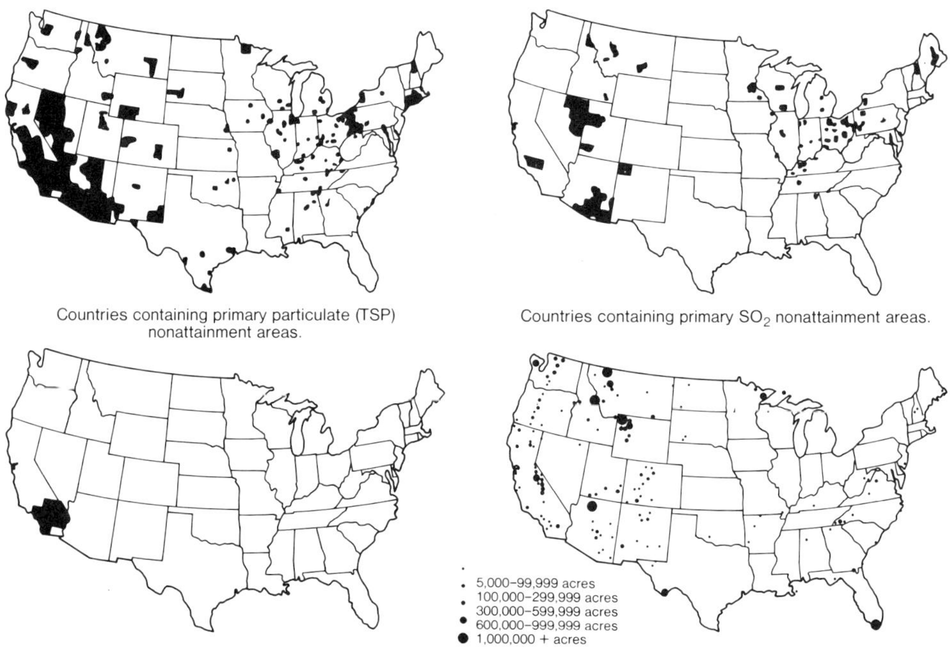

Countries containing primary particulate (TSP) nonattainment areas.

Countries containing primary SO_2 nonattainment areas.

Countries containing primary NO_2 nonattainment areas.

Clean Air Act: Class I Areas

Source: EPA, Office of Air Quality Planning and Standards

Figure 4. Potential air quality siting constraint areas for coal-fired power plants.

As more coal-fired plants with more effective air quality controls are built, emissions previously expelled into the air become solid waste on the ground. Solid waste disposal is becoming a more important siting factor. Land requirements for storage and environmental effects of hazardous substances that may be leaching from the solid-waste disposal site are two principal concerns. The nominal solid-waste disposal requirement is 300-1000 acres for a 1000-MWe plant depending on type of coal, control hardware, and disposal technique. Water availability is another siting consideration, but it is not unique to coal-fired power plants.

4.2. Petroleum, Natural Gas, and Coal-Derived Fuels

In 1980, installed capacity for petroleum and natural-gas-fired units was 209 GWe located at 1,268 sites. A further breakdown by technology type is: oil/gas steam = 262, combined-cycle = 6, combustion turbine = 220, internal combustion = 622, and a combination of oil/gas steam, combustion turbine, and internal combustion = 158 (U.S. Department of Energy 1982).

Many of these existing sites have less than 10 MWe of installed capacity with internal combustion machines accounting for the majority of the sites. Figure 5 shows existing oil and natural gas steam electric capacity by county but it excludes combustion turbine and internal combustion capacity. Oil and natural gas electric capacity are shown together because these fuels are generally interchangeable.

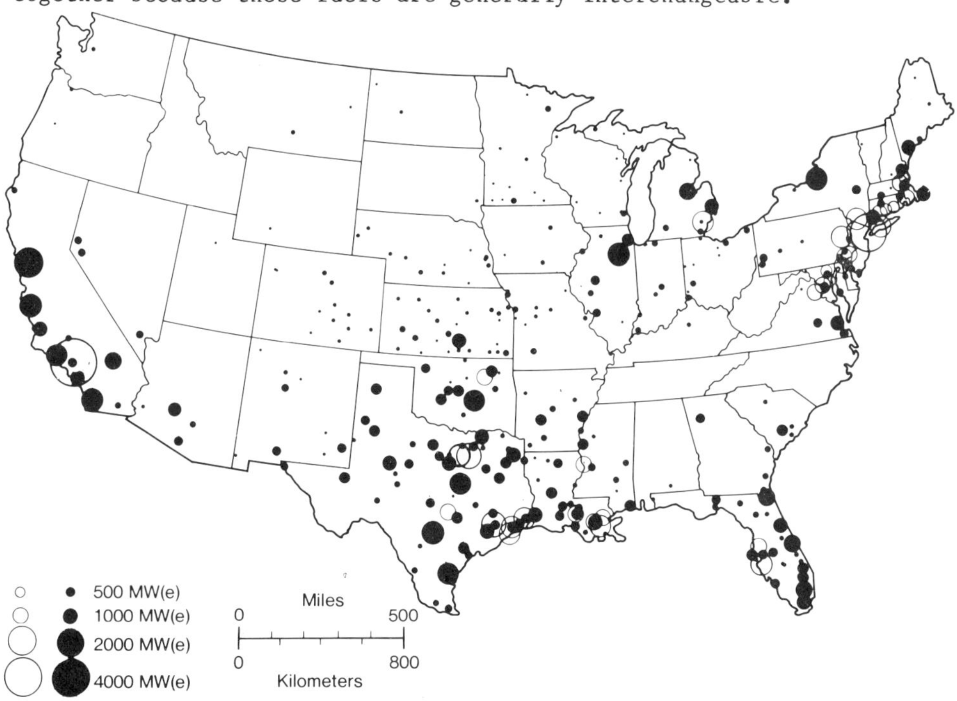

Source: Oak Ridge National Laboratory, Regional and Urban Studies Section

Figure 5. Oil and natural gas steam electric capacity in 1980 at the county level.

The total number of sites and the spatial pattern for petroleum, natural gas, and coal-derived fuel (which include coal liquids and coal gases) power plants may differ from the present number of sites only slightly by year 2000. Total petroleum- and natural gas-fired capacity will probably be less in 2000 because uneconomic oil- and gas-fired steam units will be replaced by other technologies. Existing sites may be recycled with new peaking units or smaller modular technologies, such as fuel cells, because of the relative ease in siting and constructing these technologies.

Currently, the overriding constraint on building new petroleum and natural gas-fired power plants is the Power Plant and Industrial Fuel Use Act of 1978 (Public Law 95-620), which prohibits the use of natural gas and petroleum as the primary energy source in new power plants that operate over 2500 hours per year. With few exceptions, this law effectively dictates that new liquid- and gas-fired power plants must operate as peaking units.

4.3. Nuclear

In 1980, total nuclear capacity was 51 GWe on 48 sites with 38 sites having only nuclear power plants. Figure 6 shows existing nuclear electric capacity at the county level. With the exception of nuclear plants already under construction, it is unlikely that additional nuclear plants will be built in the foreseeable future because of cost, regulatory, and public perception considerations. Projected requirements for both scenarios are 30 new sites for approximately 56 GWe of new nuclear capacity. These sites are based on plants that are currently under construction and likely to be completed and licensed by the year 2000.

The following is the projected regional nuclear plant siting pattern. The Southeast region has seven, the Northeast and South Central regions each have six projected sites. The remaining projected sites are in the West Central region (5) and in the East Central and West regions (3 each).

Radiological health concerns are the primary siting factors unique to nuclear power plants. Regulations require nuclear plants to be located away from population concentrations to reduce the risk of exposure in the event of a reactor accident. Seismic considerations include: fault rupture hazard, dynamic soil stability (liquefaction), and strong ground motion (vibratory). A detailed site study requires the consideration of all three factors. Nuclear power plant siting requirements are specified by the Nuclear Regulatory Commission in the Code of Federal Regulations (10CFR 100 Reactor Site Criteria). Cooling water availability and transportation access for moving large plant components during construction are also important siting factors for nuclear plants.

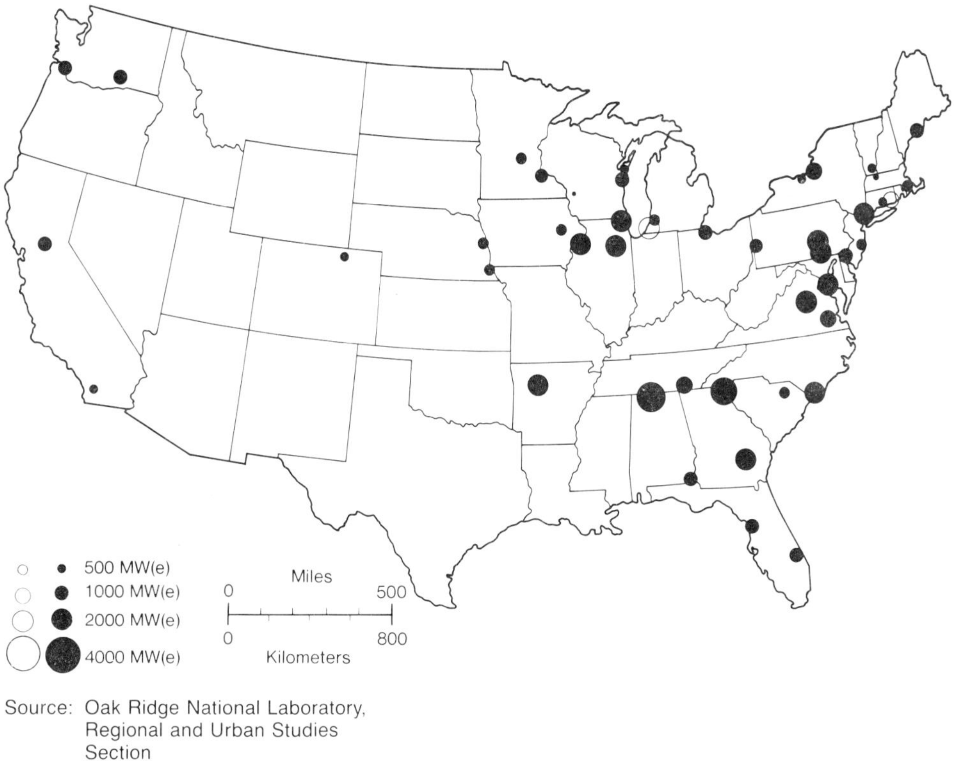

Figure 6. Nuclear steam electric capacity in 1980 at the county level.

4.4. Hydroelectricity

Hydroelectricity is currently the most abundantly used renewable resource for generating electricity. Conventional hydroelectric capacity in 1980 was approximately 65 GWe on some 1,211 sites in the continental United States. The breakdown of hydro sites is: small-size hydro plants (.05-15 MWe) 67%, intermediate-size hydro plants (15-25 MWe) 6%, and large-size hydro plants (>25 MWe) 26% (U.S. Army Corps of Engineers 1981). In 1980, pumped-hydroelectric storage capacity was 12 GWe on 32 sites.

Projected siting requirements for an additional 27 GWe of conventional hydroelectricity by 2000 entail 1000 sites for the large-plant scenario and 1400 sites for the small plant scenario. Conventional hydro sites are limited by the availability of hydroelectric resources.

Approximately 19 GWe of the 27 GWe of projected capacity may come from increases in capacity of existing hydroelectric sites as well as 600 new sites with existing reservoirs but no hydroelectric facilities (U.S. Army Corps of Engineers 1981). The remaining capacity (approximately 8 GWe) was allocated to undeveloped sites: 400 sites (average 20 MWe plant size) for the large-plant scenario and 800 sites (average 10 MWe plant size) for the small-plant scenario. The development of small-scale hydro facilities may result in an even larger number of new sites. Figure 7 portrays existing and projected hydroelectric capacity by region. Projected siting requirements for pumped-hydroelectric storage are based primarily on size. For 12 GWe the large-plant scenario may require approximately 5 sites, averaging 2,400 MWe, while the small plant scenario may require 10 sites averaging 1200 MWe.

Regional percentages of development potential for projected regional hydroelectric siting patterns were calculated from the National Hydroelectric Power Resources Study (U.S. Army Corps of Engineers 1981) and applied to the large- and small-plant total siting requirements. The West and Northeast regions had the highest percentage of feasible development potential 35% and 32% respectively. The other regional development potential percentages were: East Central = 9%, South and West Central each = approximately 7%, and Southeast = 6%.

Other siting factors for hydroelectric plants involve environmental concerns and questions of public acceptance. Environmental and public acceptance issues include low-flow augmentation for aquatic and ecological concerns, reservoir fluctuations, wild and scenic rivers, and land inundation, competition for land and water use including recreation, fish and wildlife protection, flood control, water supply, irrigation, and preservation of wilderness areas. Some of these issues may reduce the likelihood that certain hydroelectric sites will be developed because mitigation measures may reduce the economic viability of a project.

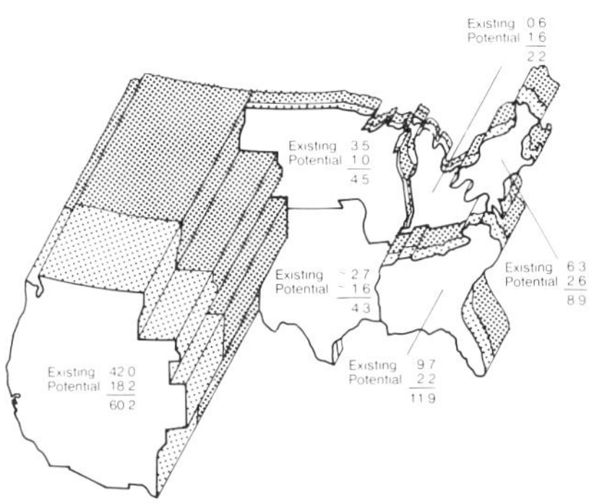

Source: Modified from EPRI, *1982–1986 Overview and Strategy*

Figure 7. Existing and reasonable potential hydroelectric capacity in the year 2000 (GW).

4.5. Other Renewable and Alternative Resources

Geothermal, wind, photovoltaics, solar-thermal electric, and biomass are grouped together because most are relatively new and their existing capacity is small compared to the other technologies discussed here.

4.5.1. Geothermal. In 1980, the principal geothermal site for electricity generation in the United States was the Geysers in Northern California with 837 MWe of natural steam cycle geothermal capacity. Other geothermal sites were being developed in Imperial Valley in Southern California, but only 20 MWe of capacity was in operation in 1980.

There are only three known dry-steam reservoirs in the United States: Yellowstone and Lassen National Parks and the Geysers. At present, only the Geysers is available for commercial development with potential capacity estimated to be 2-3 GWe for 30 years. Approximately 24 GWe of electricity-grade hydrothermal resources have been identified in 51 known geothermal resource areas (KGRA) in the western United States (Muffler 1979). Total geothermal capacity is expected to be approximately 8 GWe by 2000. An arbitrary estimate of 35 KGRAs was used in the large-plant scenario and 45 KGRAs in the small-plant scenario because of the difficulty of specifying the number of power plants to be located on a geothermal field. The rationale for these estimates is an average of 225 MWe/KGRA for the large-plant scenario and an average of 175 MWe/KGRA for the small-plant scenario. In either scenario a majority of capacity will probably come from the Geysers and Imperial Valley in California.

Location of the resource is one of the principal siting factors related to geothermal development. All geothermal (hydrothermal) sites are allocated to the West region. Figure 8 shows the distribution of hydrothermal resources in the western United States. Other siting considerations include ground subsidence resulting from withdrawal of geothermal brines and disposal of spent geothermal fluids and hydrogen sulfide emissions which are regulated. Water availability may be an important siting factor for geothermal power plants, particularly for a binary-cycle plant that has about four times the cooling requirements of a comparable fossil plant.

4.5.2. Wind. Approximately 20% of the United States has a potential annual average wind resource of 150 watts per square meter (W/m^2) or more. If this resource can be captured, the United States wind resource has been estimated at 284,300 GWe (Pacific Northwest Laboratory 1981). However, extraction limits posed by siting constraints, machine spacing requirements, and power conversion losses sharply reduce the achievable potential. Although good individual sites may be found in low wind-density zones, the best wind resources are found in the Northeast and Northwest, along the West Coast, in or near the Appalachian and Rocky Mountains, in the Sierra Nevada, and in the southern plains states.

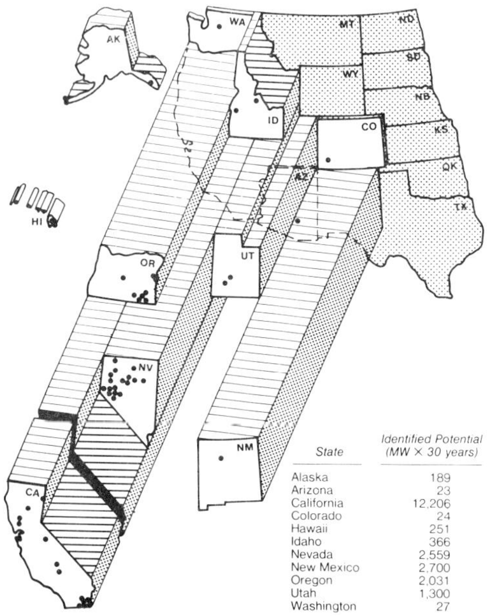

State	Identified Potential (MW × 30 years)
Alaska	189
Arizona	23
California	12,206
Colorado	24
Hawaii	251
Idaho	366
Nevada	2,559
New Mexico	2,700
Oregon	2,031
Utah	1,300
Washington	27

States without identified potential: Kansas, Montana, Nebraska, North Dakota, Oklahoma, South Dakota, and Texas. Wyoming has identified potential, but area withdrawn from commercial exploration or development because of national park status.

● Locations of identified hydrothermal convection systems with reservoir temperatures >150 C

Source: Muffler, L.J.P., *Assessment of Geothermal Resources of the United States—1978*

Figure 8. Potential capacity of major known hydrothermal resources.

In 1980 there were five large utility-operated wind turbine machines in the continental United States located in Clayton, NM, Goldendale, WA, Block Island, RI, Boone, NC, and Cuttyhunk Island, MA. These five sites had a total capacity of approximately 10 MWe, and most are used for research, testing, and demonstration purposes. More recently, large numbers of small wind machines (25-50 kWe) have been installed, primarily in California, by developers who sell electricity to electric utilities. Altamont, Tehachapi, and San Gorgonio Passes in California are the primary sites for these wind machines.

Projected siting requirements for 4 GWe of windfarms by the year 2000 vary from 55 (large plant) to 160 (small plant), based on an average utility windfarm size of 75 MWe for the large-plant scenario and 25 MWe for the small-plant scenario. Because of the nature of the wind resource, it is difficult to make an absolute distinction between windfarm sites.

Regional projected siting patterns for windfarms are based on wind power density and oil-fired capacity (which wind is likely to displace). The West and Northeast regions were each allocated approximately 33% of the total wind sites for both scenarios. The remaining regions have the following percentages of total wind sites: South Central = 17%, both West and East Central = approximately 6%, and Southeast = 5%.

Location of a reliable wind resource is the primary siting factor for wind. Land requirements are another important factor--from 5,600 to 11,200 acres may be required for 100 MWe (Electric Power Research Institute 1981), resulting in land-use conflicts. Other siting factors include aesthetics, noise, and safety.

4.5.3. Solar: Photovotaics and Solar-thermal Electric. The size of the available solar resource in the United States is estimated to be 16×10^{15} kWh per year. While solar energy is ubiquitous, there are many environmental, technological, and sociopolitical obstacles to using even a small fraction of this resource. Two of the more promising methods for generating electricity from solar energy are photovoltaics and solar-thermal. In 1980 photovoltaic (PV) applications were primarily in remote areas and small R&D demonstrations. More recently, two PV installations over 1 MWe in size have been built in California. Approximately 1 GWe of utility photovoltaic capacity is projected by 2000. This capacity may be located on 20 to 200 sites based on an average utility PV installation of 5 MWe for the small-plant scenario and 50 MWe for the large-plant scenario.

Projected siting patterns for PV are based primarily on solar insolation. Figure 9 portrays annual average insolation in the United States. The West region was allocated approximately 33% of the total PV sites for both scenarios. Other regions have the following percentages: South Central and Southeast each have 17%, while East Central, West Central, and Northeast each have 11%.

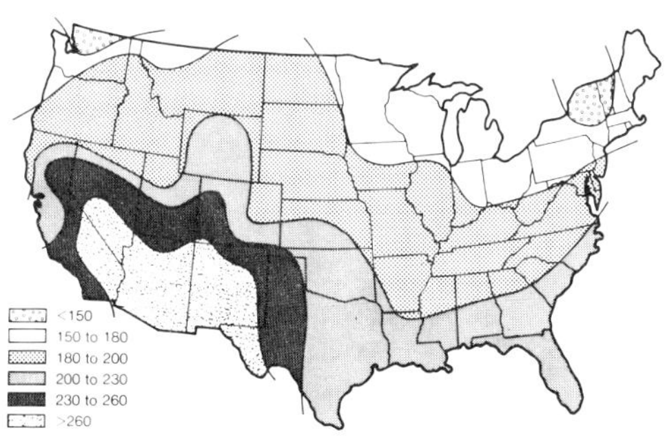

Source: *EPRI, 1982–1986 Overview and Strategy*

Figure 9. Annual average insolation in the United States(W/m^2).

The primary siting requirement for photovoltaics is a large land area with access to solar radiation. Land requirements from 1,400 to 2,800 acres for 100 MWe, may cause potential land-use conflicts (Electric Power Research Institute 1981). Since more solar radiation is required for concentrated than for flat plate systems, concentrated systems are more likely to be located in the southwestern United States.

In 1980, there were no solar-thermal central receiver power plants in the United States. Today, the only utility solar-thermal central receiver in the United States is the 10-MWe "Solar One" pilot plant in Barstow, California, which came on-line in 1982. By 2000 about 1 GWe is projected on 15 to 30 sites, based on an average 35-MWe plant for the small-plant scenario and a 75-MWe plant for the average large-plant scenario. All projected sites are located in the West region. The primary siting factors are very similar to photovoltaics, but land requirements may be less depending on the technology. Water availability is also a concern if plant cooling water is required, because most of these plants are likely to be located in the desert Southwest.

4.5.4. Biomass. In 1980 there were approximately 8 utility biomass sites having some 80 MWe of installed capacity. These sites use a variety of fuels, including wood, animal, and municipal solid waste. By the year 2000, approximately 2.0 GWe of installed capacity could be located on 40-80 sites. The small-plant scenario is based on an average site of 25 MWe and the large-plant scenario on an average site of 50 MWe. Regional projected siting patterns are primarily based on the woody biomass resource. Of the total projected biomass sites, the West and Northeast regions each have 33%, the Southeast and East Central regions each have 9%, and the West Central and South Central each have 8%.

The primary siting factor is availability of the biomass resource. Fuel transportation costs strongly influence biomass economics--generally the fuel must be within a 50-mile radius of the plant in order to be economic. While the land requirements of the generating facility are not great, the amount of land to produce biomass feedstock can be substantial, resulting in a requirement of 40,000 to 90,000 acres for a 100 MWe plant (Electric Power Research 1981). Other siting factors include air quality and water availability.

4.6. Combined-Fuel Types

In 1980 there were 281 mixed fuel sites, a category allowing sites containing any mixture of fuel types. By 2000, projections for additional sites in the large- and small-plant scenarios are 150 and 100, respectively. The rationale for the larger number of sites is based on the premise that more new sites will be mixing fuel types in the large-plant scenario than in the small-plant scenario. Projected combined-fuel type sites were allocated by region, based on the percent of total existing (1980) combined-fuel sites as follows: West Central = 28%, East Central = 21%, Northeast = 17%, Southeast = 16%, South Central = 10%, and West = 8% (U.S. Department of Energy 1982).

5. IMPLICATIONS AND RESEARCH NEEDS

The small-plant scenario results in significantly more sites which may affect more people from a socioeconomic and environmental perspective. This has been shown in Kaman Tempo (1984) that examines the environmental effects of four 550-MWe coal-fired units located at one site compared to four sites. Results from the study indicated that environmental and social costs were greater for the dispersed siting option than the consolidated siting option.

Some hypotheses that need to be examined further include: While renewables and smaller-sized technologies may be easier to site individually, the cumulative effort to site these units is probably greater than locating the same amount of capacity at fewer sites. Public perception seems to favor the siting of smaller technologies, but what will happen when people begin to discern the siting and environmental magnitude of the smaller units. For example, the number of wind machines needed to provide a nominal amount of capacity. Other factors that need to be studied are the projected amount of cogeneration sites: Will they reduce the number of utility sites? If so, by how much? Also, what effect will extending the life of existing units have on future siting requirements?

Further research is needed to better understand siting factors for newer technologies, particularly their relative importance and overall influence in the siting process. Power plant siting patterns need to be further examined within the construct of economic and resource geography and location theory. While fossil-fuel plant location has been the subject of study for a quarter century (early examples include Deasy and Greiss 1960 and Smith 1973) location patterns of other types of generating plants need to be systematically analyzed. Location models could also be improved as discussed by Hobbs (1984).

6. SUMMARY

Power plant siting is a complex spatial decision-making process. Total siting requirements for large- and small-plant scenarios are projected to be between 1500 and 2530 new sites, respectively, in the year 2000. Coal and nuclear technologies already provide and are projected to provide a substantial amount of capacity on a relatively small number of sites. A majority of projected sites, particularly for coal technologies, may be in the South Central and Southeast regions. Petroleum, natural gas, and coal-derived fuels furnish a moderate amount of capacity on a large number of sites, many with capacities less than 10 MWe. No new sites are projected for petroleum and natural gas-fired capacity because of prohibitions on new petroleum and natural gas-fired power plants and reuse of existing sites for replacement capacity. Hydroelectricity, from existing and projected capacity, contributes a modest amount of electricity on a very large number of sites. Most of the projected hydroelectric sites will be in the West and Northeast regions. As with petroleum and natural gas, many of the hydroelectric sites will probably have capacities less than 10 MWe. Other renewable resources currently generate very little electricity and occupy only a few sites. A small fraction of electricity produced in 2000 is projected to be generated from these resources on a moderate to large number of sites. Virtually all of the renewable resource technologies are constrained to locations where the specific resources are naturally available, primarily in the West and Northeast regions. The primary implications of the two scenarios relate to public perception and the magnitude of socioeconomic and environmental effects. Further research is needed to better understand the two scenarios and their implications.

7. ACKNOWLEDGMENTS

The author would like to express appreciation to Frank S. Young, Benjamin F. Hobbs, Jerome E. Dobson, Richard W. Zeren, and some anonymous reviewers for their help and guidance in preparing this paper. The views presented in this paper are solely those of the author and do not necessarily represent the official views of EPRI or its constituents.

8. REFERENCES

Deasy, G., and Greiss, P. 1960 Factors influencing the distribution of steam electric plants. Professional Geographer 12:1.

Dobson, J.E. 1977. The Maryland Power Plant Siting Project: An Application of the ORNL-Land use screening procedure. ORNL/NUREG/TM-79. Oak Ridge,TN.: Oak Ridge National Laboratory.

Eagles, T.W., Cohon, J. L., and ReVelle, C. 1980. Modeling plant location patterns: Applications. EPRI EA-1375. Palo Alto,CA.: Electric Power Research Institute.

Electric Power Research Institute. 1981. 1982-1986 overview and strategy. P-2156SR. Palo Alto, CA.: Electric Power Research Institute.

Hanham, R. Q. and Calzonetti, F.J. 1983. Regional and temporal trends in power plant unit siting, 1912-1978. The Professional Geographer. 35:416-426.

Hillsman, E. L., Alvic, D. R., and Church, R. L. 1983. BUILD(Baseload Utility Integrated Location Decisions): A model of future spatial distribution of electric power production. ORNL 5969. Oak Ridge, TN.: Oak Ridge National Laboratory.

Hobbs, B. 1984. Regional energy facility location models for power system planning and policy analysis. In Analytic Techniques for Energy Planning, ed. B. Lev et al. Amsterdam, Netherlands: North-Holland.

Honea, R.B., Hillsman, E.L. and Mader, R.F. 1979. Oak Ridge siting analysis: A baseline assessment focusing on the National Energy Plan. ORNL/TM-6816. Oak Ridge, TN.: Oak Ridge National Laboratory.

Kaman Tempo. 1984. Environmental impacts of dispersed and concentrated siting of coal-fired power plants. EPRI EA-3484. Palo Alto, CA.: Electric Power Research Institute.

Meier, P., Hobbs, B., McCoy, M. and Stern, R. 1979. The Brookhaven Regional Energy Facility Siting Model (REFS): Model development and application. BNL 51006. Upton, NY: Brookhaven National Laboratory.

Muffler, L.J.P. (ed.). 1979. Assessment of geothermal resources of the United States - 1978. Circular 790. Arlington, VA: U.S. Geological Survey.

Pacific Northwest Laboratory. 1981. Wind energy resource atlas. PNL-3195 WERA 1-12. (12 Vols). Richland, WA: Pacific Northwest Laboratory.

Smith, B.W. 1973. Analysis of the location of coal-fired power plants in the eastern United States. Economic Geography 49: 243.

U.S. Army Corps of Engineers. 1981. National hydroelectric power resources study. (Volumes 1-23). Ft. Belvoir, VA: U.S. Army Corps of Engineers, Institute of Water Resources.

U.S. Department of Energy. 1982. Inventory of power plants in the United States. Washington, DC: Energy Information Administration, Department of Energy.

CHAPTER 11

MODELING IMPERFECT SPATIAL ENERGY MARKETS

Benjamin F. Hobbs
Department of Systems Engineering
Case Western Reserve University
Cleveland, OH 44106

ABSTRACT. Coal, electricity, natural gas, and oil markets are imperfect spatial markets. In such markets, consumers, haulers, and/or producers possess market power which derives from the insulation that transport cost provides from competition with distant rivals. Policy analysis tools have been recently developed for simulating such markets. In this chapter, models for predicting prices in imperfect spatial markets are summarized and their application to energy problems is critiqued. It is concluded that much research remains to be done, and that regional scientists, geographers, operations researchers, and economists all must contribute.

1. INTRODUCTION

A major focus of policy analysis is the projection of responses of markets to changes in public policy. Microeconomic models are frequently used in policy analysis to predict equilibrium prices in energy, agricultural, and other markets and to assess their efficiency and equity. The purpose of this chapter is to critically examine the formulation of models of imperfect spatial markets and their use in energy policy analysis.

Imperfect markets are defined here as those in which firms or consumers perceive that they are able to affect price. This can be contrasted with pure competition, in which firms and consumers take prices as given and equilibrium prices equal marginal cost. Spatial markets are those in which significant transport costs and spatial dispersion of firms and customers allow prices to differ from place to place. In imperfect spatial markets, the source of market power is the cost of transport which insulates firms and/or consumers from competition with their rivals. Markets for coal, oil, electricity, natural gas, and other forms of energy are of this type.

As geographers and regional scientists have long recognized, energy markets are spatial. Most energy forms are costly to transport, and centers of production are often located far from consumers. As a

F. J. Calzonetti and B. D. Solomon (eds.), Geographical Dimensions of Energy, 179–199.
© 1985 by D. Reidel Publishing Company.

result, energy prices vary over space. For example, the delivered price of coal is much lower in West Virginia, where there are many mines, than in New Hampshire, which has no coal seams. Policy analysts have modeled such markets by assuming that pure competition prevails so that price equals marginal cost everywhere. Equilibrium prices are usually calculated by applying Samuelson's (1952) result that a pure competition equilibrium can be obtained from a mathematical program by maximizing the sum of consumer surplus and profit (see Takayama and Judge 1971). Takayama (1978), for example, calculates equilibrium prices for the world oil market using this approach. Labys (1982) lists 20 studies which modeled spatial energy markets using variants of Samuelson's principle.

The shortcoming of the pure competition approach to modeling spatial energy markets is that such markets are not purely competitive. Firms and consumers often possess great market power; this, together with pervasive government regulation, often causes prices to diverge dramatically from marginal cost. Monopoly or oligopoly occur when there are one or several firms, respectively, in a market and they realize that, by altering their outputs, they can influence the prices at which they sell their products. In such markets, firms are able to raise prices above marginal cost. Monopsony or oligopsony, in turn, prevails when there exist one or several consumers, respectively, in a market and they can affect prices by changing how much they purchase. Consumers, in that case, are able to force price below the marginal worth to them of the product. As examples, OPEC exerts oligopoly/monopoly power in the oil market, while some electric utilities are oligopsonies in the U.S. coal market.

Samuelson's principle cannot be used to simulate such imperfect markets. Prices in this type of market are the result of a game where firms and consumers set prices or outputs/ purchases based on how they expect existing and potential rivals to react. Because firms and consumers can hold many different expectations regarding how rivals will react to their decisions, there is no single set of equilibrium prices and no single model that is applicable. One type of model is appropriate for monopolists/monopsonists who lack competition, and another type for oligopolists/oligopsonists who might compete relatively intensely among each other. OPEC's behavior, for example, is modeled by some workers using a Hotelling (1931) type model of a monopolist who controls an exhaustible resource, and by others who use a Stackleberg model in which OPEC maximizes profits subject to the reactions it expects from a "competitive fringe" of producers who believe they cannot affect price. Labys (1982) gives 24 examples of the use of imperfect market models in oil and other energy markets.

But none of the energy applications surveyed by Labys (1982) recognized the importance of market power conferrred by transport costs. In some markets, space is the main source of market power. For example, in the U.S. natural gas market, there are many pipeline companies, but most gas distribution utilities can only buy from one or a very few of them because the others are too distant. In the case of electricity, the existence of approximately 350 private electric utilities in the U.S. does not imply that a deregulated market for bulk power would necessarily be competitive; the cost of transmission would likely limit certain markets to a handful of firms. In coal markets, even though there are many railroads in the U.S., coal producers in some regions face a monopoly or duopoly for transport services. Projections of prices for such markets can be realistic only if the spatial origins of market power are recognized.

Models that explicitly consider the spatial interrelationships of firms and customers in imperfect markets have been proposed in the regional science and economics literature over the last half-century (e.g., Hotelling 1929; Hoover 1937), but are only beginning to be applied to actual markets. The purpose of this paper is to summarize how models of spatial imperfect markets calculate equilibrium prices and to discuss how they have been used to analyze energy markets. Emphasis is placed on market imperfections that result from market power, rather than those that are the outcome of government intervention. Additional applications of imperfect spatial market models to important energy problems are recommended, and modeling issues which require additional attention are discussed.

2. MODELS OF IMPERFECT SPATIAL MARKETS

Models of imperfect markets, both spatial and spaceless, usually focus on the market power of producers while assuming that consumers have none, or vice versa. There are few models for the case in which both producers and consumers exercise market power. This is because of the additional complexity of such situations. Spatial imperfect markets can be even more complicated if, in addition, providers of transport services possess market power. No theory has been developed for the situation in which producers, haulers, and consumers all wield market power; indeed, most spatial models consider only monopoly or oligopoly on the part of producers, and assume that haulers and consumers are competitive. In this section, some concepts in the theory of spatial imperfect markets are summarized, and areas in which further theoretic work is desirable are indicated.

Models of imperfect spatial markets have been formulated for the problems of equilibrium prices and locations. Emphasis is placed here on pricing models, for pricing has been the focus of energy policy analysis. Like models of spaceless imperfect markets, spatial models differ from one another in terms of the conjectures that firms and consumers hold regarding the reactions of rivals. In the case of oligopoly, relatively intense competition results, for example, from the Bertrand conjecture, in which each party believes that rivals will hold prices fixed. In contrast, much higher prices can occur if firms collude. Spatial models based on a range of conjectures are summarized <u>infra</u>.

Imperfect spatial market models also differ from one another in the spatial structure of prices they assume. For example, each firm might charge a single price at its plant while letting customers bear the cost of transport (f.o.b. mill pricing), or each may charge a single delivered price that is the same everywhere (uniform delivered pricing). Alternatively, the delivered prices a firm charges might differ from place to place, but not by the difference in transport cost from the plant (spatial price discrimination). Spatial models also differ in terms of their representation of space. The simplist models have firms and customers uniformly distributed over an infinite or circular one dimensional market; others represent the market as a two dimensional plane and allow transport to take place in any direction. These types of models are called continuous models. Yet another type represents space as a set of nodes interconnected by a set of arcs which define a transport network. Such models are referred to as network models. Because spatial models can differ in the price structure they assume and in their representation of space, and because producers, haulers, and consumers can possess market power, the possible number of spatial imperfect market models greatly exceeds the possible number of spaceless models. For this reason, it is impossible to survey all available spatial models. Instead, a few are presented to illustrate the range of models available. Mention is also made of applications to non-energy markets.

2.1. One-Dimensional Models

To demonstrate the nature of price equilibria in imperfect spatial markets, several simple models are presented in this section. More realistic models are summarized in a following section. Models in which producers have market power are discussed first.

The models presented here make the following assumptions:

1. Each firm owns a single plant. Firms are uniformly distributed along an infinite one dimensional market, each d units from its two neighboring rivals.
2. Goods from different firms are indistinguishable, marginal production cost per unit equals m, and each firm has a fixed cost of F.
3. Demand functions are the same everywhere, and are linear:

$$Q = a - bP \tag{1}$$

   where Q is quantity demanded per unit length and P is delivered price. Consumers buy from the firm offering the lowest price.
4. Transport cost per unit distance equals t, implying that the marginal cost at a point r units from a firm is m + tr.
5. Each firm picks values of P(r) and R to maximize profit:

$$\text{Max}\pi = 2 \int_0^R [a - bP(r)]\,[P(r) - (m + tr)]dr - F \tag{2}$$

Quantity Demanded Price-Marginal Cost Fixed Cost

where R is maximum distance served and P(r) is the delivered price at r. This expression assumes that R and P(r) are symmetric around the plant (r=0).

The polar extremes of competition in spatial markets are pure competition and monopoly. Under pure competition, price equals marginal cost (P(r) = m + tr). In this simple model, this maximizes social welfare, if, as is often done in analyses of imperfect spatial markets (e.g., Villegas 1982), welfare is defined narrowly as the sum of producer and consumer surplus. But if firms are spatially separated, as they are here, competitive forces are unlikely to force prices that low. At the other extreme, there is collusion, where firms split the market, each serving out to d=R/2 and charging net revenue maximizing prices. The results of this model, also called the Loschian model, are well known (Beckmann 1968). For spatial price discriminators, the collusive price at each point equals 0.5((a/b) + m + tr). Colluders who adopt uniform delivered pricing charge 0.5((a/b) + m + 0.25td), while for mill pricers, P(r) = 0.5((a/b) + m - 0.5td) + tr. Prices in each case are well above marginal cost, and social welfare is smaller than that under marginal cost pricing. Beckmann (1968) and others have explored the relative welfare and prices under the different price structures; it is sufficient here to note that different structures do not, in general, yield the same mean prices, welfare, or profit. The Loschian model has been used in qualitative analyses of the behavior of competing banks (Benson 1980) and supermarkets (Benson and Faminow 1985).

Neither of these extremes characterize energy markets; generally, producer competition is neither as intense as pure competition nor entirely absent, as under collusion. A variety of models representing intermediate levels of competition have been proposed. Several, for example, are based on the Bertrand conjecture, where each firm believes that rivals will not change their prices if the firm changes its price (e.g., Hotelling 1929; Schuler and Hobbs 1982). Holahan (1977), for instance, generalizes Hoover's (1937) Bertrand model of spatial price discriminators. Under it, each firm conjectures that rivals will not alter their prices at each point in space. As a result, firms whose marginal costs are lower than the prevailing price at a point in space. As a result, firms whose marginal costs are lower than the prevailing price at a point will drop their prices slightly to steal customers. This happens repeatedly until only the firm with the lowest marginal cost at the point is willing to sell there. Prices will then equal the lower of: (1) the second lowest marginal cost among firms (m + t(d - r)); and (2) the collusive price 0.5((a/b) + m + tr). In either case, the oligopolists keep price above marginal cost. For most energy markets, where transport costs constitute a large but not overwhelming proportion of the costs, the second lowest marginal cost would be the price. Because lower equilibrium prices are only possible if firms believe that rivals will respond to price cut with price hikes, and *vice versa* (Eaton and Lipsey 1978), Bertrand prices represent a reasonable lower bound to prices in spatial oligopoly.

Cournot equilibria represent a less intensive level of competition; in contrast to the Bertrand case, Cournot firms believe that rivals will hold outputs rather than prices fixed. Hobbs (1984a) presents a Cournot mill pricing model, while Greenhut and Greenhut (1975) develop a spatial price discrimination version in which each firm assumes that neighbors will hold their sales at each point fixed. In the latter model, the equilibrium P(r) at each point equals $mc^*(1-(1/\varepsilon n))$, where mc^* is the mean marginal cost, including transport cost, among the n firms serving that point, and ε is the price elasticity of demand at P(r). This price is at least as high as the Bertrand price, and usually higher (Hobbs 1984a). Neven and Phlips (1984) explore the properties of the duopolistic version of this equilibrium and suggest that it may explain the observed pattern of automobile prices in Europe.

Yet another type of spatial oligopoly model results if firms collude, but face competition from new firms who may enter and steal customers. A rational strategy in that case might be limit pricing, where prices are held low to dissuade energy. For example, in the mill pricing case, a limit mill price P_L can be calculated by assuming that existing firms charge a price that would allow entrants to earn, at most, zero profit. If entrants have a maximum capacity of CAP_e, marginal cost m_e, and fixed cost F_e, and believe that existing firms will not alter their prices after entry, then P_L solves:

$$P_L = (F_e/CAP_e) + t[(d/2) + R_e] + (bt^2R_e^3/3) \tag{3}$$

$$P_L = (a/b) + [t(d + R_e)/2] = [CAP_e/2bR_e)] \tag{4}$$

where R_e is the maximum distance served by the entrant who is producing at full capacity. This presumes (1) that entrants place their plants halfway between those of existing firms in an effort to maximize the probability of survival (Eaton 1976), and (2) that collusive prices are even higher than limit prices. Norman and Nichols (1982) presents a more sophisticated dynamic limit pricing model for spatial price discriminators. In that model, entry takes place at every point at a rate that is a function of delivered price. Firms choose profit maximizing prices, which in general are a function of time, so that entry takes place at an optimal rate. A mill pricing version of such a model might be appropriate for the world oil market.

A number of spatial oligopoly models based on other conjectural variations have also been proposed (e.g., Eaton and Lipsey 1976; Greenhut and Ohta 1975). Like the Bertrand, Cournot, and limit pricing models, they yield prices that fall between the extremes of pure competition and collusion.

In contrast to the bounty of available spatial oligopoly models, there exist few spatial models in which haulers or consumers have market power. Schuler and Holahan (1978) propose two such one-dimensional models in which the provider of transport services to producers in a monopolist, producers collude, and new producers enter if producers are earning excess profits. In the Cournot version, the hauler assumes that producers will not alter their prices or service radius R, while producers assume that the hauler won't change its transport fee t. In the Stackleberg version, the hauler realizes that the producers are Cournot players, and takes advantage of that fact. The result of the latter model is that the hauler extracts all of the producers' excess profits; this result may not surprise those who are familiar with the behavior of railroads in the western U.S. towards coal producers.

2.2. Two-Dimensional and Network Models

Although the models described above can serve as tools for understanding the general behavior of imperfect spatial markets, they are too simplistic to be used to simulate actual energy markets. For this reason, more realistic two-dimensional and network based models have been developed.

Two-dimensional continuous models are fairly straightforward generalizations of the one-dimensional models. Additional features are usually added to increase their realism, such as transmission losses (for electricity market models), limited capacity for existing firms, and nonlinear demand functions. For example, Hobbs (1983) and Hobbs and Schuler (1984) derive such models for the Bertrand and limit pricing cases using both straight-line (Euclidean) and city-block (rectilinear) distance metrics.

Continuous models become unwieldy to use, however, if demand functions are not the same everywhere, if firms are not equally sized and uniformly distributed over space, or if production costs differ between firms. Network models, solved using mathematical programming rather than the calculus, can incorporate such spatial heterogeneity and be practical to solve. Network models of perfectly competitive spatial markets have frequently been used to model energy and other markets (Labys 1982). Many are formulated as quadratic programs in which linear demand and supply curves lead to an objective function consisting of the sum of producer and consumer surplus (Takayama and Judge 1971). Recently, the competitive spatial price equilibrium problems has been reformulated as both variational inequality and nonlinear complementarity problem, which enables solutions to be found for more general conditions, including those in which demand curves are not integrable (e.g., Friesz *et al.* 1983). But network models have been developed for only a small subset of imperfect market types for which continuous models are available. Nevertheless, because network models of imperfect markets represent an area of active research, their applicability should be much wider in the near future.

Most available network models of imperfect spatial markets have been developed specifically for application to energy markets. Many are the network analogues of one-dimensional models that are presented *supra*. For example, Harker (1984a) presents an oligopoly and two monopoly network models in which producers have market power, but haulers and consumers have none. The oligopoly model calculates a Cournot equilibrium at each node of the network. It is a generalization of Greenhut and Greenhut's (1975) model to the network case and presents a practical means of calculating such equilibria. In essence, the model has a submodel of each firm with an objective function equaling the profit of the firm. Each submodel, consistent with the Cournot conjecture, takes the quantities supplied to each node by the other firms as fixed. An equilibrium solution is obtained by first solving the submodels using a trial set of values for quantities supplied, and then iteratively resolving the submodels until convergence is achieved. Hashimoto (1984) formulates a similar model, which uses quadratic programming to obtain Cournot equilibria for the linear demand and supply curve case.

Harker's (1984a) monopoly models are variations of a spatial price discriminating monopoly model developed by Takayama and Judge (1971). One model represents the situation in which the monopolist takes transport costs as given and the other simulates the case in which the monpolist is also a monopsonist with respect to transport services. Song and Hallberg (1983) use a similar model to calculate equilibrium world wheat prices under the assumption that several wheat producing nations form a cartel and behave as a monopolist with a competitive fringe. The cartel is assumed to take transport costs as given.

Falk and McCormick (1982) present a mill pricing collusive network model which is based on the concept of a cooperative Nash equilibrium (Nash 1950; Friedman 1977). The nonlinear objective function which is used to obtain the equilibrium is $\pi_k(\pi_k - \pi_k^*)$, where π_k is the profit of firm k in the solution and π_k^* is the "threat point" --the profit that would occur if the cartel collapsed. The Nash cooperative equilibrium is the only one that satisfies the following five reasonable conditions in the case of a two player game (Friedman 1977). Let Z be the set of feasible values of π_k such that $\pi_k \geqslant \pi_k^*$, and assume that Z is compact and convex. The five conditions are: (1) the solution is unique; (2) it is undominated, in the sense that no other solution exist which yields higher profit for at least one firm without lowering the profits or the others; (3) if the π_k are subjected to positive linear transformations, then the new solution is related to the old one by the same transformations; (4) if (a) two problems A and B have the same threat points π_k^*, (b) Z for problem A is contained within Z for problem B, and (c) the solution for B is contained within the Z for problem A, then both problems have the same solution; and (5) arbitrary reordering of the firms' index i will not change the solution.

Unfortunately, because Falk and McCormick's (1982) model assumes a mill pricing structure rather than spatial price discrimination, their problem is non-convex and very difficult to solve. In contrast, Harker's (1984a) model and the other network models discussed <u>infra</u> allow spatial price discrimination, which in most cases yields a convex problem. Mill pricing yields nonconvexities because if a firm raises its mill price, its revenue may inititially increase--but only until its delivered price at a node exceeds another firm's. Revenues then plummet, only to perhaps grow again with further increases in price (Hobbs 1983; Falk and McComrick 182). Work is needed on algorithms that can solve the mill pricing problem more efficiently.

Hobbs(1984b) also develops an oligopoly network model which calculates a Bertrand equilibrium analogous to Holahan's (1977) continuous model. Similar to Harker's (1984a) model, there are implicit subproblems, each of which represents one firm's profit maximizing problem. In this case, firms maximize profit subject to a belief that rivals will hold their prices fixed. Again, an interative procedure is needed to obtain an equilibrium solution. The model was developed particularly for the electric utility industry, and power generation and transmission is modeled in a manner similar to the spatial generation planning models described by Anderson (1972) and Hobbs (1984c). Hobbs (1984b) also presents a limit pricing network model for spatial price discriminators that makes the same basic assumptions as the one-dimensional mill pricing limit pricing model discussed _supra_.

A few network models have been developed that recognize market power wielded by haulers and consumers. Krohm _et al_. (1978) formulate a model in which consumers collude and behave as a monopsonist, and producers and haulers are competitive. Song and Hallberg (1983) present a similar model in which a subset of consumers form a cartel and the remaining consumers behave competitively. They applied their model to the world wheat market.

Brooks (1984) develops a network model for a market in which producers and haulers are a bilateral monopoly, similar to the situation analyzed by Schuler and Holahan (1978). Haulers purchase the product at the plant (in Brook's 1984 application, the natural gas wellhead) and resell it downstream to the consumer. The model allows some of the hauler's services to be regulated so that the price of transport equals average cost, while permitting others to be unregulated. This reflects the situation faced today by many natural gas pipelines in the U.S. The producer prices and transport charges that result depend on the relative strength of the two parties, as reflected in weights upon their profits in the model's objective function. Although the model has some intuitive appeal, it is not based on any recognizable concept of economic equilibrium. It may be possible, however, to provide a theoretic basis for the model by reformulating the objective function so that the model yields a cooperative Nash equilibrium.

Friesz and Harker (1984) draw a distinction between "spatial price equilibrium" models, which yield perfect competition solutions, and "freight network equilibrium" models, which allow haulers and producers/shippers to exercise market power. They survey a number of freight network models, most of which (1) focus on the behavior of railroads and (2) obtain solutions by first predicting freight origins and destinations and then predicting the hauler's decisions. Such sequential solution strategies cannot yield true equilibria between producer and haulers. Only recently have models been proposed that obtain simultaneous equilibria. Harker's (1984b) model is the first that allows rival haulers to behave noncompetitively, although producers are assumed to be competitive. Types of hauler behavior that can be modeled include marginal cost pricing, cost-plus pricing, and rate ceilings. The rate ceilings version could be used to obtain a limit pricing solution in which rates are set at levels to discourage entry by rival transport firms. Although the cost-plus version does not represent a true oligopolistic equilibrium, further developments in that direction may yield practical and realistic models of oligopolistic hauler behavior.

Fisk and Boyce (1983) present a conceptual model of oligopolistic haulers and shippers. Depending on the specific formulation, each hauler can be a Bertrand player, which takes the prices of rival haulers as fixed, a Cournot player, taking the amount and location of traffic hauled by rivals as constant, or a Stackleberg leader, who knows how rivals will react to price or traffic changes. At the same time, each hauler is a Stackleberg leader with respect to shippers who independently minimize their cost of transport. Although the model is elegant, its mathematical structure is very complex and it is not at all clear that it can be used for problems of realistic size (Harker 1984b). According to Friesz and Harker (1984), Harker (1981) formulated a similar model, but even for small test examples, it proved very difficult to solve. There is an obvious need here for practical yet theoretically defensible models that will account for the market power not only of producers, but of haulers and consumers as well. With continued theoretical and algorithmic research, such models may soon be forthcoming.

3. APPLICATIONS TO ENERGY MARKETS

Models of imperfect spatial markets are just beginning to be used by policy analysts to project the impacts of alternative policies upon energy prices, market efficiency, and income distribution. Past and possible future applications to four different energy markets are discussed in this section: electricity, coal, natural gas, and oil.

3.1. Electricity Markets

Electricity markets in the U.S. are, at the moment, heavily regulated by federal, state, and local governments. This in contrast to some European countries, such as Sweden, in which private power producers are subject to less regulation (Bentz and Associates 1982). In the U.S., facility expansion decisions are subject to review, retail rates are strictly controlled so that utilities earn only a 'fair' rate of return (i.e., zero economic profit), and bulk power exchanges between different utilities are usually subject to state and/or federal oversight. Nevertheless, electricity markets exist and there is significant competition among utilities for bulk power contracts and sometimes even retail sales (U.S. Department of Energy 1980; Hobbs 1983). The Public Utilities Regulatory Policy Act of 1978 now allows nonutility entitites to participate in the market by forcing utilities to buy power form independently owned cogeneration and renewable generation sources.

Many economists and regulators believe that the U.S. electricity market should be opened up even more by lifting the restraining hand of regulation from bulk and/or retail power sales (Plummer et al. 1984). For example, the transmission grid might be operated as a common carrier, allowing local distribution companies and large customers to solicit bids for bulk power contracts from many different power generation companies. The argument that deregulation advocates make in the case of bulk power markets is that economies of scale in generation are now small relative to the size of the market. This is in large measure due to advances in transmission technology which have dropped transport costs, thus lessening the market power bequeathed by spaces (see Weiss 1975). Recent studies have examined that argument by using models of imperfect spatial markets to guage the ability of deregulated power producers to exploit consumers.

Schmalensee and Golub (1984) attempt to estimate the likely competitiveness of deregulated bulk power markets by calculating indices of effective market concentration for each of 170 geographic market areas in the U.S. One index is the difference between the quantity sold in a market under pure competition (marginal cost pricing) and that sold under a Cournot equilibrium. The two quantities are calculated for each of the markets one at a time (not simultaneously as in true spatial price equilibrium models, such as Greenhut and Greenhut 1975 or Neven and Phlips 1984). This is done by first calculating the approximate marginal transmission and generation cost for each large power plant within a certain arbitrary radius of the market and then treating, in seperate cases, either each plant, each operating company, or each holding company as a separate firm and calculating the resulting equilibria. An arbitrary distances cutoff is used because insufficient data is available on transmission capacities and costs. They conclude that many markets would be characterized by high effective concentration, implying that deregulated firms would have considerable market power. However, if publicly owned plants behaved competitively by expanding output until their marginal cost equaled price, a strong check would be exerted upon the ability of profit-seeking private utilities to exert market power. But these and the other results of that study are difficult to evaluate because the analysis considered neither actual transmission constraints nor the effect of generation capacity commitment in one market upon prices in other markets. As a result, both the competitive and Cournot prices they calculate are probably far lower than those that would be yielded by a true spatial price equilibrium model.

Hobbs (1983, 1984b) and Hobbs and Schuler (1984) take a different tack by estimating true spatial price equilibria for hypothetical deregulated bulk power markets in upstate New York and the southwestern states of Utah, Arizona, Colorado, and New Mexico. They attempt to define the likely range of deregulated prices by calculating Bertrand prices as a lower bound and limit prices as an upper bound. Continuous two-dimensional models are applied to both regions using region-wide averages for customer densitites, generation and transmission capacities and marginal costs, and regulated prices (used to calibrate linear demand functions). Both mill and spatial discriminatory pricing

Table 1. Summary of Results of Deregulated Power Market Simulations[a]

Region/Year/Model Type	Average-Cost Based Regulation		Marginal Cost Pricing		Bertrand Equilibrium		Limit Pricing	
	Mean Price[b]	Social Welfare[c]	Mean Price[b]	Social Welfare[c]	Mean Price[b]	Social Welfare[c]	Mean Price[b]	Social Welfare[c]
Upstate New York								
1980: Continuous[d]	0.0319	0	0.0468	23	0.0476	23	0.0514[f]	21
1980: Network	0.0328	0	—[e]	—	0.0478	30	0.0587[g]	19
2000: Network	0.0416	0	—[e]	—	0.0521	357	0.0606[h]	197
Long Run: Continuous[d]	0.0508	0	0.0508	0.1	0.516	-0.3	0.516[i]	-0.3
Southwestern U.S.								
1979: Continuous[d]	0.269	0	0.269	0	0.269	0	0.393[j]	-61
Long Run: Continuous[d]	0.381	0	0.0381	0	0.0402	-3	0.0406[k]	-4

a. Sources: Hobbs (1983), Hobbs and Schuler (1984).
b. Average price, weighted by quantity demanded, over the study area.
c. Differece in the sum of producer and consumer surplus compared to price regulation.
d. F.o.b. mill pricing solution.
e. Not calculated directly. Welfare insignificantly above that in the Bertrand solution.
f. Entrant with a 500 MW coal fired steam unit plus peaking gas turbines assumed. 1980 coal prices used.
g. Entrant with a 600 MW coal unit plus turbines assumed. 1990 coal prices used.
h. Entrant with a 600 MW coal unit plus turbines assumed. 2000 coal prices used.
i. Entrant with a 750 MW coal unit plus turbines assumed. 1980 coal prices used.
j. Entrant with a 500 MW coal unit plus turbines assumed. 1979 coal prices used.
k. Entrant with a 750 MW coal unit plus turbines assumed. 1979 coal prices used.

models are applied, since different deregulation proposals would lead to different price structures. Continuous model equilibria are obtained both for the short run (fixed capacity) and long run cases. Linear programming network models are also applied, but only to the upstate New York study area. The network models consist of 55 nodes, one per county, and represent the actual distribution of existing plants, their ownership by each of six utilities, customer densities, and transmission facilities. Both peak and baseload power markets are modeled by including separate transshipment networks for each. In year 2000 solutions, utilities could build additional coal fired power plants in their service areas.

Table 1 summarizes the average results of the deregulated market simulations. The short run analyses show that if deregulated firms compete intensely (Bertrand competition), social welfare would either be the same or even better than under average cost based price regulation. Welfare improvements result in New York because Bertrand prices were closer to marginal cost than average cost based prices. Under Bertrand competition, prices would be, on average, 2% to 5% above marginal cost. Limit prices, however, would be roughly 50% higher than regulated prices, and yield welfare losses in the Southwest. Limit prices are so much higher primarily because the cost of new coal plants (the assumed technology for entrants) are much higher than the imbedded cost of existing facilities. In the long run, however, the bounds upon deregulated prices are much tighter, and are not much higher than regulated prices. The reason for this is that marginal and average costs of existing firms in the year 2000 and beyond are closer to those of entrants.

The network and continuous model results for New York State are surprisingly consistent, given the basic differences in the methodologies. The mean Bertrand results for 1980 are almost identical, and the differences between the limit prices can be explained by the different entrant cost assumptions used. What the network models show in addition is the geographical distribution of price changes across the state. Whereas both models show that _mean_ prices would rise under deregulation and that _mean_ profit would increase, the network model solutions reveal that in areas of the state where utililty prices are now relatively high, prices would actually fall, and some utilities would be threatened by bankruptcy. In essence, under deregulation, power would be diverted from counties in which power prices are presently low to counties where the power is more highly valued.

But as Schmalensee and Golub (1984) point out, these analyses are limited in that they examine only two regions of the country and, in the case of the continuous models, assume a uniformity in demand and cost conditions that doesn't exist. Fortunately, the New York analyses indicate that the average results of continuous and network models are similar. What the continuous models fail to show is how different firms and groups of consumers are affected. Another shortcoming of the analyses is that the assumed upper bound may be too optimistic; entrants might be unable to build large power plants cheaply and quickly enough to provide a check on collusion. Analyses that consider the range of technologies that entrants are likely to adopt (including small-scale plants and cogeneration) and the rate at which entry occurs would be more realistic. The dynamic limit pricing model of Norman and Nichols (1982) may provide a useful framework for such an approach.

3.2. Coal Markets

The coal mining industry in the U.S. is a relatively competitive one, even considering the spatial dispersion of production. But the transport of coal is not. Railroads have considerable spatial market power, particularly in the west, and have thus far been adept at fending off challenges by coal slurrly pipelines to their markets. The ownership by some railroads of coal reserves further complicates the picture. Thus far, no one has used spatial imperfect market models to explicitly analyze the market power possessed by railroads.

But recent developments in freight transportation modeling, summarized *supra*, indicate that the capability to do such analyses may be available soon. For example, the marginal cost pricing version of Harker's (1984b) model was used to solve a large scale problem (approximately 3500 nodes and 15,000 arcs). With it, the rates that 16 U.S. railroads would charge for the transport of steam coal were obtained and the effect of closing three U.S. ports to coal exports was estimated. The cost-plus pricing or rate ceiling versions might be used to estimate the effect of railroad market power upon use of U.S. coal resources. The Fisk and Boyce (1983) approach, should it prove computationally practical, would provide a theoretically rigorous means of accomplishing the same goal.

A few analyses of the U.S. coal market have been made using a spatial oligopolistic framework. Hwang (1979) econometrically estimates spatial price schedules for coal and showed that producers use spatial price discrimination. This betrays the presence of at least some market power on the part of coal mines (see Mertens and Ginsburgh (1984) for a similar analysis of the European car market). Krohm *et al.* (1978) solve two versions of the Argonne Coal Market model, one representing pure competition and the other assuming that consumers collude and are monopsonists. Shifts in production among regions resulted. But as neither type of market is credible, those results are of minor interest.

Although U.S. coal producers may behave fairly competitively, the international coal market in strongly oligopolistic (Abbey and Kolstad 1983). South Africa, Australia, and Poland have significant market power as producers, and Japan and the EEC have power has consumers. The expense of coal transport is the source of much of this market power. Falk and McCormick (1982) formulate a Bertrand network mill pricing model, but reject its use because of computational difficulties posed by its nonconvexities. Instead, they develop a spatial Nash cooperative equilibrium model in which the producer's cartel has three members (east and west Australia and South Africa). Consumers have no market power. Even this simple model proved difficult to solve, primarily because of the imposition of a mill pricing structure. Formulations based on spatial price discrimination would permit the solution of more realistically sized models.

3.3. Natural Gas Markets

The U.S. natural gas market is not a purely competitive one. Pipelines possess market power both in buying gas from producers, as there are at most a handful of pipelines who buy gas from a given field, and in selling it downstream, as there are few pipelines who deliver to any given market. Pervasive and rapidly changing government regulation makes the picture even more complicated. Transport costs and government regulation cause gas prices to very significantly across the country, yielding differences of up to 50% between regions (Montgomery and Gottlieb 1983). Broadman et al. (1983) conclude that our knowledge of the degree of competition among pipelines is very limited, and that research is needed on the boundaries of spatial markets. Application of models of spatial oligopoly and oligopsony using data on gas transmission costs and locations of consumers and producers could help assess the competiveness of the pipeline industry under different kinds of regulation. This is now an especially important topic, as more of the pipelines' sales to industry are becoming unregulated. Such a modeling exercise would have to explicitly consider the complicated nature of government regulation of production and prices.

Brooks (1984) has developed the only spatial oligopolistic model of the natural gas sector. As mentioned _supra_, it is not based on a rigorous concept of economic equilibrium. It represents a form of collusion, which is a 'worst case', in terms of prices, for consumers. A 'best case' could easily be calculated by solving a pure competion version of the same model. Intermediate levels of competition might be simulated by reformulating the model so that pipelines are Cournot or Bertrand players and producers are marginal cost pricers. But before this can be done, more understanding of the present and likely future dynamics of competition in local gas production and consuming markets is needed. This would help guide the choice and structuring of an appropriate imperfect spatial market model.

The model of Brooks (1984) also treats government regulation in a simplified manner. Murphy _et al_. (1980) developed means for including realistic regulatory price and quantity adjustments into the natural gas sector of the U.S. Department of Energy's PIES model; incorporation of those features into the Brooks (1984) model might make it a more useful tool for policy analysis.

3.4. Oil Markets

Although the degree of collusion among U.S. oil companies is the subject of intense debate, what market power they may exercise does not seem to derive from the spatial nature of the market. But this may change if, as has been proposed by the Presidential Task Force on Regulatory Relief, oil pipelines are deregulated. Love (1983) asserts that pipeline owners would be able to exploit upstream oil producers who would have only one or a very few different pipelines to sell to. Models that are appropriate for modeling the market power of natural gas pipelines would also be useful for assessing the possible consequence of oil pipeline deregulation.

The international oil market was the first energy market to be modeled as an oligopolistic market. Indeed, most of the studies cited by Labys (1982) focus on the world oil market. Surprisingly, despite the significant expense involved in shipping oil, no effort has been made to analyze the market as a spatial oligopoly. This is probably because OPEC's market power does not have its origins in the cost of transport and the spatial separation of firms. But as OPEC continues to be weakened by competitive forces in the forms of conservation and substitution of non-OPEC for OPEC oil, spatial oligopoly is likely to become more important. Producing centers that are closer to consuming countries might be able to demand a premium, while those further away or located in conflict-prone areas of the world may have to cut their prices. Large consumers may be able to exercise market power; market power, after all, is one of the raisons d'etre of our Strategic Petroleum Reserve.

Spatial imperfect market models of the world oil market would have to take several characteristics of the market into account in order to be realistic. One aspect is the finite nature of the resource, which affects the pricing policy of some OPEC countries. Another is the dynamic nature of entry; exploration and development efforts in non-OPEC countries both depend upon and affect the price of oil. These two characteristics imply that a hybrid of Hotelling's (1931) exhaustable resource model and Norman and Nichol's (1982) spatial limit pricing model might be appropriate. The possibility of buyer market power should also be considered, which, by making the problem a bilateral oligopoly, significantly compounds its complexity.

4. CONCLUSION

Models of imperfect spatial markets can capture important aspects of energy markets missed by spaceless oligopoly models and traditional spatial price equilibrium models. Recent developments in the field have been many and exciting. Both continuous and network models have been developed for a variety of producer conjectural variations, such as Bertrand and Cournot competition, limit pricing, and collusion, and for both mill pricing and spatial price discriminating producers. A few models have also been formulated for the cases where consumers or haulers wield market power. Models involving imperfect competition among haulers present particularly difficult theoretical and computational problems, but recently there have been some encouraging developments.

Most applications of imperfect spatial models have been to energy markets, and for good reason: energy is a critical component of the economy, most energy forms have significant transport costs, and producers and transporters of energy are spatially dispersed. In some applications, the models are used to analyze the extent of market power that would be exercised under alternative market structures, and in others, they have been used to analyze other economic and environmental issues. Successful application of these models in the future will depend on: (1) theoretical developments, such as the definition of price equilibria when both producers and haulers have market power; (2) algorithmic developments, such as the creation of computer programs that can solve realistically sized nonlinear network models with arc losses and capacity constraints; and (3) empirical investigations of energy markets, which can reveal the extent of spatial barriers to competition and how they actually affect behavior of producers, haulers, and consumers. Contributions will be needed from geographers, regional scientists, operations researchers, and economists.

5. ACKNOWLEDGMENT

The author thanks an anonymous referee for several helpful suggestions.

6. REFERENCES

Abbey, D.S., and Kolstad, C.D. 1983. The structure of international steam coal markets. Natural Resources Journal 23: 859-91.

Anderson, D. 1972. Models for determining least cost investments in electricity supply. Bell Journal of Economics and Management Science 3: 267-99.

Beckmann, M. 1968. Location theory. New York: Random House.

Benson, B.L. 1980. An examination of U.S. vs. Philadelphia National Bank in the context of spatial microeconomics. Industrial Organization Review 8: 27-65.

______, and Faminow, M.D. 1985. An alternative view of pricing in retail food markets. American Journal of Agricultural Economics 67.

Bentz, E.J., & Associates and Burns & Roe Industrial Services Group. 1982. Characterization of foreign electric utility experiences for potential useful applications in the U.S. Washington, D.C.: U.S. Department of Energy, Division of Electric Utility Policy, Report DOE/PE/70225-1.

Broadman, H.G., Montgomery, W.D., and Zimmerman, M.B. 1983. Natural gas markets after deregulation: Methods of analysis and research needs. Baltimore: Johns Hopkins University Press.

Brooks, R.E. 1984. The TENRAC gas pipeline competition model. In Analytic techniques for energy planning, eds. B. Lev, F. Murphy, J. Bloom, and A. Gleit. Amsterdam: North-Holland.

Eaton, B.C. 1976. Free entry in one-dimensional models: pure profits and multiplier equilibria. Journal of Regional Science 16: 21-23.

______, and Lipsey, R.G., 1978. Freedom of entry and the existence of pure profit. Economic Journal 88: 455-69.

Falk, J.E., and McCormick, G.P. 1982. Mathematical Structure of the International Coal Trade Model. Washington, D.C.: U.S. Department of Energy, Report DOE/NBB-0025.

Fisk, C.S., and Boyce, D.E. 1983. Game Theory and Transportation Systems Modeling. Urbana, Ill.: Transportation Planning Group, Department of Civil Engineering, University of Illinois, Publication 18.

Friedman, J.W. 1977. Oligopoly and the Theory of Games. Amsterdam: North-Holland.

Friesz, T.L., Gottfried, J., Brooks, R.E., Zielen, A.J., Tobin, R., and Meleski, S.A. 1981. The Northeast Regional Environmental Impact Study: Theory, Validation, and Application of Freight Network Equilibrium Model. Argonne, Ill.: Argonne National Laboratory, Report ANL/ES-120.

Friesz, T.L., and Harker, P.T. 1984. Freight Network Equilibrium: A Review of the State of the Art. Philadelphia, Penn.: University of Pennsylvania, unpublished paper.

Friesz, T.L., Tobin, R.L., Smith, T., and Harket, P.T. 1983. A nonlinear complementarity formulation and solution procedure for the general derived demand network equilibrium problem. Journal of Regional Science 23: 337-59.

Greenhut, J.G., and Greenhut, M.L. 1975. Spatial price discrimination, competition, and locational effects. Economica 42: 401-19.

Greenhut, M.L., and Ohta, H. 1975. Theory of Spatial Pricing and Market Areas. Durham, N.C.: Duke University Press.

Harker, P.T. 1981. A Simultaneous Freight Network Equilibrium Model with Application to the Network Design Problem, M.S.E. Thesis. Philadelphia, Penn.: University of Pennsylvania.

______. 1984b. A generalized spatial price equilibrium model. Papers of the Regional Science Association, forthcoming.

Hashimoto, H. 1984. A Spatial Nash Equilibrium Model. Washington, D.C.: The World Bank, Commodity Studies and Projections Division, Working Paper 1984-2.

Hobbs, B.F. 1983. A Location Theoretic Analysis of Competition in Electrical Generation. Unpublished Ph.D. dissertation. Cornell University, Department of Civil Engineering.

______. 1984a. Bertrand and Cournot models of spatial competition under mill pricing and spatial price discrimination. Unpublished manuscript.

______. 1984b. A spatial linear programming analysis of the deregulation of electricity generation. Unpublished manuscript.

______. 1984c. Regional Energy Facility Location Models for Power System Planning and Policy Analysis. In Analytic Techniques for Energy Planning, eds. B. Lev, F. Murphy, J. Bloom, and A. Gleit. Amsterdam: North-Holland.

______, and Schuler, R.E. 1984. Spatial competition among deregulated electricity producers in the southwestern United States. Modeling and Simulation 15: 295-302.

Holahan, W.L. 1977. Competitive Spatial Price Policies and Consumer Welfare. Milwaukee, Wisc.: Department of Economics, University of Wisconsin, unpublished paper.

Hoover, E.M. 1937. Spatial price discrimination. Review of Economic Studies 4: 182-91.

Hotelling, H. 1929. Stability in competition. Economic Journal 39: 41-57.

______. 1931. The economics of exhaustible resources. Journal of Political Economy 39: 137-75.

Hwang, M.J. 1979. A model of spatial price discrimination for the pricing schedule of coal. Journal of Regional Science 19: 231-43.

Krohm, G.C., Mehring, J., and Van Kuiken, J.C. 1978. The representation of markets in optimization models. In Coal in Transition: 1980-2000. Stanford, CA: Energy Modeling Forum, Stanford University.

Labys, W.C. 1982. Evaluation of the International Energy Evaluation System: Report of Phase I. Cambridge, Mass.: Energy Model Analysis Program, Energy Laboratory, Massachusetts Institute of Technology.

Love, J. 1983. Quoted in Energy Daily, September 28.

Mertens, Y., and Gisburgh, V. 1984. Product differentiation and price discrimination in the European community. Louvain-la-Neuve, Belgium: Universite Catholique de Louvain, Center for Operations Research and Econometrics, Discussion Paper No. 8424.

Montgomery, W.D., and Gottlieb, M. 1983. Transport costs and the spatial distribution of natural gas prices. Washington, D.C.: Resources for the Future, Unpublished Discussion Paper D-92C.

Murphy, F.H., et al. 1980. Modeling natural gas regulatory proposals using the Project Independence Evaluation System. Operations Research 29: 876-902.

Nash, J.F. 1950. The bargaining problem. Econometrica 18: 155-62.

Neven, D., and Phlips, L. 1984. Discriminating oligopolists and common markets. Louvain-la-Neuve, Belgium: Universite Catholique de Louvain, Center for Operations Research and Economics, Discussion Paper 8422.

Norman, G., and Nichols, N.K. 1982. Dynamic market strategy under threat of competitive entry: an analysis of the pricing and production policies open to the multinational company. Journal of Industrial Economics 31: 152-74.

Plummer, J., Ferrar, T., and Hughes, W., eds. 1984. Electric Power Strategic Issues. Arlington, Va.: Public Utilities Reports.

Samuelson, P.A. 1952. Spatial price equilibrium and linear programming. American Economic Review 42: 283-303.

Schmalensee, R., and Golub, B.W. 1984. Estimating effective concentration in deregulated wholesale electricity markets. Rand Journal of Economics 15.

Schuler, R.E., and Hobbs, B.F. 1982. Spatial price duopoly under uniform delivered pricing. Journal of Industrial Economics 31: 175-88.

______., and Holahan, W.L. 1978. Competition vs. vertical integration of transportation and production in a spatial economy. Papers of the Regional Science Association 41: 209-25.

Song, D.H., and Hallberg, M.C. 1983. An analytic model of the imperfect world trade market: a case study of world wheat cartels. Canadian Journal of Agricultural Economics 31: 389-400.

Takayama, T. 1978. An application of spatial and temporal price equilibrium model to world energy modeling. Papers of the Regional Science Association 41: 43-58.

______., and Judge, G. 1971. Spatial and temporal price and allocation models. Amsterdam: North-Holland.

U.S. Department of Energy 1980. The national power grid study. Report DOE/ERA-0056-2. Washington, D.C.: U.S. Government Printing Office.

Villegas, D.J. 1982. Comparative performance of spatial models of economic markets with linear household demands. Southern Economic Journal 48: 893-908.

Weiss, L.W. 1975. Antitrust in the Electric Power Industry. In Promoting competition in regulated markets, ed. A. Phillips. Washington, D.C.: The Brookings Institution.

ut th
nication

CHAPTER 12

NUCLEAR POWER IN THE US AND UK: THE ROLE OF SITING IN SAFETY PHILOSOPHY

John Fernie
Huddersfield Polytechnic
Department of Geographical Sciences
Queensgate
Huddersfield HD1 3DH
England

Stanley Openshaw
University of Newcastle
Department of Geography
Newcastle upon Tyne NE1 7RU
England

ABSTRACT. Reactor safety has featured prominently in the nuclear debate in the US and UK throughout the 1970's and 1980's. Public acceptability of nuclear power will largely be related to the resolution of safety issues. In view of the proposal by the UK Central Electricity Generating Board to build PWRs, this paper discusses the safety philosophies which have evolved in both countries. Whereas UK safety authorities have consistently advocated siting as a means to minimise the residual risks of nuclear generation, their UK counterparts consider siting to be only of marginal relevance as a safety measure. This is borne out from a cluster analysis of 109 sites using 12 demographic variables. A sixfold classification illustrates the predominance of remote sites in the US and high population density sites in the UK. A search for more remote, safer sites in the UK is advocated.

1. INTRODUCTION

The late 1970's and early 1980's has been a period of crisis for the nuclear industry in the USA and UK. The euphoria which heralded nuclear power in the 1950's and 1960's as a safer, environmentally benign and cheaper form of electricity generation than alternative sources has given way to a phase of consolidation as attempts are made to revive the industry's fortunes.

F.J. Calzonetti and B.D. Solomon (eds.), Geographical Dimensions of Energy, 201–223.
© 1985 by D. Reidel Publishing Company.

In the US, massive ordering programmes in the late 1960's and early 1970's were never fully realised as a combination of high prices, conservation and recession depressed demand while nuclear power costs escalated because of high interest rates, construction delays and stringent regulatory modifications. As a result, cancellation of orders began to exceed those licensed or under construction. By January 1984, 82 reactors were licensed, 49 were under construction and 116 had been cancelled (UCS, 1984; USDOE, 1984). In the UK, similar market conditions were experienced. The state power utilities, the Central Electricity Generating Board (CEGB) and the South of Scotland Electricity Board (SSEB) had a strong commitment to nuclear power but uncertainty over reactor choice prevented any large ordering programme. Nevertheless, previous investment plans were sufficiently optimistic in forecasting demand that it is unlikely that any new plant will need to be commissioned until the 1990's (MacKerron, 1984; CEGB, 1982). The present proposal by the CEGB to construct the first of a series of Pressurized Water Reactors (PWRs) at Sizewell is therefore justified on the grounds of economics as well as system costs and diversification of fuel usage (CEGB, 1982).

Ordering in advance of need will also prevent the run down of the British nuclear manufacturing and construction industries in order that they can meet demand in more buoyant times later in the century. One of the problems in ordering PWRs, however, is that they may not benefit British industry for some time, as much of the equipment and materials will be imported. Franklin (1984), former managing director of the National Nuclear Corporation (NNC) which builds British plants, argues that the ordering of a PWR will diminish the prospects of the Advanced Gas-cooled Reactor (AGR) and will lead to the abandonment of teams with considerable experience in designing, manufacturing, and constructing AGRs in the UK. Furthermore, he draws parallels between the disillusionment of US utilities with the PWR at the present time and the CEGB's problems with the AGR a decade earlier. Indeed, Starr and Braun (1984) claim that many of the problems associated with the PWR in the US can be attributed to the fragmentary nature of the nuclear industry. Hundreds of US utilities have ordered reactors from the four reactor vendors, but the large number of competing architect-engineering firms and associated manufacturing concerns have produced custom built plants which have not facilitated 'learning curve' experience for utilities and have delayed Nuclear Regulatory Commission (NRC) review procedures. It is not suprising that NRC advocates the standardisation of plant designs to expedite this process (NRC, 1984).

In the UK, this problem of fragmentation does not exist. Since 1973 when the NCC was created, the two state utilities have liaised with one architect-engineer-construction company and a few plant and component manufacturers. Knowledge of the AGR after early technical and construction difficulties is now at the top of the 'learning curve' while knowledge of PWRs is only at an early stage. This is reflected in the involvement of Westinghouse and Bechtel as consultants to the CEGB and NNC for the Sizewell PWR.

The uncertainty over PWRs in the US and AGRs in the UK is a reflection of a crisis of confidence over these particular nuclear technologies. This uncertainty has increased public concern about nuclear power. In the US, the accident at Three Mile Island (TMI) and the associated adverse publicity did little to allay fears over nuclear safety. In the UK, some of the criticisms of the nuclear industry at the Sizewell inquiry have been overshadowed by incidents related to the back end of the nuclear fuel cycle; for example, radioactive leaks at Sellafield (formerly Seascale, formerly Windscale) and the publicity associated with the site selection procedure for radioactive waste disposal. Overall, reassuring statements by the industry and government officials have been met with increased scepticism (Fells, 1984). In essence, those responsible for promoting nuclear energy are doing a poor job of selling their product.

It is clear that public acceptability of nuclear power is related to safety issues. Reactor safety has dominated the nuclear debate in the US from the public hearing on the Emergency Core Cooling System in the early 1970's to the revised safety goals produced in the wake of the TMI accident. Indeed, several nuclear enthusiasts such as Lilienthal (1980) have maintained that the only way to achieve public acceptance of nuclear power is to design an ultra safe reactor which would avert an accident such as that at TMI. Although the industry is far from convinced by this argument, the Congressional Office of Technology Assessment (1984) endorses this view. It argues that fundamental changes in technology, management and regulation will be necessary to restore faith in nuclear power. A new approach to safety is advocated through modifications of existing designs, including the Process Inherent Ultimately Safe (PIUS) concept and the development of advanced reactor systems. Paradoxically, one of the 'safe' designs under consideration is the Modular High Temperature Gas Reactor; a development of current gas-cooled reactor technology. Meanwhile, the UK is contemplating a switch to light water reactors.

The CEGB has flirted with light water technology since the early 1960's but successive governments, on the advice of the UK Atomic Energy Authority (UKAEA), have favored the British AGR design. It was in 1973, however, when the CEGB first announced its commitment to the PWR that the question of reactor safety became a prominent issue. In the ensuing decade a former Chief Scientific Advisor to the Government expressed his doubts about the integrity of the reactor pressure vessel, the UKAEA endorsed the design for licensing in the UK, and the safety case of the CEGB became a prominent feature of the Sizewell inquiry. Moreover, the inquiry inspector has hinted that his final decisions may depend upon the resolution of safety issues between the CEGB and the Nuclear Installations Inspectorate (NII), the nuclear regulatory equivalent of the NRC in the UK.

2. THE ROLE OF SITING IN SAFETY PHILOSOPHY

In view of the commitment by the CEGB to a US designed reactor and the current emphasis on safety issues as a means of achieving public credibility of nuclear power, it is appropriate to discuss the safety philosophies which have evolved in the US and UK.

The safety of nuclear plant depends on a number of factors: the basic design and engineering safeguards to reduce the risk of failure of any part of the system, the quality of equipment and materials, the standard of construction and maintenance work, the ability of operators and the skill of operators, the siting of reactors to minimise the risks of radiological exposure in the event of an accident, and the absence of unexpected events. In the US these safety criteria are met by utilities complying with licensing regulations which were originally derived from the Atomic Energy Act of 1954 and which have subsequently been modified with regulatory experience. Utilities have to submit a safety evaluation report for agency and public comment prior to a mandatory public hearing before the issuance of a construction permit. Further evidence is required after construction is complete before an operating licence is granted and the NRC maintain surveillance of the plant throughout its operating life.

In practice, the fragmentation of the nuclear industry has led to differential standards throughout the country. Indeed, 'bad' publicity associated with utilities which have violated NRC regulatory codes has not enhanced the industry's image in the eyes of the public. Nevertheless, the NRC has considerable experience of light water reactor operations and from a wealth of case law has produced more stringent safety standards in the wake of the TMI accident. Since this accident in 1979 utilities have had to consider in their safety review accidents beyond the design basis. The design objective for the frequency of occurrence of a large uncontrolled release of radioactivity into the environment is less than 10^{-5} per reactor year of operation.

This greater consideration for possible but unlikely events is also reflected in the revised regulations for emergency planning. At the time of the TMI accident, evacuation plans only covered a low population zone of 2 to 3 miles from the reactor. Since 1980 new regulations were introduced to expedite evacuation within 10 miles of the reactor (NRC, 1981).

Even this planning zone has been criticised as too narrow in view of the response by residents to a real, rather than a hypothetical accident (Zeigler and Johnson, 1984).

In the UK, the state utilities have absolute responsibility for the safety of their nuclear plants. The regulation of nuclear installations is carried out by the NII which grants site licenses and monitors the operations of licensed plants. Both the utilities and the NII share common safety objectives. They are: that the risk presented by nuclear plant should be as low as reasonably practicable and that all reasonably practicable steps should be taken to minimize the radiological consequences of an accident (HSE, 1979).

In many ways the task of the NII is much easier than that of the NRC with regard to the quality control aspects of nuclear safety because of the more centralized institutional structure of the British nuclear industry. The main differences in safety philosophy between the NRC and the NII, however, lie in the realm of design safeguards and siting considerations. The more conservative approach to safety standards adopted in the US is not being followed in the UK. Although the CEGB has no operational experience with PWRs, it feels that by spending £100 million on additional safety measures, it can guarantee safety to nearby residents.

The design target for a credible accident for the British PWR and the AGR is 10^{-7} per reactor year (CEGB, 1982; Dale, 1982). This implies that British reactors are two orders of magnitude safer than their US counterparts. The CEGB has only undertaken limited studies of the possible consequences of accidents beyond the design basis and even then it has assumed that the built-in engineered safeguards would either prevent the release of radioactivity completely or else contain it to a very small geographical area (George and Hilsley, 1982). This optimism is reflected in UK emergency planning procedures which only envisage the worst case evacuation of people downwind from the reactor to a distance of 2 miles (HSE, 1982).

These guidelines were formulated in 1968 when the use of a concrete pressure vessel in the AGR design was considered to be sufficiently safe to allow plants to be located in urban and industrial areas. Since then, the choice of the PWR as the favoured option by the CEGB has posed questions about whether the PWR's pressure vessel can be built to the same standards of safety and whether microprocessor control systems can make up for the far shorter response times of the PWR system. In essence, should this 'relaxed' siting policy be applicable to the PWR?

The entire question of accidents beyond the design basis and the nature of emergency plans to accommodate these unlikely events would be relatively insignificant if the predicted reactor accident frequencies proposed by the CEGB on the basis of zero experience with the PWR were not a matter of international scientific dispute. Objectors at the Sizewell inquiry have been quick to point out the more pessimistic forecasts emanating from the US (Minarick and Kickielka, 1982) and West Germany (Armstrong, 1984). Even the inspector at Sizewell asked the CEGB to consider a 5 mile evacuation zone rather than the statutory 2 miles assumed in their evidence about emergency plans.

It is clear from this discussion that geographical location is important in the safety debate because the consequences of a major release of radiation depend on: the size and nature of the release, the prevailing atmospheric conditions, the response of the emergency services, and the distribution of population downwind of the reactor. On these variables, only reactor location can be manipulated and this is under the direct control of the safety authorities and the site selection teams representing the utilities.

Safety, however, is only one of the many considerations in site selection. Once a utility has determined that a nuclear plant offers the most lucrative return on its capital investment to meet future demand in part of its service area, alternative sites will be evaluated in relation to the variables shown in Table 1. Figures 1 and 2 show the geographical

Table 1. Factors Influencing the Siting of Nuclear Power Plant

Variables	Description
Physical	adequate and unfailing supply of cooling water geological conditions suitable for heavy loads availability of a flat site with development potential
Economic	proximity to transmission lines costs of any special site related features
Environmental	avoidance of high amenity areas effects of cooling water on local aquatic ecology landscaping requirements
Political	regional or local attitudes to nuclear power job creation benefits
Safety	demographic criteria feasibility of local evacuation plan

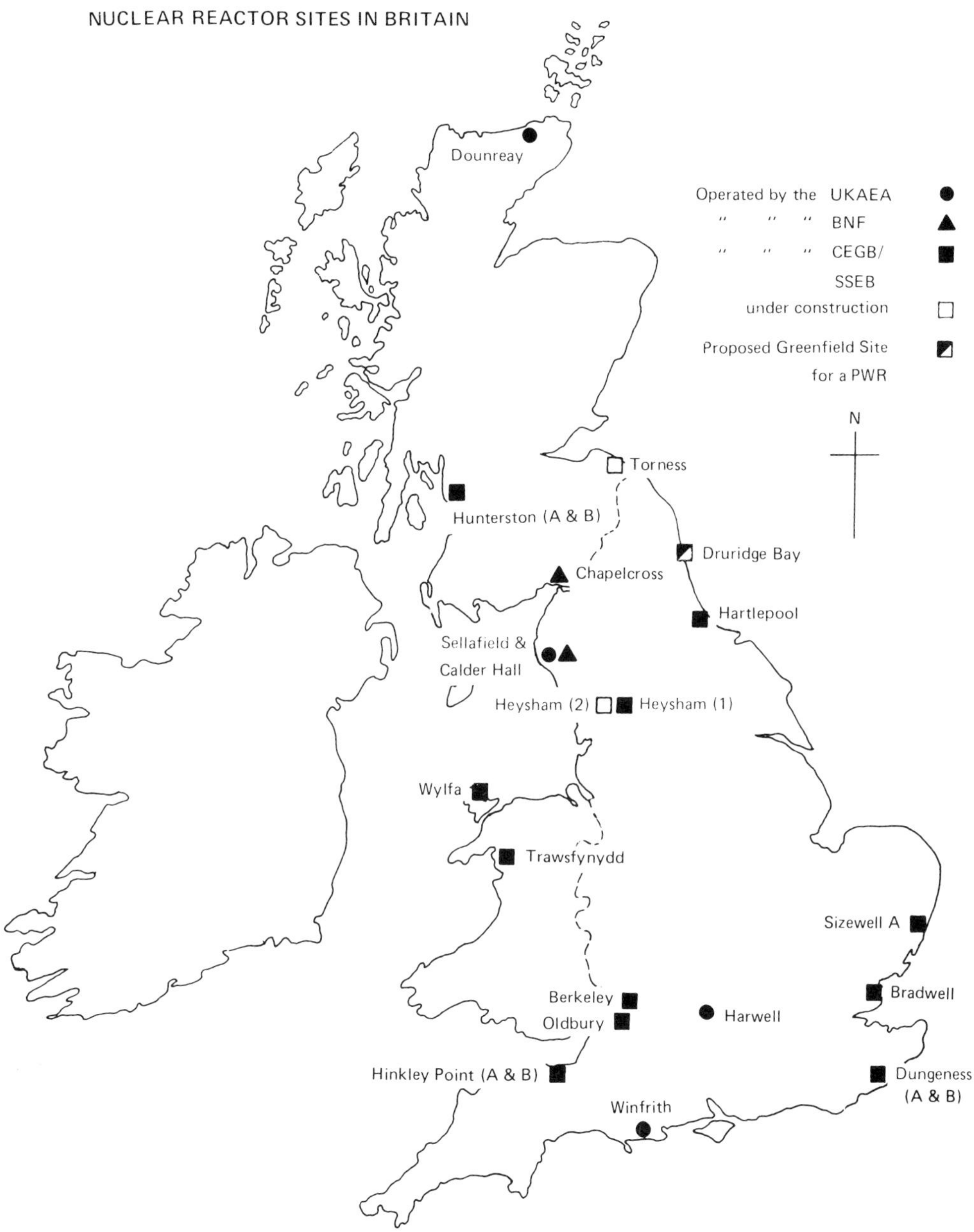

Figure 1. Nuclear reactor sites in Britain.

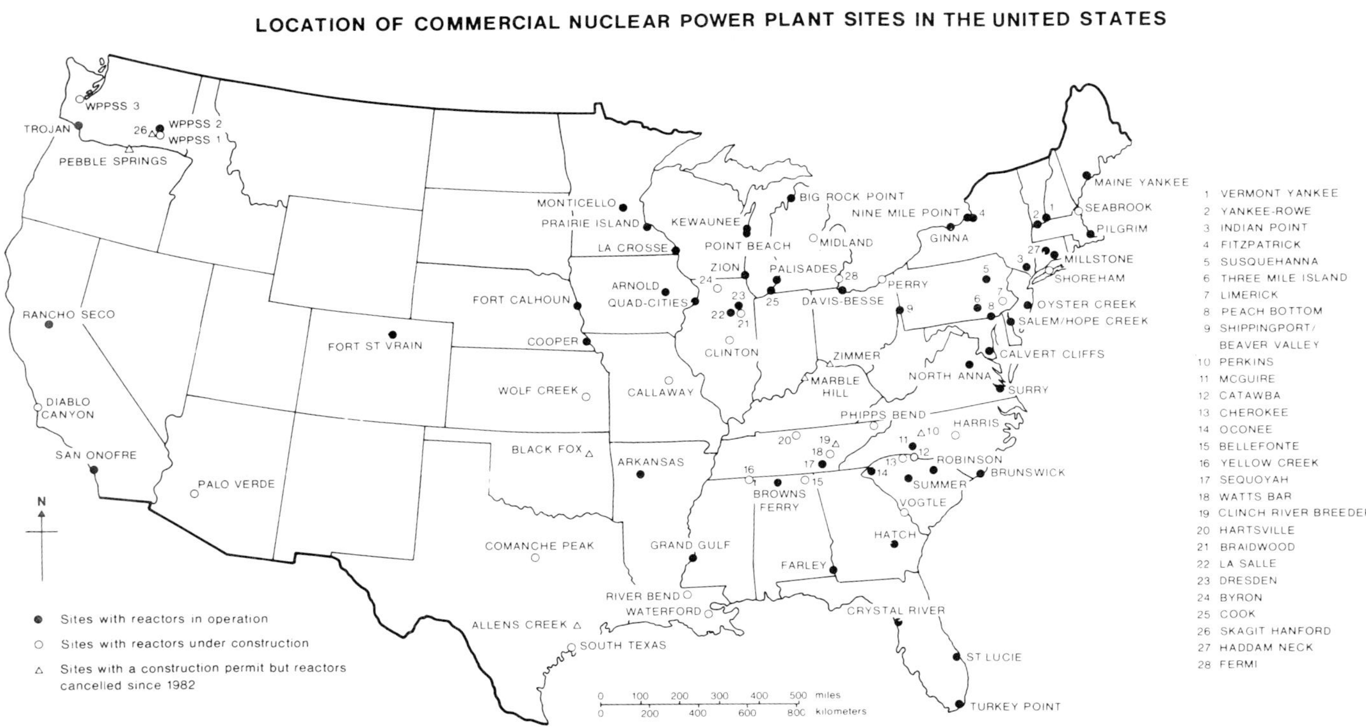

Figure 2. Location of commercial nuclear power plant sites in the United States.

distribution of sites in the UK and USA where in general it seems that nuclear electricity has a competitive advantage over other forms of power generation and where environmentally acceptable and 'safe' sites have been found.

At the micro level it seems that US utilities have been faced with many more constraints upon site selection than their UK counterparts. In the UK, valued environments have never been a major obstacle to developers if national interest arguments can be substantiated. One site (Trawsfynydd) is actually in a national park, some others are located in areas of high landscape or amenity value. By contrast, the US utility is not only excluded from designated areas but has to comply with a plethora of environmental legislation at all administrative levels.

Overall, the site selection process in the UK is primarily a technical exercise (Hunt, 1970) with most emphasis placed on the desirability of avoiding new long transmission lines by locating new stations to utilise unused capacity on the National Grid near centers of unsatisfied demand (Gammon and Pedgrift, 1983). Although US utilities share these objectives, the emphasis placed on siting as a safety measure by US safety authorities has prevented the optimisation of this siting strategy to the same extent (Kelly et al, 1984). In view of the importance attached to safety at the Sizewell inquiry, it is useful to evaluate the differences in approach adopted by the agencies responsible for safety in the US and UK.

3. SITING CRITERIA

The earliest civilian nuclear reactors, like their military predecessors, were located in 'remote' sites in order to minimise the risk to the public from a new technology. This siting and safety philosophy was embodied in the UKAEA's siting criteria of lq55 (Marley and Fry, 1955) and 1962 (Farmer, 1962) and the US Atomic Energy Commission's (USAEC) guidelines of 1962; see Tables 2 and 3.

In principle, both sets of siting criteria embraced similar concepts but a key difference in approach relates to the demographic and distance separation factors used in site evaluation studies. More importantly, the US guidelines were devised not only to reduce the risk of a maximum credible accident but also to account for Class 9 accidents (accidents beyond the design basis). This was achieved through a minimum distance separation between reactor and the nearest population center of 25,000 or more persons. The other components of the US siting criteria were formulated to ensure that the radiation exposure from a specified maximum credible accident were within certain explicit limits. By contrast, in the UK the nature of various maximum credible accident scenarios was not specified and the siting policies were derived with reference to completely arbitrary radiation releases under unfavourable weather conditions. For

Table 2. UK demographic siting criteria.

1. Remote siting policy.
a. Marley and Fry (1955). For any 10 degree sector around a site: few within 0.5 miles, 500 within 1.5 miles, 10,000 within 5 miles and 100,000 within 10 miles. For any 60 degree sector, 10 times these values
b. Farmer (1962). Based on 20 one mile distance bands divided into 30 degree sectors. An index value is derived for the maximum populated sector based on estimated ground level concentration of iodine 131 following a 1 curie release under unfavourable weather conditions.

2. Relaxed siting policy, 1968.
a. Initially no population restrictions were specified other than that people within 2 miles could be evacuated within 2 hours
b. NII devised a series of sector and radial population limits covering a 20 mile radial region. The population limits were derived from studying the risk of thyroid cancer resulting from a 1000 curie release of iodine 131 under unfavourable weather conditions. These limits were again applied to 30 degree sectors.

Table 3. US demographic siting criteria.

1. Reactor site criteria, 1962. This specifies: an exclusion zone around the reactor, a low population zone and a minimum population center distance to the nearest population center. The size of these variables were expressed as functions of reactor power and then reduced for safety credits due to engineering safeguards.

2. Regulatory Guide 4.7, 1975. This specifies population density guidelines to trigger a more detailed evaluation of alternative sites.

3. Emergency planning zones, 1981. A plume exposure evacuation zone of 10 miles radius and a food ingestion pathway zone of 50 miles specified.

4. Possible new regulations. A standard population exclusion zone to minimize the risks from design basis accidents and population/density limits to restrict the consequences of Class 9 accidents.

example, the 1962 Farmer criteria are based on a 1 curie release of iodine 131.

In practice both regulatory bodies began to apply a more 'relaxed' siting philosophy by the late 1960's. In the US this reflected the development of reactor designs with improved engineering safeguards to compensate for site deficiencies. In the UK the AGR reactors were considered to be almost completely safe (Charlesworth and Gronow, 1967). However, in the US the minimum distance separation rule acted as a deterrant that precluded the development of metropolitan sites (Ashley, 1965; Bunch, 1978; NRC, 1979). Nevertheless, there was growing concern that with the increasing size of reactors even this distance factor would not offer much protection from a Class 9 accident. In the wake of the National Environmental Policy Act, the NRC began to cajole applicants to choose a remote site if a genuine choice was available and introduced population density guidelines to facilitate site evaluation (see Table 3). Since then the whole question of siting policy has been under investigation. In 1978 a siting task force was established to accomplish the following goals: (1) to strengthen siting as a factor in defence in depth by establishing requirements for site approval that are independent of design considerations; (2) in site assessments, to take into consideration the risk associated with accidents beyond the design basis through the use of population density and distribution criteria; and (3) to require that sites selected minimise the risk from energy generation (NRC, 1979).

The task force reported in 1979 and made a number of recommendations, some of which were adopted, others which were subjected to further investigation. An emergency planning zone of 10 miles was established within which evacuation had to be possible and a 50 mile food ingestion planning zone was added for the control of food and water (NRC, 1981). The question of specific population density and distribution criteria was postponed pending further detailed studies of the effects of different criteria on site availability. Furthermore, the NRC was awaiting the results of research into a categorization of potential radioactive releases from a nuclear plant accident before producing definitive regulations (NRC, 1983).

Much of the commissioned research for the NRC on demographic criteria has already been published (Aldrich et al, 1982; Durfee and Coleman, 1983; Kelly et al, 1984). The latter research proposes six alternative demographic criteria with alternative four being the likely favourite to be incorporated into any new revised regulation (Table 4). In essence, this approach moves away from the recommendations of the task force as it seeks to justify all but the most marginal of existing sites and it does not deliberately promote a search for more remote sites.

Table 4. Six alternative US demographic criteria.

Nature of restriction	Alternative criteria 1	2	3	4	5	6
Exclusion zone radius (miles)	0.5	0.5	0.5	0.5	0.5	0.5
Population density of 0.5 to 2 mile zone	250	250	250	250	250	250
Average density of 2 to 30 mile zone	500	500	500	750	750	750
Average density of any two adjacent 22½ degree sectors	1000	1500	2000	1500	2250	3000

Source: Kelly et al (1984).

Table 5. Comparison of US and UK demographic siting criteria.

	UK demographic criteria			US demographic criteria		
distance band	1955 remote	1962 remote	1968 relaxed	1975	1984 criteria 1	1984 criteria 6
	RADIAL POPULATION DENSITIES					
0 -0.3	1061	none	none	500	250	250
-1.5	424	none	2062	500	250	250
-5	763	none	2248	500	450	670
-10	1910	none	2134	500	490	730
-20	none	none	2005	500	497	747
-30	none	none	none	500	499	748
	SECTOR DENSITIES					
0 -0.3	2122	2122	none	500	250	250
-1.5	5093	4558	10438	500	250	250
-5	1528	4558	12576	500	920	2680
-10	3820	4558	12766	500	979	2919
-20	none	4558	12782	500	995	2980
-30	none	none	none	500	997	2991

The value of the current US siting work, however, is that an extensive effort is being made to build computerized data bases which will facilitate site evaluation in the future. The recent work by the national energy laboratories augments previous commissioned research which has evaluated existing and proposed sites for their development potential according to the variables outlined in Table 1; see Burwell et al (1979); Burwell and Lane (1980), Cope and Bauman (1977). The lower 48 states have been thoroughly screened for their site suitability for nuclear plants. Concurrently, research at the Brookhaven National Laboratory has monitored demographic and land use changes around existing sites to assess whether the original remote siting intentions of the USAEC and NRC have been respected (Metz et al, 1984; Greenberg et al, 1984).

By contrast, the NII has never been actively involved in the site evaluation process. It only has to respond to those sites already 'adopted' by the state utilities when they actually apply for a nuclear site license. The onus is with the prospective licensee to identify an acceptable site and they have traditionally relied on non-automated methods and subjective 'expert' opinion in their search for sites. Moreover, only a very limited number of alternative sites are evaluated and these evaluations together with the search for alternative locations have never been subjected to public scrutiny (Openshaw, 1985). This was borne out at the Sizewell inquiry when the CEGB admitted that Sizewell was chosen as the first PWR site in 1978 without any evaluation of environmental impacts, capital costs, or ecological impacts and without even a complete reference reactor design (Day 64 transcript, Sizewell Power Station Inquiry). It seems that the decision was first made in principle and then justified retrospectively.

At this point it is interesting to compare the demographic criteria that have evolved in both the US and the UK as a result of these differences in safety and siting philosophies. A simple comparison is made difficult by the use of different sizes of radial zone, different sector widths, and the use of densities in the US and population totals in the UK. Another problem is that the 1962 US guidelines are incapable of a precise numeric representation as they are design-dependent, although it could be reasoned that the 1975 population density guideline of 500 persons per square mile is a reasonable representation of what was originally intended. Likewise, the 1962 UK criteria can only be given a population limit interpretation under the special conditions of an uniformly distributed population. Finally, two of the six alternative 1984 US criteria are selected to represent current US intentions; alternative one which is the most restrictive and alternative six which is the least binding. Table 5 compares the various criteria in terms of implied radial and sectoral population densities.

The differences are quite remarkable and provide a good reflection of the impact of the different national safety philosophies. The UK so-called 'remote' siting criteria are in US terms far more relaxed than any of the alternative US criteria. There seems to be a factor of two difference in radial densities. The UK's relaxed criteria are between 3 and 4 times more relaxed than current US criteria in terms of radial densities and between 4 and 24 times more relaxed in terms of sectoral densities.

4. DEMOGRAPHIC CHARACTERISTICS OF UK AND US SITES

Although neither the regulatory agency nor the utilities have undertaken computerized searches of the UK space for potential future sites, the evaluative research by Openshaw (1982a, 1982b, 1985) on the 1971 and 1981 census provide demographic data for UK sites which can be compared directly with similar work on US sites commissioned by the NRC. The best way to evaluate the demographic characteristics of both sets of sites is to perform a multivariate classification using the full range of comparable demographic data (see Table 6 for a list of these variables). The

Table 6. Demographic Variables Used to Classify Reactor Sites.

Variable	Description
1	population density within 0 to 5 miles
2	population density within 5 to 10 miles
3	population density within 10 to 20 miles
4	population density within 20 to 30 miles
5	population density within 0 to 5 miles in maximum populated 22½ degree sector
6	population density within 5 to 10 miles in maximum populated 22½ degree sector
7	population density within 10 to 20 miles in maximum populated 22½ degree sector
8	population density within 20 to 30 miles in maximum populated 22½ degree sector
9	site population factor for 5 miles
10	site population factor for 10 miles
11	site population factor for 20 miles
12	site population factor for 30 miles

results may depend on the choice of these variables which were in turn dictated by what was available for the US sites. Thus, population densities were calculated for radial zones and 22½ degree sectors in distance bands out to 30 miles for all UK sites because comparable estimates had already been published for US sites (Aldrich et al, 1982). In addition, a site population factor was computed for radial distances of 5, 10, 20, and 30 miles. This statistic is used in US siting guidelines to control population densities; values equivalent to 500 and 1,000 persons per square mile have a particular significance (NRC, 1974).

The sites selected for analysis reflect sites in both countries with operating reactors, construction permits or preliminary approval. A final aspect concerned the choice of an appropriate number of clusters. Table 7 shows the within cluster sum of squares for a range of classifications produced by the non-hierarchical iterative relocation procedure that was used. A value of zero would indicate a perfect classification which never occurs with real data while a value of 35 per cent within cluster sum of squares might be typical of general usage. Here a sixfold typology is selected as the most parsimonious choice and one that reflects a significant jump in the sum of squares function between 5 and 6 clusters.

Table 7. Within cluster sum of squares for alternative site classifications.

Number of clusters	Percentage within cluster sum of squares
2	57.2
3	47.7
4	44.0
5	44.2
6	28.4
7	28.2
8	23.5
9	19.1
10	18.2
11	16.8
12	16.3

A number of sites were assigned to each of the six types and the cluster numbers were arranged in such a way that type 1 represents the most populated sites and type 6 the most remote. As can be seen from Table 8, most of the US sites (60 per cent) are located in areas with

Table 8. A classification of reactor sites.

Site type	Total number	Percentage of US sites	Percentage of UK sites
1	3	2.1	5.5
2	7	1.0	33.3
3	4	3.2	5.5
4	7	5.4	11.1
5	29	27.4	22.2
6	59	60.4	22.2
Total sites	109	91	18

very low population densities with a further 27 per cent in semi-remote areas. By contrast only 44 per cent of UK sites came into the remote or semi-remote categories. These differences can largely be explained by the rejection of applications by the USAEC for metropolitan and periurban sites in the US at a time when similar applications were being approved in the UK. Furthermore, the narrowness of the UK 'remote' siting criteria, which only considered distances up to 10 miles from a reactor, meant that several of the early so-called remote sites were within 11 to 20 miles of highly populated cities. Figures 1 and 2 combined with Appendix 1 can be used to identify further details of the various site types. It would appear that the only 3 US sites that occur in the most heavily populated site types (Limerick, Indian Point, and Zion) slipped through the "regulatory" net. It is now highly unlikely that these sites will be developed any further and, therefore, by implication that many existing UK sites would be rejected were they to be located in the US. It is interesting to note also that one of the UK sites in this most densely population type of area (Druridge Bay) is currently earmarked for a PWR development.

5. CONCLUSIONS

It would be wrong to dismiss these results with the assertion that the more urban sites of the UK are a reflection of higher population densities. It has been shown elsewhere (Openshaw, 1985) that parts of the US have comparable population densities to that of the UK and yet still have sites which are more remote from population centers than currently found in Britain. Nor are the differences due to an assumed shortage of sites in the UK. Quite simply there is no evidence that either current demographic criteria or concern for the environment cause any such problems.

The main reason for the different demographic characteristics of UK and US sites is the application of different siting policies. The US safety agencies have always planned for residual risk minimisation. At present, siting and safety goals are under review to strengthen this philosophy and it is unlikely that metropolitan or peri-urban sites will be approved in future licensing decisions.

In the UK, by contrast, siting is considered to be of only the most marginal relevance to the safety debate. The relaxed siting policy of 1968 is still in force. Emergency planning arrangements only cover a 2 mile radial zone. The lessons of TMI have been declared of no relevance to the UK. This philosophy may have been understandable with gas-cooled reactors but it is also being applied to PWRs. The UK safety authorities have complete confidence that a PWR can be designed to British engineering standards and constructed, operated and managed by British staff to be inherently two orders of magnitude safer than the same design in the US, despite a total lack of operating experience. This is a potentially dangerous philosophy, primarily because of the industrial inertia inherent in nuclear power station planning. A 'bad' locational decision now could affect future generations of reactors and prove impossibly expensive to correct. The CEGB and the NII have such faith in engineering safeguards and the ability of operators and management that they totally discount the need for any sites more remote than those currently in use. As the CEGB's financial liability to third parties in the event of a reactor accident remains at £5 million, there is no financial incentive to minimise residual risks by incurring real additional costs. A switch to remote sites would incur cost penalties in terms of infrastructure investment and cause credibility problems with those existing sites which are located close to population centers. It is now too late to start again and there is every indication that the existing UK sites will be expanded and redeveloped as need occasions. Any change to this current philosophy will require far more central government intervention than in the past and it is to be hoped that a major accident on a marginal site is not the only catalyst that could trigger a long overdue review of siting and safety philosophies in the UK.

6. REFERENCES

Aldrich, D. C. et al. 1982. Technical guidance for siting criteria development. NUREG/GR 2339, SAND 81 - 1549, Albuquerque, New Mexico: Sandia National Laboratories.

Armstrong, J. 1984. Town and Country Planning. April: 129.

Ashley, R. L. (ed). 1965. Nuclear power reactor siting. Proceedings of a National Topical Meeting. Los Angeles: American Nuclear Society, Division of Technical Information.

Bunch, D. F. 1978. Metropolitan siting: a historical perspective. NUREG 0478, Washington DC: USNRC.

Burwell, C. C, Ohanian, M. J, and Weinberg, A. M. 1979. A siting policy for an acceptable nuclear future. Science **207**: 1043-51.

Burwell, C. C, and Lane, J. A. 1980. Nuclear site planning to 2025. ORAU/IEA-80-5A(M), Oak Ridge, Tn: Oak Ridge Associated Universities, Institute for Energy Analysis.

CEGB. 1982. CEGB statement of case: volumes 1-2, Sizewell Power Station Public Inquiry. London: CEGB.

Charlesworth, F. R, and Gronow, W. S. 1967. A summary of experience in the practical application of siting policy in the United Kingdom. In Containment and siting of nuclear power plant. Vienna: IAEA, 143-70.

Cope, D. F, and Bauman, H. F. 1977. Expansion potential for existing nuclear power station sites. ORNL/TM-5927, Oak Ridge, Tn: Oak Ridge National Laboratory.

Dale, G. C. 1982. The safety of the AGR. London: CEGB.

Durfee, R. C, Coleman, P. R. 1983. Population distribution analyses for nuclear power plant siting. NUREG/CR 3056, ORNL CSD TM-197, Oak Ridge Tn: Oak Ridge National Laboratory.

Farmer, F. R. 1962. The evaluation of power reactor sites. FPR/INF/266, Harwell: UKAEA.

Fells, I. 1984. The world nuclear power scene and UK energy policy in 1984. Energy Policy **12**: 3: 306-09.

Franklin, L. 1984. What Sizewell means for the UK power plant industry. Energy Policy **12**: 3: 271-75.

Gammon, K. M, Pedgrift, G. F. 1983. Changes in the investigation and selection of sites for nuclear power stations. Nuclear engineering **22**: 41-45.

George, B. V, Hilsley, D. C. 1982. Britain's approach to the PWR stresses safety and reliability. Nuclear Engineering International **21**: 34-46.

Greenberg, M, Krueckeberg, D. A, Kaltman, M. 1984. Population trends around nuclear power plants. In M J Pasqualetti and K D Pijawka (eds), Nuclear power. Boulder, Westview: 189-212.

Health and Safety Executive. 1979. Safety assessment principles for Nuclear power reactors. London: HMSO.

Health and Safety Executive. 1982. Emergency plans for civil nuclear installations. London: HMSO.

Hunt, F. R. 1970. Power station site selection in England and Wales. In Environmental aspects of nuclear power stations. Vienna: IAEA: 647-59.

Kelly, M. J. et al. 1984. Analysis of availability of previously identified sites under alternative demographic criteria. NUREG/CR 3057, ORNL 5936, Oak Ridge Tn: Oak Ridge National Laboratory.

Lilienthal, D. 1980. Atomic energy: a new start. New York: Harper and Row.

Marley, W. G, and Fry, T. M. 1955. Radiological hazards from an escape of fission products and implications in power reactor locations. In Proceedings of the international conference on peaceful uses of atomic energy. New York: United Nations: 102-05.

MacKerron, G. 1984. Sizewell: good value for consumers money. Energy Policy **12**: 3: 296-301.

Metz, W. C, Daum, M. L, Pearlman, K. T and Waite, N .D. 1984. Land-use controls and nuclear power plants. In M J Pasqualetti and K D Pijawaka (eds), Nuclear power. Boulder: Westview: 137-64.

Minarick, T.W and Kickialka, C. A. 1982. Precursors to potential severe core damage accidents. 1969-1979, a status report. NUREG/CR-2497, Oak Ridge Tn: Oak Ridge National Laboratory.

Nuclear Regulatory Commission. 1974. The site population factor. WASH 1235, Washington DC: USNRC.

Nuclear Regulatory Commission. 1979. Report of the siting policy task Force. NUREG 0625, Washington DC: USNRC.

Nuclear Regulatory Commission. 1981. Scoping summary report - environmental impact statement on the siting of nuclear power plants. Washington DC: USNRC.

Nuclear Regulatory Commission. 1982. 1981 annual report. Washington DC: USNRC.

Nuclear Regulatory Commission. 1983. 1982 annual report. Washington DC: USNRC.

Nuclear Regulatory Commission. 1984. 1983 annual report. Washington DC: USNRC.

Openshaw, S. 1982a. The siting of nuclear power stations and public safety in the UK. Regional Studies **16**: 183-98.

Openshaw, S. 1982b. The geography of reactor siting policies in the UK', Transactions of the Institute of British Geographers, New Series **7**: 150-62.

Openshaw, S. 1985. The siting of nuclear power stations and public safety. London: Routledge Kegan and Paul.

Starr, C, and Braun, C. 1983. US nuclear power performance. Energy policy **12**: 3: 253-58.

Union of Concerned Scientists. 1984. Nuclear power plants in the United States: current status and statistical history. Boston: UCS.

US Atomic Energy Commission. 1962. Reactor siting criteria 10CFR Part 100. Washington DC: USAEC.

US Department of Energy. 1984. Nuclear power program information and data. Washington DC: USDOE.

US Office of Technology Assessment. 1984. Nuclear power in an age of uncertainty. OTA-E-216, Washington DC.

Weinberg, A M. 1984. Emerging trends in nuclear energy. Energy Policy **12**: 3: 247-52.

Zeigler, D. J, and Johnson, J. H. 1984. Evacuation behaviour in response to nuclear power plant accidents. The Professional Geographer **36**: 207-15.

7. APPENDIX

CLASSIFICATION OF US AND UK REACTOR SITES

Site type 1

This is based on only three sites; Indian Point, Limerick, and Hartlepool. These sites have population densities at least eight times greater than the average for the total set of 109 sites. In essence these are metropolitan sites which in the US 'slipped through' before people realised their poor locations. By contrast the Hartlepool site in the UK was quite deliberate and reflects the apparently superior safety characteristics of the AGR system. In many respects Hartlepool is the most heavily populated of all 109 sites. It was in fact never properly evaluated but merely declared acceptable by Cabinet fiat.

Site type 2

This cluster has seven members all of which are located in peri-urban situations which are basically rural areas in close proximity to large cities. In the US the only site representing this type is Zion. In the UK, the sites are: Bradwell, Berkeley, Oldbury, Heysham, Druridge Bay (adopted but no permit as yet), and Harwell. Typical demographic feature are high population densities in the 10 to 20 mile distance band which are three times greater than the average. The UK sites included here are a mixture of early remote sites (pre 1963) and later (post 1968) relaxed semi-urban sites. It is very interesting that both groups are considered to have similar demographic characteristics and this clearly shows the distorted picture that can be given by such a misleading terminology as 'remote' and 'relaxed' when applied to UK reactor sites.

Site type 3

These are mainly peri-urban sites located near to small or medium sized towns rather than large cities. The US sites are: Midland, Millstone, and Three Mile Island. The only equivalent UK site is Winfrith, a location which is also considered suitable for a PWR and possibly a fast breeder reactor. The main feature of this cluster are the high population densities within 5 miles which are generally three times greater than the average for all 109 sites.

Site type 4

The US sites included here are: Bailly, Beaver Valley, Fermi, Haddam Neck, and Waterford. The UK sites are: Hunterston and Hinkley Point. The distinguishing feature is the presence of high population densities in the 20 to 30 mile distance band which are between two and five times higher than average. The UK sites included in this cluster are all designated 'remote' although they are within 30 miles of major cities.

Site type 5

The members of this cluster are characterised by less than average population densities in all distance bands. The US sites are: Byron, Catawaba, Cook, Dresden, Arnold, Forked River, Fort Calhoun, Ginna, Marble Hill, McGuire, Oyster Creek, Peach Bottom, Perkins, Perry, Pilgrim, Quad Cities, Salem, San Onofre, Sesquoyah, Shoreham, Skagit, Surry, Susquehanna, Turkey Point, and Zimmer. The only UK sites are Chapelcross, Windscale, Dungeness, and Sizewell.

Site type 6

This is a collection of very remote sites with population densities between 2 and 4 times lower than the average for all distance bands. It is also the most popular US site type and the least popular UK site type. The US sites are: Allens Creek, Arkansas, Bellefonte, Big Rock Point, Black Fox, Braidwood, Browns Ferry, Brunswick, Callaway, Calvert Cliffs, Cherokee, Clinton, Commanche Peak, Cooper, Crystal River, Davis-Besse , Diablo Canyon, Fairley, Fitzpatrick, Fort St Vrain, Grand Gulf, Hartsville, Hatch, Kewaunee, La Salle, La Crosse, Maine Yankee, Monticello, Nine Mile Point, North Anna, Oconee, Palisades, Palo Verde, Pebble Springs, Phipps Bend, Point Beach, Praire Island, Rancho Seco, River Bend, Robinson, Saint Lucie, Seabrook, Harris, South Texas, Summer, Trojan, Vermont Yankee, Vogtle, Watts Bar, WPSS 1-2-3, Wolf Creek, Yankee Rowe and Yellow Creek. The only UK sites are Dounreay, Trawsfynydd, Wylfa, and Torness.

CHAPTER 13

ENERGY, PARTICIPATION AND PLANNING: THE CASE OF ELECTRICITY GENERATION IN GREAT BRITAIN

Susan Owens
Department of Geography
University of Cambridge
Downing Place
Cambridge CB2 3EN
United Kingdom

ABSTRACT. This paper shows how issues and conflicts over power station proposals in the United Kingdom have changed and developed during the past three decades, then looks to the future to consider whether planning and democratic procedures will be able to cope with anticipated development pressures. It is argued that current planning 'delays' result from a long established trend towards greater concern for environmental quality, superimposed on the emergence of 'generic' issues since the mid-1970's. There is consensus that the present situation is unsatisfactory, but reforms proposed by development interests and environmental groups (broadly defined) reflect fundamentally different views about policy and participation. Several scenarios are explored in the final section and it is concluded that in spite of a 'remission' from development pressures due largely to the current recession, there is little room for complacency about conflicts over power station developments in the future.

1. INTRODUCTION

In Britain, as in other advanced economies, opposition to major development proposals has become a notable feature of the land use planning process. Planning applications often cause intense controversy and are subject to long delays before any decision is made. Electricity generating stations have all the characteristics of the most contentious developments: they are large, intrusive, technically complex and are perceived to have serious and possibly irreversible environmental impacts. Nuclear power in particular has been found to embody all those dimensions of perceived risk which are most unacceptable to the public (see, for example, Fischoff _et al_. 1981; Royal Society 1983). Opposition to the siting of power stations is not a new phenomenon, but it has undergone profound changes in nature and intensity. Ministerial decisions on power station applications, which sometimes took only a matter of weeks in the 1950's, can now take upwards of five years. Delays of this kind have significant

F. J. Calzonetti and B. D. Solomon (eds.), Geographical Dimensions of Energy, 225–253.
© 1985 by D. Reidel Publishing Company.

implications for national energy policy and concern has frequently been expressed about the ability of existing statutory procedures to cope with major applications.

This chapter explores how issues and conflicts have changed over the past three decades, culminating in the present experience of lengthy planning delays. An analysis of 68 applications for major power stations (44 for fossil fueled and 24 for nuclear plant) (Figure 1) in Great Britain since 1955 shows that there is no simple explanation for the major changes which have taken place; they have arisen from a combination of trends and circumstances, some long-standing and others more recent in origin. Nor is there any simple solution to the present difficulties, since different participants in the debate hold fundamentally different views about policy and participation.

The present situation raises important questions about the implementation of energy policy and the role of public participation in the future. The current economic recession is providing something of a remission from development pressures, but the electricity supply industry can still find grounds to justify a power station construction programme, whose precise nature awaits the outcome of the current public inquiry into Britain's first pressurized water reactor (PWR) at Sizewell in Suffolk.

Conflict is inherent in both the national and local issues surrounding the siting of power stations, but it is difficult to predict how it will develop in a situation of great uncertainty about the future health of the British economy and the direction of energy policy. In the final section of this chapter several scenarios are used to explore the prospects for the power station programme and for democratic procedures in the future.

2. THE POLICY BACKGROUND

Reaction to power station proposals cannot be considered in isolation from the national policy context, so it is useful to review developments in the British electricity supply industry during the past 30 years and to outline the relevant statutory consents procedures. This background is necessarily brief, but there are many detailed accounts elsewhere (Dobry 1965; Drapkin 1974; Forster 1974; Hunt 1975; Pearce et al. 1979; Williams 1980).

2.1. Electricity Supply

Throughout the period considered in this chapter, British electricity has been generated primarily from coal. In the 1960's there was a growing commitment to oil burning (and some natural gas), but this became uneconomic in the 1970's and has been cut back. Nuclear power has slowly increased its contribution, which will be about 16% when plants presently under construction are commissioned. Fueling of power plants has often been a controversial issue. The electricity supply

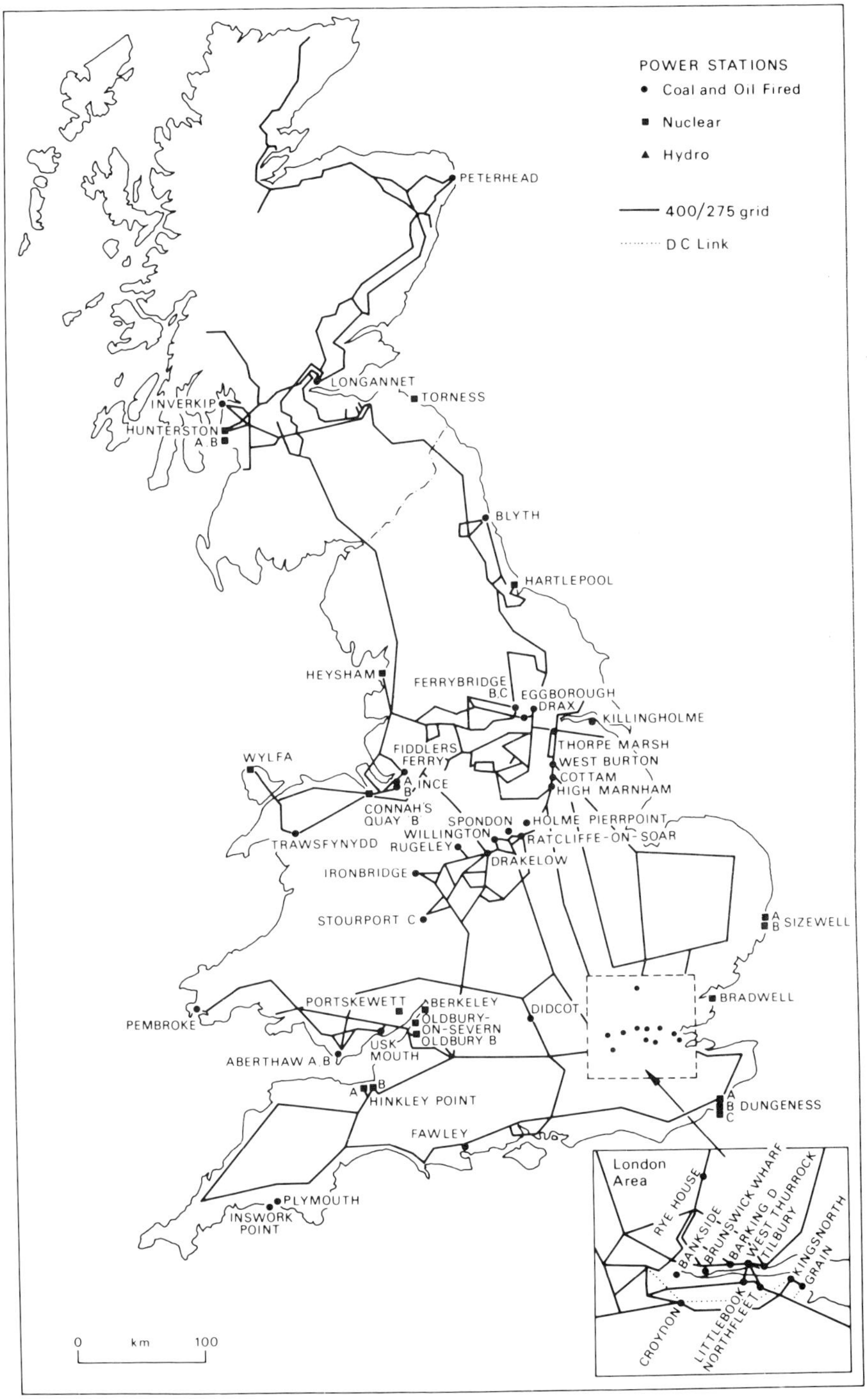

Figure 1. Power station sites for which consents applications have been made since 1955.

industry has an interest in using the cheapest fuel, but as a nationalized industry it has been subject to the wishes of successive governments to use fuel mix as a policy lever for other purposes, for example to mitigate the effects of decline in the coal industry in the 1960's (Forster 1974).

The early part of the period considered here was characterized by a supply shortfall in the face of rapidly growing electricity demand, resulting in frequent power cuts and voltage reductions. The situation became critical in the abnormally severe winter of 1962-3. After this, the Central Electricity Generating Board (CEGB)[1] increased the 'planning margin' and embarked on a huge ordering and construction programme, assuming a 4% annual increase in gross national product. At the same time they began to order new and much larger generating sets than had previously been in operation. Development pressures were therefore considerable.

By the late 1960's, it was obvious that demand had been greatly overestimated, but this was offset by the fact that the CEGB was facing problems in getting plants commissioned, and were also running into difficulties with the nuclear programme. In the 1970's, demand for electricity has stagnated, so that there is now a 'temporary' over-capacity in the electricity supply system (Electricity Council 1984). This has a significant bearing on attitudes towards development proposals.

The first civil nuclear power programme was announced in 1955, amidst great euphoria about its potential. It was trebled after the Suez Crisis of 1956, but thereafter progressively cut back because of continued criticism by the CEGB, which thought it too ambitious a programme for a new technology. Nine magnox stations were eventually constructed. A second programme of advanced gas cooled reactors (AGR's) was announced in 1964, after much debate about reactor choice, and construction of several AGR's began; public optimism about nuclear power was still at a high level.

In the late 1960's many problems beset the nuclear industry. There was trouble with corrosion in the magnox reactors, the AGR's were proving extremely difficult to construct and commission and there was increasing public hostility towards large scale technology in general and nuclear power in particular. Problems intensified throughout the 1970's, with growing opposition, indecision about reactor choice and economic difficulties in the power plant industry, and were exacerbated by uncertainties about future electricity demand. Nevertheless, in December 1979 a third, ten station nuclear power programme was announced by the Secretary of State for Energy, though there is no commitment to reactor type beyond the first station which, if the current inquiry at Sizewell finds in its favour, is to be a PWR. The public debate about nuclear power greatly intensified in the 1970's and was fuelled in Britain by a number of significant events, including publication of an influential report by the Royal Commission on Environmental Pollution (RCEP 1976), the Windscale inquiry into British Nuclear Fuels Ltd.'s proposed reprocessing plant in Cumbria, and the Three Mile Island near meltdown in 1979.

The outcome of the Sizewell Inquiry is now awaited, and until then the future of the power station programme is uncertain. If consent for a PWR is given, further applications are likely to follow; if it is refused, the AGR and coal fired generation are the main alternatives considered feasible by the electricity supply industry (Electricity Council 1984).

2.2. Statutory Consents Procedures

Within the electricity supply industry, power station planning involves demand forecasting, identification of system requirements, site searches and detailed investigation, obtaining consents and licenses and design, construction and commissioning of the plant. The whole process, which at one time could be completed in 6 or 7 years (Electricity Council 1984) can now take at least 15 years, though delays must be blamed upon long construction times as well as on planning and participation procedures (Table 1).

Table 1. Anticipated Completion Dates for Plant Under Construction

Power Station	No. of Units and Rating (MW)	Year of Start On Site To Expected Completion Of Commissioning
Dungeness B (AGR)	2 x 600	1966-1984
Hartlepool (AGR)	2 x 626	1969-1984
Heysham I (AGR)	2 x 626	1969-1984
Heysham II (AGR)	2 x 615	1970-1984
Drax, Units 4-6 (Coal fired)	3 x 625	1978-1986
Grain Unit 4 (Oil fired)	1 x 631	1971-1984
Littlebrook D (Oil fired)	1 x 631	1974-1984

Source: Electricity Council. 1984.

Consent to construct a power station must be obtained from the Secretary of State for Energy under Section 2 of the 1909 Electric Lighting Act and the 1957 Electricity Act. This differs from normal development control procedures where consent is obtained from the local planning authority (under the Town and Country Planning Act 1971) and the responsible Cabinet Minister is the Secretary of State for the Environment. It could be construed as less democratic, since the role of the local planning authority is diminished. But in practice the local authority and the public are involved just as much--or some would argue just as little--in power station siting as in applications for other major developments which are dealt with under the normal planning legislation. The local planning authority is formally consulted when a Section 2 application is made, but this formality is normally preceded by extensive pre-application consultations, so that the CEGB will be applying for a site to which the local planning authority is agreeable or will be well aware of the kind of opposition it is likely to face. In the past it was often at this stage that the first public notice of the proposal was given, but more recently the nature and scope of pre-application consultations have been greatly extended. Ironically, the effect of this may well be quite the opposite to that intended--it is just as likely to stimulate controversy as to smooth the path towards an acceptable proposal.

After carrying out its own consultations, the planning authority returns comments to the Department of Energy via the CEGB. If it objects to the proposal an inquiry <u>must</u> be held and nowadays at this stage the Department of the Environment is formally brought into the proceedings. If, however, the planning authority does not object to the proposal, the holding of an inquiry is at the Minister's discretion, even if there is a large volume of objection from the public. It is unlikely now that an inquiry would not be held in this situation, but it has happened in the past, for example at Didcot, Oxfordshire, where a large coal-fired power station was constructed in the 1960's, with no inquiry, in spite of vigorous public protest.

At a public inquiry both the proponents and opponents have a right to be heard before an inspector appointed by the Secretary of State and both may call and cross examine witnesses. Procedures are similar to those for inquiries held under the Town and Country Planning legislation (Department of the Environment 1976).

Recently, procedures for major public inquiries have included a number of innovations, such as pre-inquiry meetings, availability of written evidence to all parties by certain dates and, at Sizewell, appointment of legal counsel to cross-examine witnesses on behalf of the inquiry itself (for details, see Purdue, Kemp and O'Riordan 1984).

If, after considering the inspector's report, the Secretary of State grants consent, it includes "deemed" planning permission. This implies that the proposed power station is acceptable in his view on general siting and environmental grounds, but separate financial sanction must be obtained before construction can begin.

For the CEGB, obtaining Section 2 consent is only one stage in the lengthy process of translating estimated needs for new supply capacity into physical developments, and by this stage there is a considerable degree of commitment. It is hardly surprising therefore that opponents of power station proposals, who have no role in the wider process of policy and planning, have been largely ineffective. Of the 68 applications for major power stations considered here, 20 have been the subject of a hearing or inquiry and only two of these have been refused consent by the Secretary of State.

After site selection, Section 2 consent is only one of many which must be obtained. For example, nuclear power stations require statutory consents under the 1965 Nuclear Installations Act, the 1960 Radioactive Substances Act, the 1963 Water Resources Act and the 1951-61 Rivers (Prevention of Pollution) Act. This causes an interesting dilemma now that major public inquiries, originally designed to hear local objections, increasingly focus on much wider "generic" issues. At the Sizewell Inquiry, for example, safety has been a major concern and it has been a point of great controversy that the Nuclear Installations Inspectorate will not be in a position to license the proposed PWR before the end of the inquiry. The separation of issues in the legislation no longer reflects the way in which these issues are perceived in public debate.

3. ISSUES AND CONFLICTS

Power stations are not popular neighbors; in Britain they have long been among the most controversial of major planning proposals. By 1959 siting of power stations had been identified by the Council for the Presevation of Rural England (CPRE), an influential amenity pressure group, as "one of the greatest problems facing the planning authorities throughout the country" (CPRE 1959). But it must be acknowledged that opposition has not always been unanimous, especially in the earlier part of the period considered here. Potential local benefits--employment, multiplier effects and prestige--have provided persuasive arguments to counter local opposition in many cases. (Socioeconomic impacts have been examined in depth by Glasson et al. 1980). Anticipation that a major new power station will "put a place on the map" has been clearly demonstrated by concern about how the station should be named. In the 1960's both the Pembroke and Hartlepool power stations were so called after the local authorities had objected to the CEGB's plans to use more obscure names.

During the past 30 years, about a quarter of the applications for coal- and oil-fired power stations and approximately half of those for nuclear power stations have been the subject of a public hearing or inquiry, though as noted above these were not the only controversial proposals (Figures 2-5).

Three trends are particularly apparent from an analysis of all applications. One is increasing public resistance; the second is a significant shift in the issues which have dominated the debate; the third, related to the other two, is the growing "delay" between application and a final decision by the Secretary of State (Figures 3 and 5). These delays apparently cause great consternation to the Government and the electricity supply industry, yet environmental and anti-nuclear groups remain convinced that proposals are the subject of adequate debate. It is therefore worth exploring in more detail the trends which have led to a situation generally agreed to be unsatis-

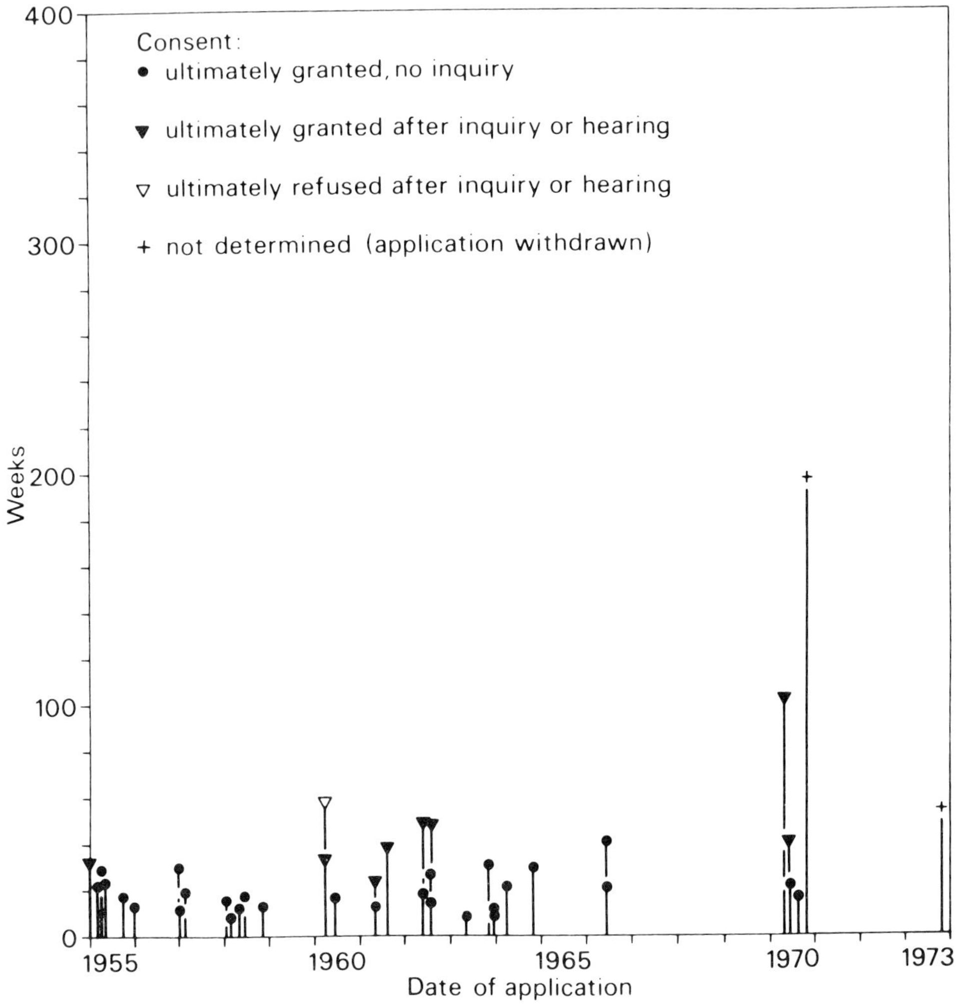

Figure 2. Timescale for local planning procedures for fossil fuel power stations (excluding Scotland.

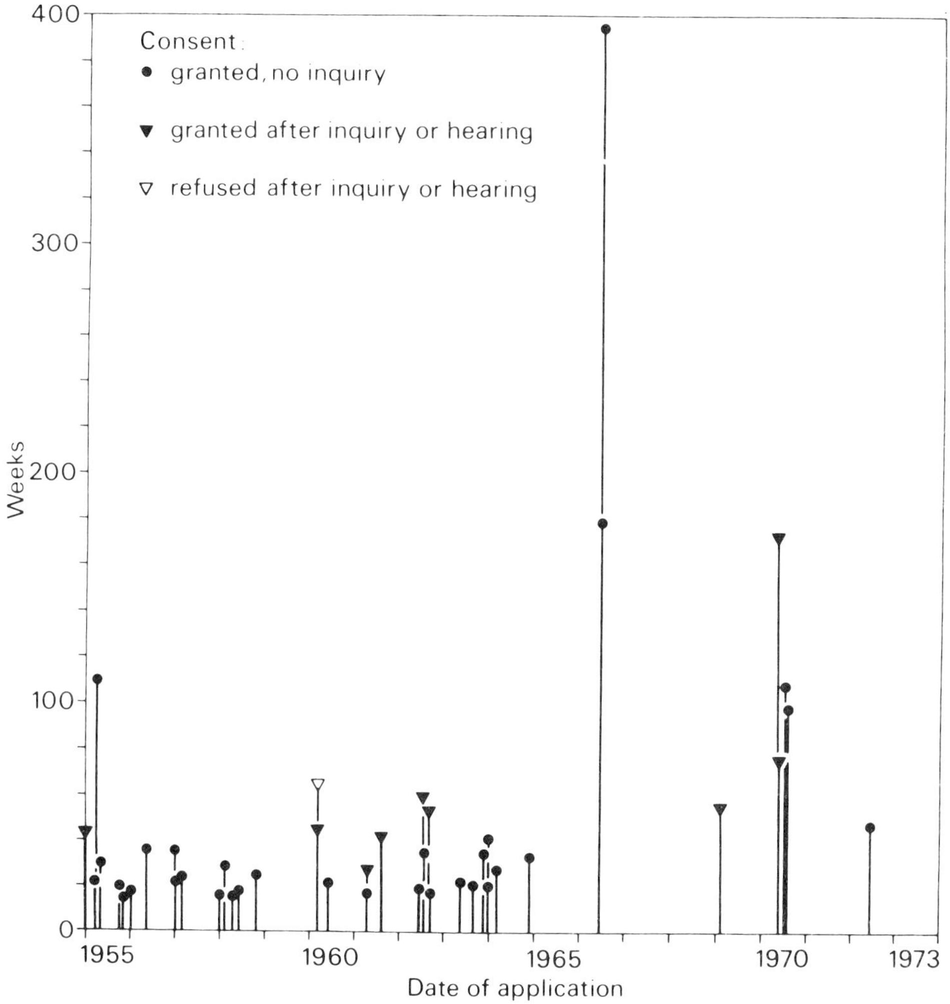

Figure 3. Time taken to obtain ministerial consent for fossil fuel power stations.

factory by all involved.

Some of the earlier power station applications seemed to pass almost unnoticed. In 1962, for example, consent was granted for a 1000 MWe oil-fired station at Kingsnorth on the Thames estuary within 18 weeks of the CEGB's application, which had not attracted a single objection. Hearings and inquiries, when held, were simple affairs tending to last for only a few days. Locational factors were probably the most important determinants of public reaction at this time; the most contentious proposals were for sites where the development would have a serious impact on the local environment, at Holme Pierrepont near Nottingham, the only site where consent for a major coal-fired power station has been refused on amenity grounds. This interesting case is used by Gregory (1971) in his well known analysis of the "amenity versus development" battles of the 1960's.

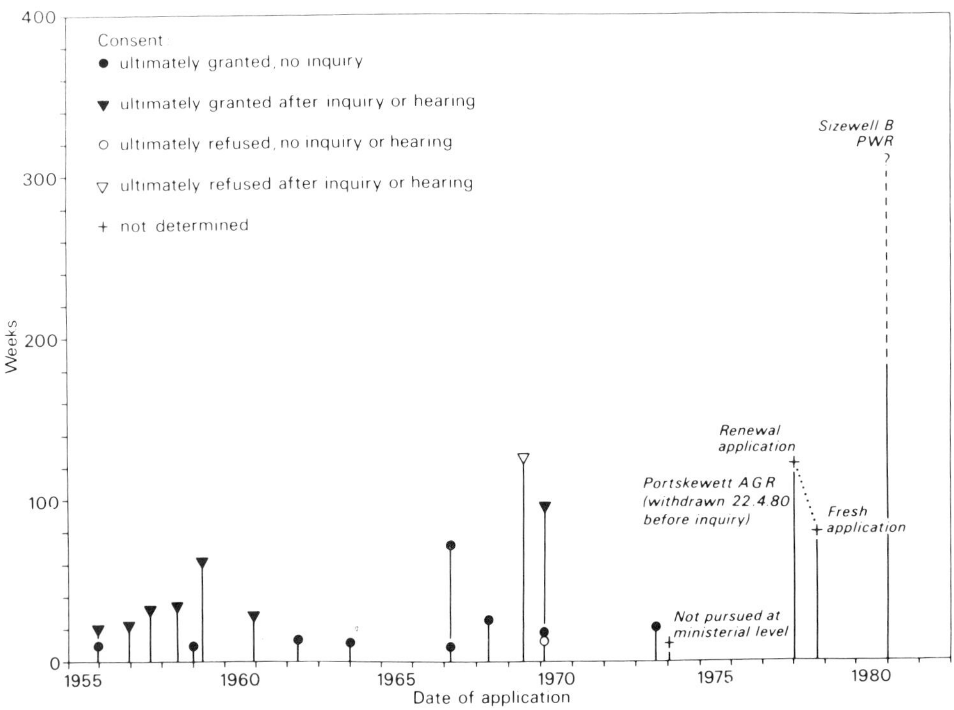

Figure 4. Timescale for local planning procedures for nuclear power stations (excluding Scotland).

Objections in the late 1950's and early 1960's were essentially local; they were made by, or on behalf of, the people who would be most directly affected (with some exceptions where national amenity was at stake in designated areas); they were relatively unsophisticated in comparison with the CEGB's case; and they were notably unsuccessful. Typically, local authorities and the public were concerned about air and water pollution, traffic generation and property devaluation. Such issues figured prominently at inquiries into the large 2000 MWe coal-fired power stations for which the CEGB was seeking consent in the early 1960's. The rather muted opposition to the first-generation nuclear power stations was also dominated by concern about loss of amenity, particularly as most were in rural areas in accordance with the Government's siting policy. It is ironic, in these days of hostility to nuclear facilities, that the advent of nuclear generation was welcomed as a way of making power stations much more acceptable to local communities. In a Parliamentary debate in 1955 the Minister of Fuel and Power (as he then was) commented on the local unpopularity of power stations, especially due to pollution and traffic problems, but informed the House that there had been "a complete change of heart":

> "in recent months I have had a queue of honourable members asking me whether they can have atomic power stations in their constituencies. This development is a very big change and I have in each case referred their requests to the Central Electricity Authority" (House of Commons Hansard 1955, 344).

The change of heart was still apparent in the mid-1960's when some objectors (at Didcot, for example) indicated that a nuclear power station might be acceptable, whereas a coal-fired station definitely was not.

The growth in prominence of environmental issues in the second half of the 1960's is now a well recognized phenomenon (see, for example, Brookes et al. 1976; Lowe and Goyder 1983; O'Riordan 1981; Sandbach 1980). It was reflected in increasing resistance to all major developments. In the case of power stations the scope of objections broadened and opponents mustered more expertise about potential environmental impacts. Few development proposals were successfully challenged, but the CEGB was forced into greater environmental consciousness and the grosser problems of atmospheric pollution and visual intrusion were mitigated. Regard for the environment became a statutory duty for the CEBG in the 1968 Countryside Act. Also at this time concern about safety began to creep into the debate about nuclear power stations, but remained relatively muted until the 1970's (Owens, Hills, and Cope 1983).

Proposals for power stations were justified by the CEGB on the grounds of national need for new capacity to meet growing electricity demand, and it was up to objectors to demonstrate if they could that national needs and local impacts would be better balanced at an alternative site. What is interesting, given subsequent experience, is that they did not question demand forecasts or the implicit definition

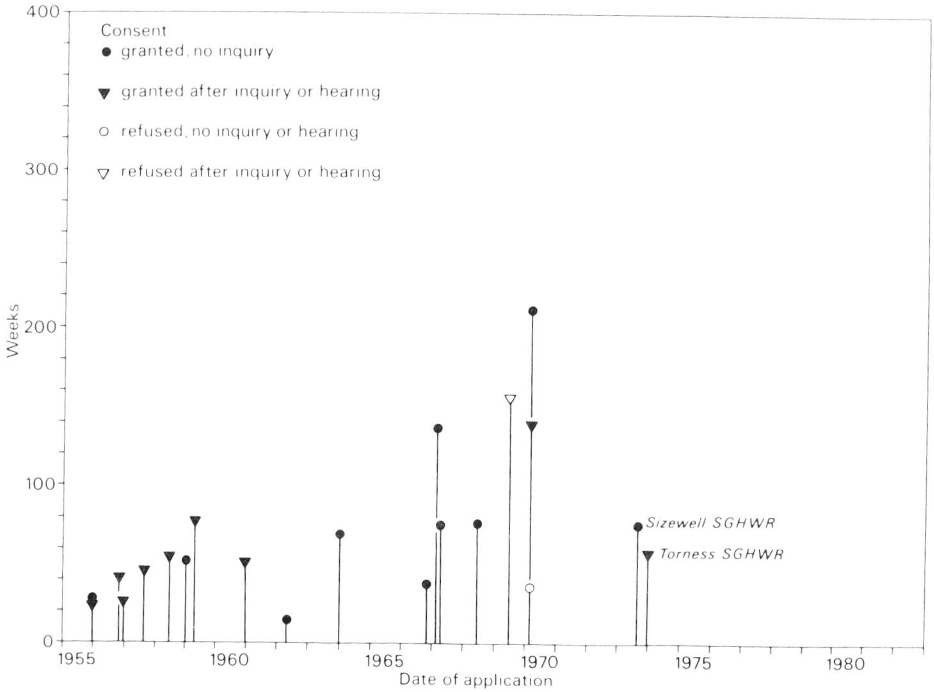

Figure 5. Time taken to obtain ministerial consent for nuclear power stations.

of "need". The arguments that demand projections are frequently no more than self-fulfilling prophecies and that in any case demand should not be confused with need were already being used in relation to other developments--notably roads and airports--by the late 1960's (see, for example, Buchanan 1981; Starkie 1982), but had yet to emerge in relation to power station proposals.

The first time that a power station application was seriously questioned on grounds of policy and "need" (within the framework of local planning procedures) was at an inquiry into a proposed 3000 MWe oil-fired station at Inswork Point in Cornwall in 1971, the longest power station inquiry that had been held up to that time. Objectors went beyond issues of local amenity to criticize the CEGB's demand forecasts, to call for a national review of power resources, and to question the commitment to oil (Hambrook 1972). (Consent for the development was granted in 1973, but due to subsequent events it has not been constructed).

The time scale for local procedures relating to power station applications (up to submission of inspector's report where there was an inquiry) actually shows little change until the late 1960's, when it begins to reflect more exhaustive scrutiny at the local level, though still not in all cases (Figures 2 and 4). Ministerial decision times show a marked increase from the mid-1960's, especially in the case of conventional power stations, where the mean decision time increased from 32 weeks before 1965 to 140 weeks afterwards (Figure 3). Ministerial decisions took on average 47 weeks for the Magnox stations in the first nuclear power programme, compared with 97 weeks for subsequent applications (Figure 5) (and the latter figure, of course, will be inflated by the Sizewell decision which is likely to take at least 6 years). Comparison of local and ministerial decision times shows that in many cases delay was not caused by local objections, but must be attributed to national fuel policy considerations including the prevailing difficulties in the electricity supply industry and controversies about fuel choice and reactor policy.

The question of 'need' for new power stations was temporarily averted by the quadrupling of oil prices in 1973/4 and the new emphasis on reducing import dependence. Projections which appeared in the wake of the oil crisis now seem incredible. The United Kingdom Atomic Energy Authority envisaged a twenty-fold increase in nuclear generating capacity by the end of the century (60 new power stations!) and a further four-fold increase, mainly from breeder reactors, in the following 30 years (RCEP 1976). Figure 6 shows sites for possible long term nuclear development under consideration by the CEGB in the late 1970's. The map is based on a confidential CEGB list obtained and published by the Suffolk Preservation Society (a branch of CPRE) during the lead-up to the Sizewell Inquiry and caused great consternation among environmental and anti-nuclear groups, though it was dismissed as being out of date by the CEGB (Suffolk Preservation Society 1982).

Ever since the oil crisis, demand projections have been progressively revised downwards. The economic recession, energy conservation, changing industrial structure and competition in the space heating market have all contributed to a much lower than anticipated rate of growth in electricity consumption. Forecasting has become more problematic in the face of economic uncertainty and the absence of smooth trends in the relationships between electricity demand and various macro-economic indicators. This is reflected in the wide range of scenarios examined in recent official projections (CEGB 1982; Department of Energy 1982).

No application has passed through all stages of the planning process since 1974, and recent applications (and pre-application consultations) have all related to nuclear power stations. These have been subject to prolonged consideration at the local level, and the debate has ranged widely over both local and generic issues. The "public" involved in the debate has also changed, with national environmental pressure groups being much more in evidence. The time scale for decisions has increased sharply (Figures 4 and 5), and it is customary to attribute this to delays in the planning and participation process. An alternative explanation could be that delays are permitted by lack of urgency rather than caused by public scrutiny; it is

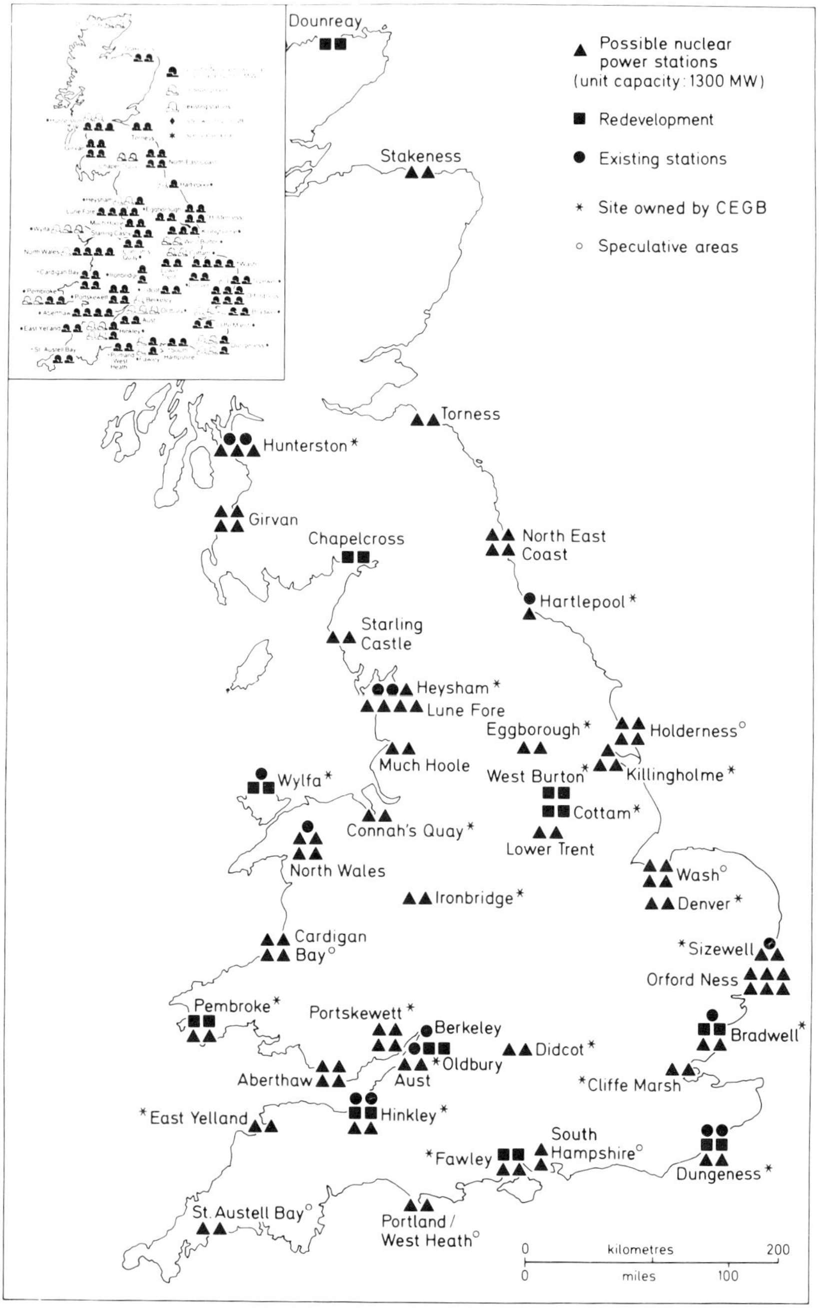

Figure 6. Sites for long term nuclear development.

politically convenient to allow debate at a time when there is no pressing need for developments to proceed. This would have serious implications for public participation in the future.

Several examples can illustrate the way in which the issues have changed. The history of the Portskewett site on the Severn Estuary serves this purpose particularly well (it is fully documented in Owens, Hills, and Cope 1983). An application for an AGR in 1970 attracted only 10 local objections. Visual intrusion, loss of amenity and disturbance during construction were the main issues at the short local hearing; safety was discussed, but did not dominate the proceedings (Baldock 1972). Consent was granted in 1972 (already a long time scale by previous standards), but no immediate start upon construction was made.[2] When the CEGB tried to renew the planning permission in 1977 and 1978, there was a dramatically different response. Well over 1000 letters of objection were received and some 3200 people signed petitions against the development. Objections, in contrast to those in 1970, came from well beyond the immediate vicinity of the site. Meetings, seminars and 'planning surgeries' were held and numerous documents produced. Need, national energy policy, reactor choice, nuclear hazards and other non-local issues virtually swamped the more parochial concerns. The CEGB faced confident and well informed opponents, some of them from national environmental groups. As the future of the AGR was also very uncertain at this time, the application was withdrawn before inquiry. At about the same time in Scotland, objections to the Torness AGR hardly mentioned amenity--all were concerned with broader issues of policy, safety and participation. A final example is provided by the proceedings at Sizewell. The public inquiry into the CEGB's PWR application has been sitting for 18 months at the time of writing (July 1984); it will probably last for two years and will cost about 10 million pounds. Overall, the decision process, from the date of application, will have taken at least 6 years. The Sizewell 'A' magnox reactor was approved within one year of the CEGB's application in 1959, with few objections and no inquiry.

In an attempt to preempt controversy and to reduce the time scale of the formal procedures, the CEGB has recently made a much greater effort to involve the public, as well as local authorities and other statutory bodies, during pre-application consultations. At prospective power station sites in the South West of England and in Northumberland they have spent much time and effort on publications, exhibitions, seminars and public meetings, though it will be several years at least before an application is submitted for any of these sites. The result, however, seems to be to stimulate awareness and to give opponents a long time in which to consider their case; some are now attending the Sizewell Inquiry. The CEGB has perhaps assumed too readily that opposition to nuclear power arises from ignorance and misunderstanding. As argued below, much of the opposition to nuclear (and other) developments is more fundamental and cannot be dispelled by "education" and information.

Evidence suggests that the present situation has arisen from the superimposition of several important trends and events, as shown in

Figure 7. Undoubtedly, one factor has been the growing tendency for local communities to reject large scale developments of any kind. This has been a long term trend, apparent even before the emergence of "environmentalism" in the late 1960's, which alone could have severe implications for national energy policy. As Pearce et al. (1979, 224) point out,

> "...a micro-political decision of this kind need never be related to the macro-political 'need' because it is always possible to say that the installation should go elsewhere."

The conflict here is essentially one of locational equity, compounded by the problem of evaluating intangible concepts like amenity--how to weigh diffuse benefits against locationally concentrated costs, and how to weigh one community's environmental quality against another's. These are profoundly difficult issues, on which there exists an extensive literature, but which have never satisfactorily been resolved.

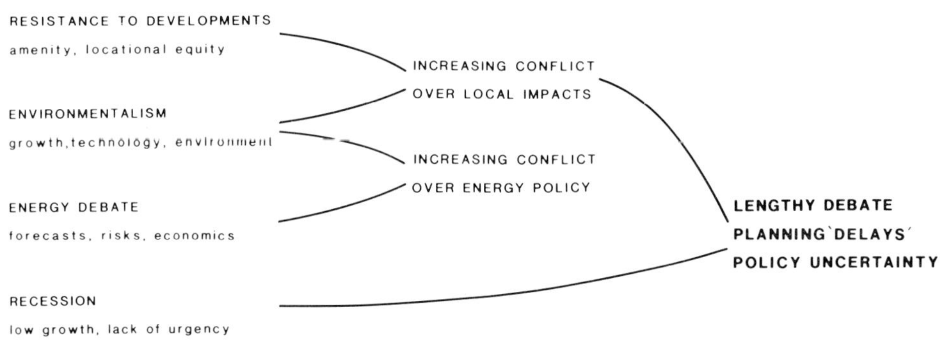

Figure 7. Social trends leading to planning 'delays' for power stations.

A second factor has been the "energy debate," stimulated by the oil crises of 1973-4 and 1979. Energy forecasts and policy have been open to public scrutiny in the United Kingdom as never before, revealing a host of value judgements and assumptions, and opening a Pandora's Box of issues for debate. In particular, this has emphasized the gulf in values between those who envisage a centralized, high technology future of material consumption and those committed to smaller scale development with emphasis on environmental values and quality of life (Cotgrove and Duff 1981; Lovins 1977; O'Riordan 1981; Wynne 1975).

Opposition to nuclear power in particular has become symbolic of this wider debate (Pearce 1978; Wynne 1983). As a result, power station proposals have been challenged on grounds of need, economics and safety, by national environmental pressure groups (including Friends of the Earth, the Anti-Nuclear Campaign and the Political Ecology Research Group) whose argument is with the principle behind individual proposals rather than with local environmental impacts at specific sites. Opportunities for public participation in the planning system are being used openly to challenge the conventional wisdom about energy policy. It is worth noting that this opportunity is available because implementation of energy policy involves planning consent for individual facilities; other controversial national policy issues, which do not involve land use changes, are not open to debate in the same way.

For want of a better forum, national and even international issues are being debated at public inquiries originally designed to hear local objections to specific proposals. This has led to considerable disaffection on all sides, though for different reasons, as argued below. The effect has been to reinforce, rather than replace, "traditional" objections; those primarily concerned with protection of local amenity can use arguments against the development _in principle_ to great advantage. But it would be a mistake to suggest, as some have done, that to question need and policy is no more than a new weapon in the old battle for local preservation.

Finally, the recession and over-capacity in the electricity supply system are important factors because they mean that lengthy debate can be tolerated. Indeed, it is probably convenient for the CEGB and the Department of Energy to be seen to be allowing "exhaustive scrutiny" at a time when speed is not crucial. In general terms, however, the "energy establishment" views the growing difficulty in obtaining consents for new power stations with considerable disquiet. Their perception of the problem, and the radically different views of other participants in the debate, will now be examined in more detail.

4. PERCEPTIONS OF THE PROBLEM--AND ALTERNATIVE "SOLUTIONS"

The CEGB and others with an interest in electricity developments, including construction companies, plant manufacturers and the relevant trades unions, argue that the system has swung too far in favour of their opponents, and that lengthy delays impose unnecessary costs on society as a whole. Typically, the argument is as follows:

> "Delays to Sizewell 'B' would mean great expense for taxpayers. We would lose 75 million a year because we would not have its cheap power and would also be faced with a severe power plant construction jam in the late 1990's when we would have to replace many existing, by then ageing, generators" (Baker 1984).

The CEGB:

> "...would like to see whether methods could be found within the framework (of current legislation) to shorten the timetable without harm to proper consideration of the project" (CEGB 1980, 15).

It is conceded (through force of circumstance!) that generic issues

> "should be subject to full examination at the first public inquiry (for a new nuclear system)...(but)...repetition at subsequent inquiries ought to be unnecessary" (CEGB 1980, 15).

A strange ambiguity has arisen on this latter point. It was widely believed that after the Sizewell Inquiry, discussion of generic issues would be curtailed at future inquiries into PWR's and the Secretary of State for Energy seemed to endorse this view (House of Commons Select Committee on Energy 1981). On the grounds that important issues would be debated "once and for all" at Sizewell, objectors requested public financial support to enable them to put their case. In a letter to the inquiry inspector explaining why he would not release funds for this purpose, the Secretary of State has suggested that there would be no grounds for curtailing discussion of generic issues at future inquiries (Lawson 1982). The issue remains confused, quite apart from the strange assumption that generic issues would remain unchanged over any lengthy period of time.

Environmental conflicts and public resistance are not openly acknowledged as potentially serious constraints in Government policy documents. The most recent discussion paper on energy policy devotes only three pages to a superficial consideration of the environment (Department of Energy 1978), and the December 1979 statement announcing a 10 station, 15GWe nuclear power programme paid scant attention to the fact that each application would be subject to vigorous public opposition. When acknowledged, opposition is represented as ignorant or misguided, rather than as a legitimate alternative view of the energy future (see Wynne 1975 for a stimulating analysis of this issue). But other sources show that there is great sensitivity in Government on this issue, and particular concern that nuclear power could become a focus for anti-Government feeling. Commitment to public debate may be largely rhetorical: Cabinet leaks reveal the feeling that, 'the Government might make more rapid progress towards its objectives by a low profile...' (Sedgemore 1981, 135) and fear of the '*danger* that a broad ranging inquiry (into the PWR) would arouse prolonged technical debate between different facets of scientific opinion' (House of Commons Select Committee on Energy 1981, 158; author's emphasis). As Wynne (1975, 125) points out:

> "...the process of social self-appraisal initiated under the aegis of technology assessment goes so far and no further:

when the light of examination reaches corporate capitalist institutions and values, the beam is hastily switched elsewhere."

Local and national pressure groups and many local planning authorities perceive the problem quite differently, and much of their criticism is directed at the conduct of public inquiries. Although some steps have been taken to improve the fairness of proceedings following the recommendations of the Dobry Report on the development control system (Dobry 1975), objectors still face the problem of defending something which has no "price", and remain at a severe disadvantage in terms of finance and other resources in comparison with the proponents, who devote large amounts of (public) money and manpower to their case. These problems are shared by objectors to all major developments and have been well documented elsewhere (see, for example, Council for Science and Society _et al_. 1979; Gregory 1971; Layfield 1982). More fundamentally, national pressure groups (and other observers) argue that a public inquiry into a site specific proposal is far too late a stage for any effective "participation" in the policy process (Council for Science and Society _et al_. 1979; Political Ecology Research Group 1978; Royal Town Planning Institute 1982; Town and Country Planning Association 1978). Some are so disillusioned that they have refused to participate in the proceedings at Sizewell. Others feel that generic issues _must_ be aired at public inquiries while there is no satisfactory alternative, and even that the adversarial nature of the proceedings at least ensures media coverage and public awareness of the issues (Grove-White 1984). The doctrine of ministerial accountability to an elected Parliament on policy questions is not accepted, since many of the issues clearly do not receive adequate debate; it is suggested instead that energy policy should be subject to scrutiny by an independent body with a broadly based and, as far as possible, impartial membership. Several variants of a multi-stage inquiry procedure have been proposed, in which national and local issues would be separately considered (Council for Science and Society _et al_. 1979; Pearce _et al_. 1979; Political Ecology Research Group 1978; Town and Country Planning Association 1978).

Provision for a two stage inquiry already exists in British planning legislation, though it has never been used (Edwards and Rowan-Robinson 1979) and similar ideas have subsequently been endorsed by ministers (Shore 1978; King 1980). In practice, extension of the terms of reference of existing inquiries, most comprehensively at Sizewell, is all that has been conceded to date. This is much "safer" than detailed independent review of policy and other generic issues because, as McAuslan (1979, 19) points out, the legal and administrative system is _constructed_ in the interests of maintaining the _status quo_ and,

"its institutional bases prevail and will always prevail over the interests of any fundamentally different points of view)"

In any case it has been too readily assumed that a multi-stage examination of policy, generic and site specific issues will improve the decision making process on the grounds that, "the better the procedure, the better (and more confidence inspiring) the likely outcome" (Council for Science and Society *et al.* 1979, 35). The implicit assumptions of a "rational comprehensive" and pluralist policy process are both questionable in the British context. An incremental model, in which agreement between groups involved in the policy process is obtained by *avoiding* clarification of objectives, is more appropriate (Richardson and Jordan 1979). Furthermore, energy policy decisions are dominated by powerful interest groups which have a close, almost symbiotic relationship with the State (Forster 1974; Richardson and Jordan 1979; Wynne 1975)--the "corporate pluralism" described by Cox (1981), Smith (1983) and others. The "amenity lobby" has become sufficiently influential to participate to some extent in this process, but environmental groups with radically different views of the energy future have not.

Much has been said in the past few years about improving decision making procedures and particularly about mitigating the problems presented by major public inquiries. But this implies a degree of consensus about the ends to be attained by reform, whereas in practice the objectives of the various participants in the planning process are quite different and cannot all be satisfied simultaneously.

It remains now to consider the prospects for the future. The following section looks first at potential development pressures based on evidence given at the Sizewell Inquiry, then considers the form which future conflicts might take and in whose interests they are likely to be resolved.

5. PROSPECTS FOR THE FUTURE

Projections of electricity demand produced for the Sizewell Inquiry are more modest and uncertain than those which have appeared previously. This is clearly demonstrated by the range of CEGB projections in Figure 8; the extremes represent continued de-industrialization and zero or negative economic growth on the one hand and early and substantial economic recovery on the other hand (CEGB 1982; Electricity Council 1984). The eight scenarios considered by the Department of Energy cover a broadly similar range of assumptions.

Implications of the higher and lower projections for the commissioning of new plants, after taking account of plant closure, are shown in Table 2. For comparison, the Department of Energy's last projections suggested that 40-50 GWe of new plant would be required by the end of the century, of which 30-36 GWe would be nuclear (Department of Energy 1979). Applications for consent for power stations will be considered 10-15 years before commissioning, so that plant to be commissioned during the 1990's and the first decade of the next century must be the subject of debate from the mid-1980's onwards. Clearly the number of *applications* will depend on the capacity for which individual consents are sought and cannot be predicted with accuracy from projected total capacity. It is assumed here that nuclear power stations will have a 1.5 GWe capacity on average and that coal-fired plant will have up to 3 GWe capacity.[3]

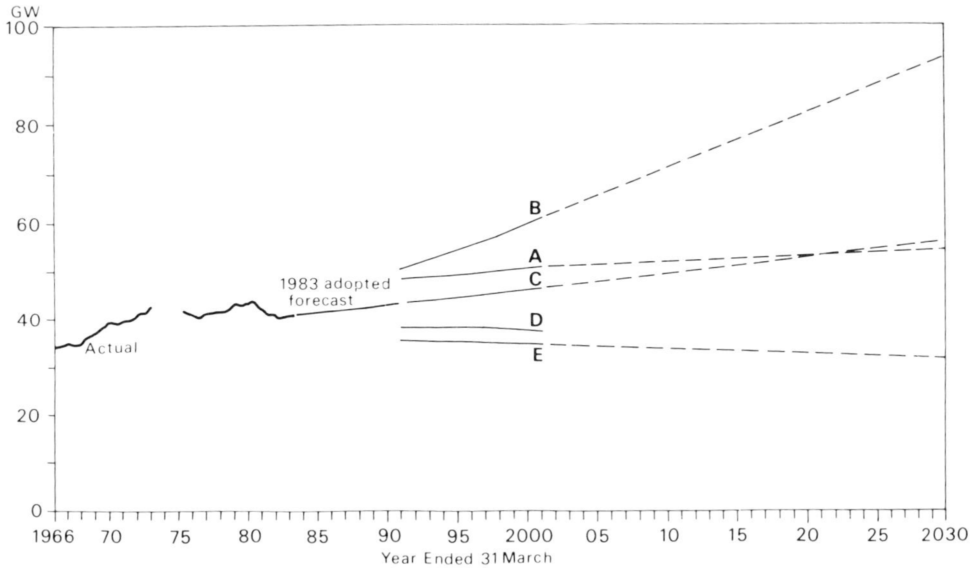

Figure 8. Electricity demand projections to 2030 (after CEGB 1982 and electricity council 1984)

The lowest projections imply that about three new nuclear power stations should be commissioned during the 1990's, together with 0.5 GWe of new coal-fired capacity (specified as a combined heat and power (CHP) scheme) and 1.7 GWe of new peaking plants (probably a hydro-electric pumped storage scheme). A further 10-15 nuclear power stations, a large coal fired plant (or perhaps smaller CHP schemes) and about 3.6 GWe of peaking plants should be ordered for commissioning during the first decade of the next century. Even these projections imply significant pressures on the planning system during the next two decades, but it is already clear that developments envisaged will not take place within this time scale. To date, only the Sizewell PWR application is being formally considered, and an application for a major pumped storage scheme has recently been deferred (Peak Park Joint Planning Board 1983). The objective of ordering one new power station every year, implicit in the 1979 nuclear programme, has clearly been abandoned, but sites for further nuclear development have been named by the CEGB. The next site will be Hinkley Point in Somerset, after which sites at Druridge Bay, Northumberland; Dungeness, Kent; Sizewell, Suffolk (a third station) and Winfrith, Dorset will be considered

(Electricity Council 1984). Only Druridge Bay is a "green field" site; Winfrith has a military reactor run by the United Kingdom Atomic Energy Authority and the other three each have one or two existing civil reactors (Figure 6). The use of existing sites is likely to reduce at least the local element of resistance to new power stations (Owens, Hills, and Cope 1983); the CEGB has recognized this and is seeking to use existing sites for nuclear stations and "to provide long term options for coal-fired development" (Electricity Council 1984, 30). (It is difficult, however, to detect any firm commitment to coal-fired development in current plans.) Observers from the sites named for possible future nuclear development have been attending the Sizewell Inquiry, fearful that discussion of the wider issues will be curtailed by the time that applications for their own sites are considered.

Table 2. Projected Need for Commissioning of New Electrical Plant (not yet ordered), GWe (upper & lower figures)

	Nuclear Plants	Coal-Fired Plants	Peaking Plants	Total
Up to 1990	-	0.2	0.0-1.1	0.2- 1.3
1991-2000	5.2-20.9	0.5- 8.4	1.3-5.1	7.4-34.5
2001-2010	18.8-42.6	2.3-17.0	3.1-6.4	25.1-65.0

Source: Department of Energy 1982.

The highest projections in Table 2 imply that some 40 new nuclear power stations, 8 large coal-fired stations, several CHP schemes and perhaps 5 major pumped storage schemes (or other peaking plants) would have to be ordered before the end of the century. Even with a rapid and unexpected turnaround in economic fortunes this is not a plausible scenario if any pretence of participation and democracy is to be maintained.

When the electricity industry itself is producing very low demand projections--two of the CEGB scenarios envisage a <u>decline</u> in electricity consumption--the projected need for new capacity seems surprisingly large. This arises partly from a redefinition of "need" as compared with the case in earlier applications, when the emphasis was on providing new capacity to meet the growing demand for electricity. Now there is more emphasis on minimizing net effective costs in the electricity supply system. New plants can replace older inefficient power stations and reduce average costs. The industry maintains that nuclear power is the most economic of the available options, but this is widely disputed and is one of the major issues at the Sizewell Inquiry (CPRE 1982; Evans 1984; Sweet 1982). In addition, the Government and the "nuclear lobby" are committed to maintaining a viable British nuclear industry. This requires the maintenance of "nuclear options from which the Board can order thermal reactor power stations with confidence" (Electricity Council 1984, 30) and a reliable stream of orders for new plants.

If the electricity supply industry attempts to pursue the programmes implied by these scenarios, conflict is potentially serious, for even the lowest projections imply a fairly steady stream of applications for new, mainly nuclear, power stations; these will come at a time when the environmental lobby (broadly defined) is more organized and better informed than in previous periods of energy development pressures.

It is clear, however, that something will have to change. If 15 applications for new power stations were submitted between now and the end of the century (in line with the "low" projections), and each took as long to determine as Sizewell, the whole process could take 90 years! While delays are being tolerated at present, it is clear that they will soon present real obstacles to implementation of the strategy endorsed by the present Government. _Reductio ad absurdum_ shows immediately that either the strategy must change or that the time scale for planning and participation procedures must be shortened.

One scenario is that the strategy will indeed be modified if the emphasis in energy policy shifts from supply to demand. The case for investment in energy conservation has become increasingly convincing, and at least one influential report has suggested that the British economy could maintain healthy economic growth over the next two or three decades with little or no increase in primary energy demand, and no need for an expanded nuclear progamme (Leach _et al._ 1979). This question is being taken seriously at a high level. For example, the Energy Select Committee:

> "...were dismayed to find that...the Department of Energy has no clear idea of whether investing around 1300 million in a single nuclear plant...is as cost effective as spending a similar sum to promote energy conservation" (House of Commons Select Committee on Energy 1981, 26).

If, for example, a decision were taken to divert civil nuclear investment into conservation, many of the 'generic' issues which have arisen in relation to recent power station applications would be defused, and if replacement capacity were, say, coal-fired plants on existing sites, local objections, while certainly still arising, might also be less intense. This scenario seems improbable, however, under the present administration (and especially in the light of the 1984 miners' strike). The nuclear industry remains a powerful lobby with much political support and clearly anticipates a resumption of regular power station orders in the near future, involving at least a "modest" programme of one nuclear unit per year (United Kingdom Atomic Energy Authority 1984, 20). The Government has done nothing to dispel this expectation.

An alternative scenario is that planning time scales will be significantly reduced in the future, but not necessarily by explicit curtailment of debate. If consent for a PWR at Sizewell is granted objectors would have to be very determined to argue the points which they had 'lost' at another PWR inquiry, and in any case may simply not have the resources to participate. There is also the possibility that applications will be submitted not in series but in parallel, for which there is ample precedent (in 1970, for example, 8 applications for Section 2 consent were made). Environmental groups could not possibly sustain the kind of case they have put at Sizewell in several inquiries at once. If consent for the PWR is refused, applications for AGR's or even major coal-fired plants may raise a new set of generic issues. In this situation, debate is still more likely to be curtailed by objectors' lack of resources, especially if applications come in parallel or in rapid succession, than by a politically embarrassing intervention by the Secretary of State.

This scenario, in which there will be some delays but also a significant shortening of the planning process, is more plausible for the immediate future than a radical revision of energy policy, because of the relative political power of the groups involved. The difficulty is in predicting how environmental groups (and the 'attentive public' who are broadly sympathetic to their views (Lowe and Goyder 1983)) will react essentially to 'losing' the battle over generic issues. Will they accept the ineffectiveness of their 'participation' in the energy debate with resignation, and turn their attention elsewhere? Will they resort to civil disobedience, like some of their European counterparts (Surrey and Huggett 1976; Taylor 1977)? Or will the whole issue of British energy policy be forced more explicitly into the party political arena where it has not traditionally been important, perhaps eventually resulting in the changes envisaged in the first scenario above? As yet it is possible only to speculate on the answers to these important questions.

6. CONCLUDING COMMENTS

Long delays in the power station planning process have often been interpreted as a temporary problem which has some satisfactory procedural solution. This paper suggests that such an interpretation is almost certainly mistaken. Intensification of conflict should be seen instead as one manifestation of a readjustment of certain values as society enters the "post-industrial" era. Underlying the generic and even the local issues are fundamentally different attitudes towards growth, technology and the environment. By definition, these issues cannot be "resolved" in the sense that all groups will be satisfied by the outcome. The next few decades, however, must see the resolution of conflict over the British power station programme broadly in the interests of one group or another, since power stations must either be ordered or not. The scenarios outlined above represent alternative possibilities. The reality will begin to unfold when the decision about the Sizewell PWR application is made.

7. FOOTNOTES

[1]The CEGB is responsible for power station planning in England and Wales. Its Scottish counterparts are the South of Scotland Electricity Board and the North of Scotland Hydro Electric Board. For simplicity, this distinction is not repeated.

[2]Section 2 consent is indefinite, but deemed planning permission lapses after five years. When the CEGB applied for renewal of planning permission, the local authority claimed that it had lapsed after three years (as for 'outline' rather than a full planning consent). Because of difficulty in reaching agreement on this matter, the CEGB submitted a fresh application in 1978.

[3]The 1979 nuclear power programme implies an average capacity of 1.5 GWe. The most recent fossil fuel plants have had total capacities of 2-4 GWe.

8. REFERENCES

Baker, J. 1984. Quoted in 'Costs row puts future of Sizewell in doubt.' The Observer 8.1.84.

Baldock, A. T. 1972. Report on a public hearing into an application by the Central Electricity Generating Board to construct a nuclear power station at Portskewett in Gwent. London: Department of Trade Industry.

Brookes, S. K. et al. 1976. The growth of the environment as a political issue in Britain. British Journal of Political Science 6: 245-55.

Buchanan, C. 1981. The Stansted controversy: No way to the airport. London: Longman.

Central Electricity Generating Board. 1980. Policy and planning procedures. Submission to Commission on Energy and the Environment's 'Coal Study.' London: CEGB.

______. 1982. Scenarios and electricity demand. Proof of Evidence P5 for Sizewell 'B' Power Station Public Inquiry. London: CEGB.

Committee of Public Accounts. 1984. *The development of nuclear power.* 32nd Report. London: HMSO.

Cotgrove, S. and Duff, A. 1981. Environmentalism, values and social change. *Britain Journal of Sociology* 32: 92-110.

Council for Science and Society, Justice and Outer Circle Policy Unit. 1979. *The big public inquiry.* London: OCPU.

Council for the Preservation of Rural England. 1959. *Annual report.* London: CPRE.

Council for the Protection of Rural England. 1982. *Statement of case for the Sizewell 'B' public inquiry.* London: CPRE.

Cox, A. 1981. Corporatism and the corporate state in Britain. *Teaching Politics* 10: 288-303.

Department of Energy. 1978. *Energy policy: A consultative document.* Command 7101. London: HMSO.

Department of the Environment. 1976. *Procedure for consultation and consent for new electricity generating stations and overhead lines.* DoE Circular 34/76. London: Department of the Environment.

Dobry, R. 1975. *Review of the development control system.* Final Report. London: HMSO.

Drapkin, D. B. 1974. Development, electricity and power stations: problems in electricity planning decisions. *Public Law* 19: 220-53.

Edwards, L. and Rowan-Robinson, J. 1979. *Whatever happened to the planning inquiry commission?* Discussion Paper 79-104. Aberdeen: Department of Political Economy, University of Aberdeen.

Electricity Council. 1984. *Medium term development plan 1984-91.* London: The Electricity Council.

Evans, N. 1984. An economic evaluation of the Sizewell decision. *Energy Policy* 12: 288-95.

Fischoff, B. *et al.* 1981. Lay foibles and expert fables in judgement about risks. In *Progress in resource management and environmental planning*, eds. O'Riordan T., and Turner, R. K. Vol. 3. London: John Wiley & Sons.

Forster, C. I. K. 1974. *Fuel policy work 1952-1972: A report.* London: Department of Energy.

Glasson, J. *et al.* 1980. *Socio-economic impacts of power stations* developments (2 Vols.). Oxford: Department of Town Planning, Oxford Polytechnic.

Gregory, R. 1971. The price of amenity. London: Macmillan.

Grove-White, R. 1979. Environmental pressure groups, energy policy and the new political climate. In council for the protection of rural England. Energy policy and local planning. Proceedings of a Symposium held in London, May 1979. London. CPRE.

______. 1984. Director of CPRE, Personal Communication.

Hambrook, L. G. 1972. Report of a Public Inquiry into an Application by the Central Electricity Generating Board for consent to the construction of a generating station at Inswork Point in Cornwall. London: Department of Trade and Industry.

House of Commons Hansard 542, Session 1955-56. London: HMSO.

House of Commons Select Committee on Energy. 1981. First report. The government's statement on the new nuclear power programme, Vol. 1, London: HMSO.

Hunt, F. R. 1975. Planning and development of major generation and transmission projects. London: Central Electricity Generating Board.

King, T. 1980. Speech to Royal Institute of Chartered Surveyors, London.

Lawson, N. 1982. Letter from the Secretary of State for Energy to Sir Frank Layfield Q.C. on the subject of funding objectors at the Sizewell Inquiry (September).

Layfield, Sir Frank. 1982. Letter to the Secretary of State for Energy on the subject of funding objectors at the Sizewell Inquiry (July 18th).

Leach, G. et al. 1979. A low energy strategy for the UK. London: Science Reviews Ltd.

Lovins, A. 1977. Soft energy paths: Towards an durable peace. San Francisco, CA: Friends of the Earth International.

Lowe, P. D. and Goyder, J. 1983. Environmental groups in politics. London: Allen and Unwin.

McAuslan, P. 1979. The ideologies of planning law. Urban Law and Policy 2: 1-23.

O'Riordan, T. 1981. Environmentalism (2nd Edition). London: Pion.

Owens, S. E., Hills, P. R. and Cope, D. R. 1983. Energy policy and land use planning: A report to the Social Science Research Council. London: Social Science Research Council.

Peak Park Joint Planning Board Press Release. 1983. CEGB shelves power station plans for Longdendale--but only temporarily. (27th July).

Pearce, D. W. 1978. The nuclear debate is about values. Nature 274: 200.

Pearce, D. W. et al. 1979. Decision making for energy futures. London: Macmillan.

Political Ecology Research Group. 1978. Public participation and energy policy. Oxford Report No. 6. Oxford: PERG.

Purdue, M., Kemp. R. and O'Riordan, T. 1984. The context and conduct of the Sizewell B Inquiry. Energy Policy 12: 277-82.

Richardson, J. J. and Jordan, A. G. 1979. Governing under pressure: The policy process in a post-parliamentary democracy. Oxford: Martin Robertson.

Royal Commission on Environmental Pollution. 1976. Sixth report on nuclear power and the environment. Cmnd. 6618. London: HMSO.

Royal Society. 1983. Risk assessment. London: The Royal Society.

Royal Town Planning Institute. The public and planning: Means to better participation. London: RTPI.

Sandbach, F. 1980. Environment, ideology and policy. Oxford: Basil Blackwell.

Sedgemore, B. 1981. The secret constitution. London: Hodder and Stoughton.

Shore, P. 1978. Speech on 'Major Public Inquiries' by the then Secretary of State for the Environment, Manchester (13th September).

Smith, G. E. 1983. Political geography, politics and the state. Social Science Research Journal 8: 30-56.

Starkie, D. 1982. The motorway age. Oxford: Pergamon.

Suffolk Preservation Society. 1982. 'Dozens of nuclear power stations: Highly confidential list published,' Press Release 16.9.82.

Lavenham, Suffolk, S.P.S.

Surrey, J. and Huggett, C. 1976. Opposition to nuclear power: A review of international experience. _Energy Policy_ 4: 288-307.

Sweet, C. 1982. _The costs of nuclear power_. London: Anti-Nuclear Campaign.

Taylor, P. 1977. The struggle against nuclear power in Central Europe. _Ecologist_ 7: 216-22.

Town and Country Planning Association. 1978. _Statement on energy policy and public inquiries_. London: TCPA.

United Kingdom Atomic Energy Authority. 1984. Nuclear industry--the road ahead. _Atom_ 333: 20.

Williams, R. 1980. _The nuclear power decisions_. London: Croom Helm.

Wynne, B. 1975. The rhetoric of consensus politics: A critical review of technology assessment. _Research Policy_ 4: 108-58.

______. 1983. _Nuclear decision-making--rationality or ritual?_ London: British Society for the History of Science.

CHAPTER 14

LAND REQUIREMENTS FOR SOLAR ELECTRICITY ALTERNATIVES

Philip R. Pryde
Department of Geography
San Diego State University
San Diego, California 92182

ABSTRACT. A major consideration prior to constructing direct or indirect solar electricity facilities is the quantity of land required. A review of existing and proposed solar electricity installations reveals a very wide range of land (or water) surface areas that would be required to produce a standardized 100 MWe. These range from under 1000 acres (400 ha.) for waste-to-energy plants, pumped storage, and parabolic dishes and troughs, to over 10,000 acres (4000 ha.) for standard hydroelectric reservoirs, OTEC, solar chimneys, and both cultivated and non-cultivated biomass. In addition, qualitative changes to the land must also be considered.

1. INTRODUCTION

With the realization that our fossil fuel resources are steadily diminishing, increased attention is being given to a variety of renewable energy alternatives. Foremost among these are the various forms of solar energy, reflecting their virtues of having a free fuel supply and of being almost infinitely renewable.

However, solar energy also has the inherent drawbacks of being both interruptible and diffuse. The spatial implications of this are that more land may be required for utilizing solar options than for nonrenewable energy options. The key questions for geographers and other planners are how much more land might be needed, and how much variation in land requirements exists among the various solar options?

This chapter investigates and quantifies the amount of area needed to provide a given quantity of energy by various solar technologies. In addition, the chapter also discusses some of the key variables or current unknowns that necessitate, in most cases, the use of a range of acreage figures. Finally, some of the qualitative aspects of the use of land for producing solar electricity is reviewed.

The subject of surface area requirements for renewable energy options has received relatively little study to date. The first to appear in the geographic literature is quite recent, and compared the

F. J. Calzonetti and B. D. Solomon (eds.), Geographical Dimensions of Energy, 255–275.
© 1985 by D. Reidel Publishing Company.

calculated land requirements for photovoltaic and heliostat facilities with those for coal (Pasqualetti and Miller 1984). This work will be expanded upon in the present chapter by examining a wide array of direct and indirect solar energy conversion technologies, and by utilizing actual operating data wherever possible. Other geographic works on the subject of energy have either been too broad to detail site requirements for land (Wagstaff 1974; Cook 1976; Pryde 1983), or treated it in total (Smil 1984), or appeared as monographs on other specific topics (Cook 1977; Calzonetti and Eckert 1981; Pryde 1984; and many others), or treated the land questions primarily from the qualitative, site planning, or environmental standpoints (Dobson 1979; Pasqualetti 1981; and numerous others). With regard to surface area requirements, only discussions relating to specific technologies or sites were encountered, and these have been used as source materials for the broader examination and comparisons provided in this chapter. Information on the facilities discussed below was gathered wherever possible by site visitation; when this was not possible, personal communication and review of project literature were utilized.

2. METHODOLOGY AND ASSUMPTIONS

Solar energy can be used to provide as an end product either heat or electrical energy. Since electricity is the most versatile form of energy, and since conventional methods of generating electricity are drawing increased criticism, this chapter focuses on solar technologies which can be used to produce electric power. These include both direct solar technologies (such as thermal electrical systems, photovoltaics, solar ponds, and ocean thermal energy) and indirect solar systems (such as wind, biomass, and hydroelectricity - see Table 1). A few proposed forms of solar conversion, such as ocean currents, have been omitted, because data on surface area requirements for these technologies are generally lacking.

For purposes of standardizing comparisons, 100 megawatts-electric (100 MWe) has been adopted as the quantity of electrical capacity to be achieved. This is an amount sufficient to service the needs of a city of about 90,000 persons. Installed capacities larger than this (e.g., 1,000 MWe) are not presently feasible in the content of renewable energy (with the exception of hydroelectricity).

In utilizing 100 MWe as the standardized facility capacity, it should be noted that electrical output, measured in kilowatt-hours per year (kWh/yr), could also be used as the unit of comparison. In some respects, kWh of output would be a more meaningful data unit, since production of electricity, rather than amount of installed hardware, is the objective of energy development. However, for purposes of this study, kWh would not represent a viable unit of comparison. This is primarily because many types of solar electricity facilities to be examined here do not yet exist on the ground, or exist only in pilot-plant scale, as opposed to being operational on a fully commercialized scale. Also, output from wind turbines is highly site specific; data

TABLE 1. Types of Electricity-producing Solar Energy Systems

A. Direct Solar

Non-thermal:
- Photovoltaic cells

Thermal:
- Heliostat arrays
- Parabolic dishes (point focusing)
- Parabolic trough (line focusing)
- Solar ponds
- Ocean thermal energy conversion (OTEC)

B. Indirect Solar

Wind (large and small systems)

Solar chimney

Hydroelectric:
- Conventional storage
- Pumped storage
- Bulb turbines

Oceanic:
- Wave energy
- Ocean currents
- Salinity gradients

Biomass:
- Cultivated feedstocks
- Non-cultivated feedstocks
- Waste material feedstocks

C. Solar-Gravitational

Tidal energy

from one windfarm might not be usable for one at a different location. Only for a few large, currently existing facilities would the use of kWh be feasible. Even here, these plants are so new that desirable long-term average annual kWh figures are not generally available. Thus, it is necessary to work instead with a standardized installed capacity figure such as 100 MWe.

There are also qualitative differences in the land requirements of the various solar options. That is, will the land used to produce solar electricity be simultaneously available for other economic uses, or will the energy production function preclude other beneficial uses on the same parcel? Also, what will be the nature of local land deterioration, if any, over the long run due to solar energy production? Although such questions are not the main focus of this study, these qualitative aspects will also be briefly discussed.

3. DIRECT SOLAR OPTIONS

Land requirements to produce electricity from direct solar sources are first considered. For some of these options, operational facilities are in place that can serve as a fairly precise measure of land requirements. For others, estimates prepared for proposed facilities will be used as the data source.

Among existing solar electricity facilities, one of the most elaborate is the "Solar One" heliostat field built by the Southern California Edison Company (SCE) near Daggett, California. Constructed on 130 acres (52.6 ha.) of land, it uses 1,818 tracking heliostats (mirrors) to focus the sun's radiation on a central boiler unit. At maximum output, the facility is capable of producing 10 MWe of electricity. Figure 1 suggests a virtually linear land-output relationship as this type of facility is scaled up. Consequently, this would suggest a maximum need for 1300 acres (526 ha.) to produce 100 MWe. However, it has been suggested (Skinrood 1984) that the full 130-acre site is not needed to generate 10 MWe, and a Sandia Laboratories study of central receiver power stations produced a theoretical calculation that only 700 acres (283 ha.) would be required for a 100 MWe system (Battleson 1981). When considering facilities of a type or size not yet built, it is not uncommon that a range of estimated land requirements will exist, since different consultants will make differing assumptions. The above range of acreages for heliostat systems, as well as those derived for the other forms of solar energy being examined in this chapter, are presented in summary form in Table 2.

A distinction should be made here between base and peaking power. Base power is that which is utilized 24 hours per day, whereas peaking power is that needed only for a few hours to meet the peak daily demand. Because most facilities that use direct solar energy have an interruptible resource base, they can be used at full output only a few hours per day. Without storage, they are usually not intrinsically suitable for either base or peak power (Holte 1984). Most solar

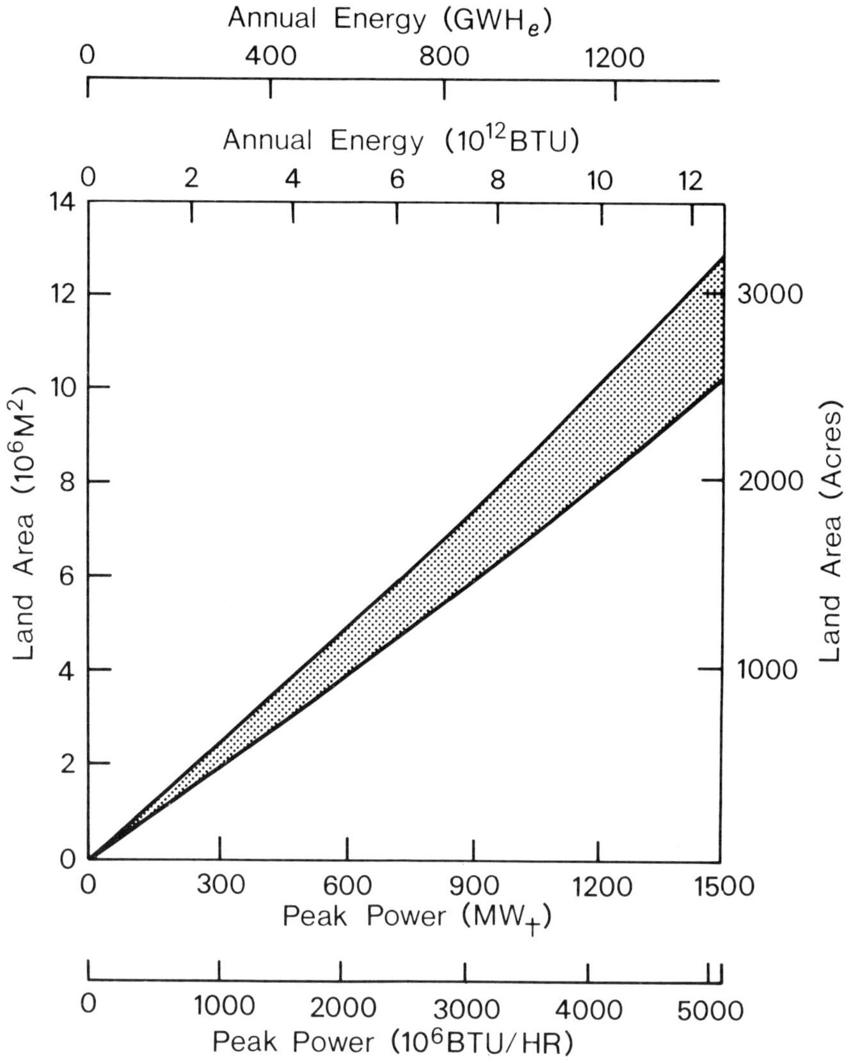

Figure 1. Land requirements for a solar central receiver plant as a function of the peak power (at the base of the tower) and annual energy output. This assumes: cavity receiver; heat transport fluid temperature, 1050°F (566°C); average heliostat field densities, 0.20 (upper curve) - 0.25 (lower curve); maximum direct normal insolation, 0.95 kW/m^2; annual direct normal insolation, 2850 kWh$_t$/m^2-yr; maximum field/receiver efficiency, 0.64; annual average efficiency 0.53. The slight curvature reflects greater atmospheric attenuation for larger heliostat fields. Peak power varies seasonally. Annual energy output excludes maintenance or equipment failure losses. A thermal-to-electric conversion efficiency of 0.39 was assumed to determine GW$_e$h/yr. Land usage when determined by an annual energy requirement for sites without 2850 kWh/m^2-yr insolation varies inversely as the annual direct normal insolation (Battleson 1981, 51).

TABLE 2. Land Requirements for Direct and Indirect 100 MWe Solar Electricity Options

Type of Process	Range of Land Required: Acres	Hectares	Assumptions and Sources (see text for elaboration)
PART A: DIRECT SOLAR OPTIONS			
Heliostat array (peaking power or direct sale)	700-1,300	283-526	Based on the existing Solar One plant near Barstow, CA, and a proposed 100 MWe facility
Photovoltaic cells (peaking power or direct sale)	2,000 3,800	810-1,540	Based on the existing 1 MWe plant at Hesperia, CA, and a 16.5 MWe plant in central California
Parabolic dish (peaking power or direct sale)	667	270	Based on the existing 4.5 MWe clustered dish facility at Warner Springs, CA
Parabolic trough (peaking power or direct sale)	667-750	270-304	Based on proposed Southern California Edison facilities of 12 and 15 MWe in the Mojave Desert
Solar pond (peak or base power)	1,200 10,400	486-4,212	Based on Ein Bokek 150 kWe plant in Israel and U.S. proposals at the Salton Sea (CA) and in Texas
OTEC (Ocean thermal energy conversion, peak or base power)	2,011-18,095	814-7,328	Larger figure based on 6-mi. technical spacing; smaller one on 2-mi. navigational clearance
PART B: INDIRECT SOLAR OPTIONS			
Wind - large turbines (peaking power or direct sale)	1,000-2,000	405-810	Based on proposed Solano County, CA, windfarm using 2.5 MWe turbines at 25-50 area turbine

TABLE 2 (continued)

Type of Process	Acres	Hectares	(see text for elaboration)
Wind - small turbines (peaking power or direct sale)	2,000- 5,050	810- 2,045	Data are from 3 existing and 1 proposed sites in California using 20-100 kWe turbines
Solar chimney (peaking power or direct sale)	24,700	10,000	Based on an existing prototype 40 kWe solar chimney in Manzanares, Spain
Hydroelectric (peak and base - see text)	762- 72,737	309 29,458	Based on a sample of 25 hydro-electric reservoirs in the U.S. (see Table 3)
Pumped storage* (peak power only)	42- 783	17- 317	Based on existing facilities at Niagra, NY, Ludington, MI, and Castaic, CA
Wave energy (base power or direct sale)	640- 1,280	259- 518	Assumes 500-100 MWe per linear mile of devices
Biomass cultivated (base power)	30,000- 59,260	12,150 24,000	Larger figure based on proposed Leucaena-burning plant in Hawaii (see text)
Biomass cultivated (base power)	138,270	56,000	Based on proposal for a multi-purpose gopher plant facility
Biomass, non-cultivated (base power)	584,000- 730,000	236,500- 295,650	Based on Burlington, VT, wood-chip plant and a proposed San Diego, CA chaparral-burning facility
Biomass, non-cultivated (base power)	1,260,000- 7,570,000	510,000- 3,070,000	These are theoretical calculations
Biomass, urban or farm waste	20- 80	8- 32	Reflects only the additional land required for power plant, storage, and buffer zones

Sources: See text for source citations

*Acres/MWe for pumped storage will vary greatly, since installed capacity at a given facility can be of varying amounts depending on how many hours a day the plant is intended to produce electricity.

facilities (direct or indirect) being built today approach this problem in one of two ways: the producers either utilize the provisions of the federal Public Utilities Regulatory Policies Act of 1978 (PURPA) to sell their energy to a local utility whenever it is produced (this is typical of wind farms and photovoltaic installations), or some form of storage can be utilized to permit the delivery of peaking power (this would be possible, for example, with thermal solar units, photovoltaic systems, and solar ponds).

The question of whether to produce base or peak power is a very fundamental problem for solar installations, and is well exemplified by Solar One. The technical description of the Solar One plant indicates that electricity can be produced at full capacity (10 MWe) for about seven hours a day near the summer solstice, but for only about four hours a day around the winter solstice (Anonymous 1982d, 3). This would present considerable problems in terms of land requirements, installed capacity, and storage capabilities if base power were to be produced. As a consequence, Solar One was constructed with storage facilities sufficient to allow a subsequent energy delivery of 7 MWe over a four hour period. This energy delivery could be in the form of either peak electrical output, or thermal energy. Energy stored at the experimental Solar One plant has to date been generally used for thermal applications (Raetz 1984). On the other hand, a few types of solar technologies, such as ocean thermal energy and biomass farms, could supply base power. For clarity, in Table 2, an indication is given as to what type of energy (base, peaking, or direct sale) will most likely be provided by each of the technologies listed.

Photovoltaic cells (sometimes called "solar cells") are another example of an existing technology. A one MWe photovoltaic installation has been built near Hesperia, California, on 20 acres (8.1 ha.) of land. Assuming similar design and equipment, this would suggest the need for 2000 acres (810 ha.) if this facility were enlarged to 100 MWe. A newer 16.5 MWe photovoltaic plant in eastern San Luis Obispo County, California, utilizes a 640 acre (259 ha.) site, which would equate to 3800 acres (1540 ha.) for 100 MWe (Tumulty 1983). This range of land requirements (2000-3800 acres or 810-1540 ha.) is presented in Table 2.

An interesting possibility exists concerning photovoltaic cells. If in the future their unit cost drops to where they are economically competitive with electricity generated from central power plants, it may become commonplace for such cells to be installed in compact arrays on the roofs of homes and businesses. If this occurs, it could substantially reduce the land requirements of photovoltaic electricity.

As another option, huge solar cell space satellites have been proposed that would beam vast amounts of energy back to earth for conversion into electricity. The receiving stations for solar satellites would require relatively little terrestrial space per 100 MWe, but for cost and security reasons it can be questioned whether they are likely to be built.

A third type of thermal solar technology involves arrays of parabolic "dishes" that concentrate the sun's energy on a boiler unit mounted at the parabola's focal point. Such a system, 4.5 MWe in size, has been constructed by the LaJet Company near Warner Springs, California, on 30 acres (12 ha.) of land (LaRue 1983). This extrapolates into a figure of 667 acres (270 ha.) for a 100 MWe facility. The Warner Springs facility does not utilize any storage, but instead sells its output directly to the local utility (Figure 2).

Figure 2. The Warner Springs facility.

A somewhat similar type of thermal solar system uses linear arrays of line-focusing parabolic troughs, instead of dishes. Two systems of this type have been proposed for the Mohave Desert by SCE, one of 12 MWe on 90 acres (36 ha.), and one of 15 MWe on 100 acres (41 ha.) (Anonymous 1982a; Anonymous 1982c). Thus, 100 MWe systems of this type would require 667 to 750 acres, or 270 to 304 hectares.

Solar ponds utilize the heating differential of saline and fresh water layers. One of their chief advantages is that these ponds can be operated 24 hours a day, thus providing base power if desired. SCE has proposed two solar pond complexes for California's Salton Sea. One would generate 5 MWe from a 210 acre (85 ha.) pond; a larger one 600 MWe from a 30,000 acre (12,150 ha.) pond (Anonymous 1982b). This would imply that 4200 to 4750 acres (1700-1924 ha.) are needed for 100 MWe. However, an existing 0.15 MWe pond in Israel (Munson 1984) needs only 1.6 acres (.6 ha), whereas a proposal for Truscott Lake in Texas would require 156 acres (63 ha.) for 1.5 MWe (LaBoeuf 1982). These latter two projects extend the range for 100 MWe from 1200 to 10,400 acres (486-4212 ha.), and this range is incorporated into Table 2. Although solar ponds could in theory provide base power, SCE has indicated that they envision using their inherent energy storage capabilities to provide peaking power (Holte 1984).

Ocean thermal energy conversion (OTEC) involves using the temperature differential between warm surface currents (such as the Gulf Stream) and deeper ocean waters to vaporize a secondary fluid which could be used to turn a low pressure turbine. A Rand Corp. research report (Anonymous 1980) envisioned a 100 MWe unit which would require a technical buffer zone (spacing) of six miles between units, or an area of 18,095 acres (7328 ha.) per unit. However, surface uses (mainly shipping) would probably not require a buffer zone this large, and a smaller navigational clearance zone could be envisioned of perhaps a one-mile radius, requiring only 2011 acres or 814 hectares.

4. INDIRECT SOLAR ELECTRICITY OPTIONS

Indirect solar systems that can be used to generate electricity include wind turbines, hydroelectricity, wave energy, and various biomass technologies. The land requirements for indirect solar systems vary over a far wider range than is the case for direct systems.

Wind systems are highly variable in terms of turbine size and configuration. Two types of systems are being built, those employing large megawatt-sized turbines, and systems typically using smaller turbines of from 20 to 100 kilowatts (kWe). For the large systems, turbines of 2.5 MWe each will be assumed (the Boeing MOD-2 model). Their spacing will depend on terrain, windflow characteristics, and safety considerations (each unit has a blade 300 feet or 91 meters in diameter). According to Pacific Gas and Electric Co., a proposed large-turbine windfarm in Solano County, California, will require 1371 acres (555 ha.) per 100 MWe (Anonymous 1981b). This equals 34.3 acres (13.9 ha.) per tower, which equates to a linear spacing between towers of 1,222 feet, or 372 meters. To allow for possible site variations elsewhere which would affect land requirements, this 34.3 acre figure will be expanded into a range of from 25 to 50 acres per turbine. Consequently, the required acreage for 100 MWe will range from 1000 to 2000 acres (405 to 810 ha.).

For the "small wind energy converison system" (SWECS) units, an average spacing of from 5 to 7 blade diameters is considered normal (Figuera 1984). This was verified on a field inspection by the author of several SWECS sites in San Diego County, where topographical considérations produced extrapolated figures of from 2000 to 5050 acres (810 to 2045 ha.) per 100 MWe (Figure 3). Printed account of three other sites in southern California (two built and one proposed) also produced figures that fell near the center of this range.

As noted above, windfarms using either small or large turbines generally sell their electricity directly to a local utility. None of the windfarms in California have to date installed any significant storage systems.

Figure 3. Small wind energy conversion system units.

An innovative variant on wind systems is called a solar chimney, in which air is heated under a surface canopy and forced up a smokestack-like tower which has a wind turbine mounted at the top. One such facility exists in Spain, generating 40 kWe from air heated under a 9.8 acre (4 ha.) transparent canopy (Hoffman 1982). This suggests a need for 24,700 acres or 10,000 ha. for an array of such units totalling 100 Mwe. Since only one solar chimney has ever been built, it is not known whether there might be spatial economies of scale resulting from larger units.

Hydroelectric facilities will inherently show considerable variability in land requirements. This is because the requisite land consists almost entirely of that needed to accommodate the reservoir behind the dam, and the size of reservoirs varies considerably depending on local topography. An added complexity is that some conventional hydroelectric facilities are built mainly for base power and others for peak power, with the latter having supplemental installed capacity.

Although other alternative energy technologies discussed in this chapter also produce peak power, there is a difference with respect to hydroelectricity. The difference is that hydroelectric reservoirs can conveniently store huge quantities (weeks or even years) of input energy, and then produce whatever level of peak MWe is desired over a short period of time without enlarging land area requirements. Thus, to a greater degree than any other solar technology, a hydro facility used predominantly for peaking power could (and probably would) have a very misleading MWe/land area ratio. For this reason, the hydroelectric data in this chapter has been divided into two sections: conventional multi-purpose hydroelectric dams, and pumped storage facilities where peak power is the sole objective. The great difference in MWe/land area will become readily apparent.

The procedure followed here has been to select 25 fairly typical multi-purpose hydroelectric stations of moderate size located in all parts of the United States (avoiding any that have excessive peaking power capabilities), convert their data to the 100 MWe standard, and present in Table 2 the full range of acreages obtained as the range of land areas required for 100 MWe of base hydroelectric power. This range calculates to be 762 to 72,737 acres (309 to 29,458 ha.), with a median size of 16,600 acres (6,723 ha.). The dams used in the hydroelectric analysis are listed in Table 3.

A variant on conventional hydroelectric power is pumped storage. Pumped storage produces peaking power (only) by using cheap nighttime power to pump water to an upper storage reservoir, and then producing high value peak electricity during the hours of maximum use the next day. Three typical pumped storage operations (Ludington MI, Niagra NY, and Castaic CA) were chosen to calculate a range of figures for this type of operation, and the range proved to be a very compact 42 to 783 acres (17 to 317 ha.) per 100 MWe. It should be noted, however, that a great range in acres/MWe for pumped storage is to be expected, since any amount of generating capacity may be installed at a given site depending on how few hours per day the generating capacity is expected to be used.

In addition to OTEC, another method of producing electrical energy from the ocean involves utilizing the vertical energy of waves. One source suggests that 50 MWe could be realized for each linear mile of wave-energy conversion devices (Schurr et al. 1979, 313); another suggests that 20-30 km could produce 1500 MWe (Ross 1979, 38). Depending on whether a one-quarter or one-half mile (400 or 800 meter) clearance would be desired in front of and behind the devices, it can be assumed that anywhere from 640 to 1280 acres (259 to 518 ha.) would be required for 100 MWe. However, the multi-kilometer stretch of wave energy conversion devices would pose a major obstacle to shipping and coastal access generally. It is difficult to factor this aspect of wave devices into the "area required" equation. A similar consideration would apply to OTEC facilities, and also to ocean current devices (not discussed here due to lack of data), and to tidal power (not classified as based on insolation). On the other hand, none of these oceanic systems use more than a token amount of terrestrial land surface, and the surface of the ocean is "rent free" (unless the federal government were to initiate some form of leasing program).

Biomass energy presents some of the most complex variables of any of

TABLE 3. Hydroelectric Reservoirs Used for Surface Area Analysis

Reservoir	Acres	MWe	Acres/100 MWe
Gunterville (AL)	70,700	97.2	72,737
Oahe (SD)	376,000	595.0	63,193
Norris (TN)	40,200	100.8	39,881
Flaming Gorge (WY,UT)	42,020	108.0	38,907
Watts Bar (TN)	43,100	150.0	28,733
Cherokee (TN)	31,100	120.0	25,917
Bull Shoals (AR)	71,240	340.0	20,953
Pickwick (TN)	46,800	216.0	21,667
Wheeler (AL)	68,300	356.4	19,164
Hartwell (GA,SC)	61,350*	330.0	18,591
Parker (AZ,CA)	20,390	120.0	16,992
Glen Canyon (AZ,UT)	161,390	950.0	16,988
Blue Ridge (GA)	3,320	20.0	16,600
Davis (NV,AZ)	28,192	225.0	12,530
Fort Loudon (TN)	15,500	128.0	12,109
Hoover (NV,AZ)	157,736	1340.0	11,771
Hungry Horse (MT)	23,750	285.0	8,333
Shasta (CA)	29,500	422.3	6,986
Fontana (NC)	10,670	238.5	4,474
McNary (OR,WA)	42,500	980.0	4,337
Grand Coulee (WA)	79,400	2280.0	3,482
Wilson (AL)	16,000	630.0	2,540
Oroville (CA)	15,500*	644.0	2,407
Lower Monumental (OR,WN)	6,590*	810.0	814
Chief Joseph (WA)	7,800	1024.0	762

Sources: Martin and Hanson 1966; Federal Power Commission 1976; Bureau of Reclamation 1971; and various agency brochures on specific dams (indicated by an asterisk).

the solar options. Much will depend on the type and source of fuel to be used (cultivated crop, non-cultivated crop, or waste products).

Land requirements for cultivated plantings of trees or other crops ("energy farms") will be considered first. The land use requirement will vary depending on (among other variables) the type of crop selected, the productivity of the site, and the assumed rotation (harvesting) period. A proposed farm of leucaena (a species of eucalyptus) in Hawaii with a 4 year harvesting cycle would require 59,260 acres (24,000 ha.) for 100 MWe (Shupe 1982a, 1196; Shupe 1982b). On the other hand, Schurr et al (1979, 375) suggest (without citing a specific crop) that for 100 MWe an area of only 30,000 acres (12,150 ha.) would be required. These figures form the range for the first biomass entry in Table 2. In another study, it was concluded that using the gopher plant (Euphorbia lathyris) for both fuel and other end-products, 138,270 acres (56,000 ha.) would be needed (Myers 1983, 175). Since the assumptions here are slightly different, a separate entry appears in Table 2 for the gopher plant.

If non-cultivated crops are to be used, acreage requirements increase dramatically. Using data for Georgia sycamores, a land requirement of as much as 7.57 million acres (3.07 million ha.) has been suggested assuming a 30 year rotation, or 1.26 million acres (.51 million ha.) if a 5 year harvesting cycle is possible (Zeimetz 1979, 13). However, analyses of the new wood chip plant in Burlington, Vermont, and of a proposed chaparral plant in San Diego, California produce much lower land use estimates. These figures are 600,000 acres (243,000 ha.) for the Burlington plant, and 584,000-730,000 acres (236,500-295,650 ha.) for the chaparral facility (Mitchell 1981, 92; Oechel 1984).

Biomass electricity may also be generated from either municipal or agricultural wastes. For agricultural wastes (which also includes timber industry wastes), the land requirements vary greatly depending on the nature of the feedstock. However, it could be reasonably argued that since it is existing wastes that are being used, no additional land is required other than that needed for storage, buffer zones, and the power plant itself. This same argument would hold for municipal solid wastes. For a proposed 21 MWe trash-to-energy plant, seven of eight candidate sites (land parcels) fell into a 10-16 acres (4-6.5 ha.) range (Anonymous 1981a). Thus, a probable range of 20-80 acres (8-32 ha.), depending on upscaling economies of land use, can be taken as the amount of additional land required for a 100 MWe waste products power plant. Comparing this figure with that for a non-cultivated biomass facility illustrates the great variance associated with biomass fuels (Table 4).

Table 4 summarizes the acreage requirements for all of the foregoing solar energy options. It presents mid-range surface area figures, derived from the data in Table 2. Table 4 provides a generalized comparison only, and the note that follows the table should be heeded.

TABLE 4. Summary of Land Areas Required by Energy Type

(based on mid-range of area figures for each type as given in Table 2)*

Type	Acres	Hectares
Biomass from wastes	50	20
Pumped storage (single-purpose peak power)	412	167
Parabolic dish	666	270
Parabolic trough	708	287
Wave energy	960	389
Heliostat field	1,000	405
Wind - large turbines (2.5 MWe)	1,500	608
Photovoltaic arrays	2,900	1,175
Wind - small turbines (20-100 kWe)	3,525	1,428
Solar ponds	5,800	2,349
Ocean Thermal Energy Conversion (OTEC)	10,053	4,071
Hydroelectric (multiple-purpose dams)	16,600	6,723
Solar chimney	24,700	10,000
Biomass (cultivated)	44,600	18,100
Biomass (non-cultivated)	580,000- 7,570,000	236,000- 3,070,000

Source: Author's calculations, based on data in Table 2.

*This table provides approximate acreage requirements. Various options, such as those using reservoirs, cannot be easily compared in terms of surface area requirements. The range of acerage requirements varies considerably. Some options permit conjunctive uses, some not. Also, some of the figures shown reflect existing facilities, whereas others reflect only design or theoretical estimates.

5. QUALITATIVE CONSIDERATIONS REGARDING LAND USE

It was noted at the outset that qualitative changes in the land being used must be considered as well as the quantity. The qualitative considerations are of three main types: the alternative use value of the land to be used, the amount of site disruption, and the possibility of conjunctive (additional beneficial) uses.

"Alternative use value of the land" refers not to its market price, but rather to its potential for other productive uses. For example, does the land to be used contain prime agricultural soils that might be better devoted to crops? Is it in a location of high aesthetic or wildlife values? Would the site more logically be developed for commercial or residential uses a few years in the future? These questions are by their nature site-specific, and will need to be investigated in the early stages of any solar electricity project.

Site disruption, if extensive, may be of concern for either aesthetic or conservation reasons. Of the non-ocean options discussed above, only the two water-based ones (hydroelectricity and solar ponds) involve total and permanent landform transformations. In both instances, existing water bodies may occasionally be utilized for the energy facility. The other options that do not involve water bodies seldom totally transform the land, but may do so to a greater or lesser extent. For example, wind farms require service roads and pads, and are sometimes built on steeply sloping land.

Conjunctive uses, if feasible, may permit greater economic returns from the site. Solar chimneys are highly compatible with potentially profitable hothouse operations. Grazing might be feasible around windfarms, but safety and security considerations would probably limit any activities involving frequent human entry. As an alternative, farmers could install wind turbines around the periphery of their fields, in which case a lower density of output would be realized, but also little additional land would be required. Banks of photovoltaic cells and thermal solar concentrators are usually placed as closely together as possible, precluding most conjunctive uses. Both solar thermal plants and photovoltaic arrays are capable of producing limited amounts of hot water as a by-product; a photovoltaic installation in Borrego Springs, California, plans to use thermally enriched cooling water in an ethanol production facility (Anonymous 1984). In the case of solar ponds, few if any conjunctive uses of the ponds will be possible. Conversely, many conjunctive uses of hydroelectric reservoirs are usually feasible.

In the case of biomass (energy farms), a variety of uses might be possible on plots where vegetative regrowth is occurring. This would depend largely on the frequency of the cutting cycle and the inclinations of the energy farm owners. At the very least, the area could provide a low level of wildlife habitat and watershed protection values, if properly managed for these ends. Also, as noted in the discussion of the gopher plant, biomass farms may sometimes be able to produce end-products other than fuel.

6. DISCUSSION AND SUMMARY

It is clear that there are a great many variables inherent in comparing the land requirements of direct and indirect solar electricity options. In viewing the data presented in Tables 2 and 4, for example, the importance of site variability should be stressed. For any direct solar or wind facility, the quality of the resource at a given site will obviously affect the amount of land needed to produce a given quantity of energy. It may be assumed that the first sites chosen for any new technology will be the most highly propitious ones. Future sites may in some cases be less advantageous, resulting in larger figures for 100 MWe than those shown in Tables 2 and 4. Land requirements for direct solar energy technologies, of course, will also vary regionally with variations in average annual insolation.

The differences between peak and base power were noted earlier. It is clear that if a photovoltaic facility or a windfarm, for example, were to be designed to produce base power, several times as much land would be needed (especially in winter) to achieve the same rating of 100 MWe. However, from a parcel of land of a given size, approximately the same amount of kWhe/year will be generated whether the facility is designed for base or peak power. This is because a fixed amount of insolation is available at any one location; the only variable in terms of annual output is the relative efficiency of the alternative technologies being considered to transform the insolation. If these efficiencies are similar, total annual output per acre (kWhe/year/acre), no matter how distributed or utilized diurnally, should not vary much regardless of technology used. As noted above, though, on a parcel of a given size, a base power plant would have a lower MWe installed capacity than would a peaking plant.

The reliability of any type of power plant can be depicted in terms of its capacity factor. "Capacity factor" is a term that is used to describe the percentage of the time (annual hours of operation/8760) that a particular energy generating plant is on line. Many types of direct solar facilities have relatively low capacity factors compared to fossil fuel plants; SCE's parabolic trough and heliostat systems, for example, have capacity factors of 30-40% (Holte 1984). Their economic performance can be enhanced by installing storage capacity so that the power could be sold at more attractive peak rates, as has been done at Solar One. To date, however, none have gone to the expense required to attempt to produce base power. In the case of wind, capacity factors are very difficult to discuss, due to the highly random nature of the wind resource being utilized, and the differing cut-in speeds and other technical variables of the individual wind turbines being used.

From the foregoing, it may be concluded that land requirements are a highly variable input into renewable energy production, as well as being intrinsically difficult to analyze. However, it is also clear that they represent considerations that must be introduced into the planning process at an early stage. In looking at the land use factor, firms or individuals producing electricity from renewable sources will of course be primarily interested in the cost of the needed land, but will also want to take into account potential land disruption and potential conjunctive uses, both of which can be translated into costs and benefits which will in some way affect total project feasibility. All these considerations can then be used to compare the land use implications of solar options with alternative ways of producing electricity, such as fossil fuel or nuclear power plants.

For a large utility, land requirements will not be the major consideration underlying a decision to build, for example, a large wind farm as opposed to a new nuclear power plant. Overall return on investment will be the basic determinant. On the other hand, if the company has already decided to explore renewable energy technologies, then land requirements may take on a fairly significant role in choosing among the many candidate technologies.

There is fertile soil here for further research by geographers. Only the cited article by Pasqualetti and Miller (1984) specifically explores the topics discussed in this chapter; other works by geographers explore the energy-land relationships from other perspectives. This subject matter is ideally suited to geographic techniques of spatial analysis, and might be particularly attractive as a research topic to those geographers who work in consulting or other applied capacities.

7. REFERENCES

Anonymous. 1980. Power from the sea. Rand Research Reports 4, 2, as reprinted in SIPIscope, 1980, July-August: 6.

______. 1981a. San Diego energy recovery project: Environmental impact report. Foster City, CA: Environmental Science Associates, Inc.

______. 1981b. World's biggest wind project. PG and E Progress June: 1

______. 1982a. Luz 15 MW power plant on target. Renewable Energy News September: 4.

______. 1982b. Salton Sea can produce electrical power, California Energy Commission News and Comment June: 3-4.

______. 1982c. So. Cal. Edison announces solar projects. Sun-Up May: 28.

______. 1982d. Solar 10-MWe pilot plant. Daggett, CA: Solar One Visitor Center.

______. 1984. The solar revolution (brochure). Foster City, CA: United Energy Corporation.

Battleson, K. W. 1981. Solar power tower design guide: Solar thermal central receiver power systems, a source of electricity and/or process heat. Livermore, CA: Sandia Laboratories.

Bureau of Reclamation. 1971. Statistical report of the Commissioner, Bureau of Reclamation (Parts I, II, & III: Statistical and Financial Appendix). Washington, DC: Bureau of Reclamation.

Burchell, R. W., and Listokin, D., eds. 1982. Energy & land use. Piscataway, NJ: Center for Urban Policy Research, Rutgers University.

Calzonetti, F. J., with Eckert, M. S. 1981. Finding a place for energy: Siting coal conversion facilities. Washington, DC: Association of American Geographers.

Cook, E. 1976. Man, energy, society. San Francisco: W. H. Freeman & Co.

______. 1977. Energy: The ultimate resource. Washington, DC: Association of American Geographers.

Dobson, J. E. 1979. A regional screening procedure for land use suitability analysis. Geographical Review 69: 224-34.

Federal Power Commission. 1976. Hydroelectric power resources of the United States, developed and undeveloped, January 1, 1976. Washington, DC: Federal Power Commission.

Figuera, A. 1984. San Diego Gas and Electric Co., personal communication, February 24.

Gustafson, M. R. 1979. Limits to wind power utilization. Science 204: 13-17.

Hoffman, D. 1982. Solar chimney. Renewable Energy News September: 1-2.

Holte, K. 1984. Southern California Edison Co., personal communication, May 16.

Knight, H., Nyhart, J. and Stein, R. 1977. Ocean Thermal Energy Conversion. Lexington, MA: D.C. Heath & Co.

LaRue, S. 1983. High-tech solar plant under way. San Diego Union December 22: B-3.

LeBoeuf, F. 1982. Truscott brine lake solar pond system. SERI in Review. Golden, CO: Solar Energy Research Institute.

Martin, R., and Hanson, R. 1966. Reservoirs in the United States

(Geological Survey Water-Supply Paper 1838). Washington, DC: U.S. Government Printing Office.

Mitchell, J. G. 1981. Whither the yankee forest. Audubon 83 No.2: 76-99.

Munson, R. 1984. Israel's solar ponds grow larger. Environment 26 No.1: 41-42.

Myers, N. 1983. A wealth of wild species. Boulder, CO: Westview Press.

Oechel, W. 1984. Director, Systems Ecology Research Group, San Diego State University, personal communication, April 5.

Pasqualetti, M. J. 1981. Energy and land use. In Land use: A spatial approach, eds. J. Lounsbury, L. Sommers, and E. Fernald. Dubuque, Iowa: Kendall/Hunt.

Pasqualetti M. J., and Miller, B. A. 1984. Land requirements for the solar and coal options. Geographical Journal 150: 192-212.

Pryde, P. R. 1983. Nonconventional energy resources. New York: John Wiley & Sons.

______. 1984. Soviet development of solar energy. Soviet Geography 25: 24-33.

Raetz, J. 1984. McDonnell-Douglas Corp., personal communication, May 22.

Ross, D. 1979. Energy from the waves. Elmsford, NY: Pergamon Press.

Schiffman, Y., and D'Alessio, G. 1983. Limits to solar and biomass energy growth. Lexington, MA: D.C. Heath & Co.

Schurr, S., Darmstadter, J., Perry, H., Ramsay, W., and Russell, M. 1979. Energy in America's future: The choices before us. Baltimore: Johns Hopkins University Press.

Shupe, J. W. 1982a. Energy self-sufficiency for Hawaii. Science 216: 1193-99.

______. 1982b. Personal communication, June 18.

Skinrood, A. C. 1984. Sandia National Laboratories. Livermore, CA. Personal communication, May 21.

Smil, V. 1984. On energy and land. American Scientist 72: 115-21.

Stobaugh, R., and Yergin, D., eds. 1980. Energy future, 2nd ed. Westminster, MD: Random House.

Tietze, W., ed. 1981. Supplementary issue on energy. GeoJournal 1981: 3.

Tumulty, K. 1983. Arco, PG&E plan solar plant. Los Angeles Times March 24: IV-1.

Wagstaff, H. R. 1974. A geography of energy. Dubuque, Iowa: William C. Brown, Co.

Zaborsky, O. R., ed. 1982. Handbook of bio-solar resources. 3 vols. Boca Raton, FL: CRC Press.

Zeimetz, K. 1979. Growing energy: Land for biomass farms. Agricultural Economic Report No. 425. Washington, DC: U.S. Department of Agriculture.

PART IV

Patterns of Energy Use

CHAPTER 15

THE URBAN GEOGRAPHY OF RESIDENTIAL ENERGY CONSUMPTION

Marilyn A. Brown
Energy Division
Oak Ridge National Laboratory
Oak Ridge, Tennessee 37831
U.S.A.

ABSTRACT. This chapter examines the link between urban ecology and residential energy consumption. An analysis of Decatur, Illinois census tract data illustrates that different types of residential areas have different kinds of heating and cooling equipment, and that the correlations are quite close. In particular, homes in areas of "low economic status" tend to have many room and window air conditioners and supplemental room heaters, but few central air conditioning systems. "Rental housing" is associated with fewer than average forced air furnaces and many hot water heating systems. Areas of "low family status" (characterized by elderly and single-person households) have a disproportionate number of room and window air conditioners. An analysis of residential energy consumption is less successful. Components of household energy use are only modestly correlated with socio-economic, demographic, and housing characteristics. These findings suggest that caution should be used when employing neighborhoods as market segments in efforts to implement energy conservation programs.

1. INTRODUCTION

Programs to conserve residential energy within metropolitan areas frequently use marketing strategies based on neighborhoods. Many conservation programs are limited to particular neighborhoods; others are implemented differently by neighborhood. Community associations are sometimes used as vehicles through which programs are advertised and are delivered. In addition to these examples of neighborhoods as explicit market segments, neighborhood market segmentation may be implicit, as

F. J. Calzonetti and B. D. Solomon (eds.), Geographical Dimensions of Energy, 279–298.
© 1985 by D. Reidel Publishing Company.

when eligibility is limited by family income or home ownership, characteristics which obviously are distributed unevenly within cities.

Given the use of neighborhoods in promoting energy conservation, knowledge about the urban geography of residential energy consumption should be of great practical value. It can assist in the effective spatial segmenting of urban populations for purposes of implementing strategies to improve the efficiency of household energy usage. It can also provide insight into how such strategies might be better tailored to the specific needs of differenct population subgroups.

These claims are based on the finding that American cities are remarkably homogeneous in their residential patterning (Rees 1979). In particular, residential areas within cities can be differentiated by standard "factors" based on socio-economic, demographic, racial, and housing characteristics. Many of these same characteristics correlate with home energy consumption.

Thus, it is surprising that there has been little effort to date to link the literatures on urban ecology and household energy use. What research there is on spatial trends in energy consumption has generally focused on the regional or national scale. At the urban scale, consumption levels of suburban versus central city households have been discussed (Grier 1978), but little attention has focussed on the energy consumption patterns of different urban neighborhoods. Fowler and Solzman (1984) have proposed a set of relationships between the urban ecology and residential energy consumption literatures, and they speculate upon the ecology of residential energy consumption in Chicago; however, no empirical tests are performed.

The lack of attention given to the residential energy/urban ecology linkage can be partially explained by the effort required to assemble the necessary data. First, data are required to determine the socio-economic, demographic, and housing characteristics of areas. Second, matching residential energy consumption data are needed. The former are relatively easy to obtain, especially for census years. The latter, however, are more difficult to collect. Residential energy consumption data are sometimes available through detailed engineering studies of individual buildings, but usually the number of buildings studied in any one metropolitan area is small. Utility company records provide a second alternative but this option is time consuming and costly. To examine an entire metropolitan area, one or more utility companies must be contacted to obtain utility bill data. Written permission authorizing disclosure of these data will likely be required of each participating household.

As a result of the effort required to measure energy consumption, much research stops short of directly measuring the residential energy consumption of individual households. It elaborates instead upon determinants of energy usage, such as household conservation practices and the thermal integrity of homes, and uses these factors to estimate consumption. One example of this is the U.S. Department of Energy's Comprehensive Community Energy Management Program, in which residential energy consumption was estimated for small areas within 16 U.S. cities, based on average regional energy consumption for different types of housing (Moore, Berger, Griggs, and Rubin 1980; 1981).

Another approach is to rely upon self-reports of consumption. Such data are problematic due to response bias and error (Seligman, Darley, and Becker 1978), and results from these studies (several of which are reviewed in this chapter) must be viewed with caution. Needless to say, energy consumption for different purposes (e.g., home heating, air conditioning, and appliance usage) has been studied with even less frequency, despite its designation as a high priority research area (Kendall 1978; Penz and Yasky 1979; Fels and Woteki 1980). The research reported here draws on a unique combination of datasets, spanning two scales (household and census tract) and a variety of measures (energy consumption estimated by end-use, socio-economic, demographic, and racial characteristics, and housing traits).

With the assistance of these datasets, this chapter seeks to capitalize on the findings and methods of urban ecology in an attempt to better understand and describe the geography of residential energy consumption in urban areas. This chapter begins with a review of relevant literature. It then describes a two-part empirical analysis based on a case study of Decatur, Illinois. First, factor analysis is applied to census tract data, combining the conventional measures of residential differentiation with descriptors of home heating and cooling equipment. Next regression analysis is used to predict residential energy consumption using surrogates of these same census measures and descriptors.

While it would be worthwhile and interesting to study the urban geography of the total household energy budget, including expenditures for transportation and the energy embodied in purchased goods, this chapter limits its attention to the energy consumed in the home. This represents approximately two-thirds of the energy consumed directly by households, the remaining supporting personal transportation needs.

As an introduction to the remainder of this paper, it is useful to present some recent statistics on residential energy consumption in the U.S. as a whole. Residential energy use comprises approximately one-fifth of the total U.S. energy budget. Average energy consumption per U.S. household has fallen almost 20 percent in the last ten years--it is now approximately 130 million BTU's per household instead of close of 160 in 1970 (Savitz 1984). This decline in average energy use has been of sufficient magnitude to counteract the increase of more than one million households each year, such that total energy consumption has remained fairly steady over the past ten years (Thompson 1984). As energy consumption per household has declined, the energy devoted to different end-uses has changed. The share of total residential energy use claimed by heating has fallen from 67 in 1970, to 63 in 1980. Air conditioning and other electric appliances have gained substantially, while water heating and cooking have shrunken as a piece of the residential energy pie (Meyers and Schipper 1984). Of course, these totals, proportions, and trends vary by region of the U.S.

2. BACKGROUND LITERATURE

The urban ecology/geography literature has successfully described and interpreted the overall social structure of residential areas within cities. It has identified the dimensions which differentiate urban subareas, the levels of residential segregation by dimension, and the spatial patterning of these dimensions (Rees 1979).

Three dimensions have been repeatedly identified as sets of characteristics along which urban neighborhoods differ. These dimensions or clusters of variables have been labelled "socio-economic," "family," and "minority or ethnic status." A fourth dimension pertaining to residential mobility is also frequently identified. High socio-economic status reflects high educational attainment, family income, and home values, and a large proportion of workers in professional and managerial occupations. High family status is associated with large households, many children, and few elderly, high female labor force participation, and newer, owner-occupied homes. High racial and ethnic status is indicated by large proportions of nonwhite and foreign born populations. High residential mobility is reflected by high rates of migration and population turnover (Rees 1979; Hamm 1982).

Of the three major dimensions, urban areas in the U.S. appear to be highly segregated by race, moderately segregated by socio-economic status, and slightly segregated by family status (Rees 1979). Concerning the latter, only the elderly and single-member households exhibit high levels of residential segregation. Socio-economic status tends to be distributed unequally across sectors or wedges emanating from the city center, with high income sectors typically buffered from low-income sectors by middle-income households. Family status tends to vary by concentric zones, increasing with distance from a city's downtown. Finally, minority and ethnic status tends to exhibit a clustered pattern in which concentrations of black and other minority and ethnic populations are segregated from others (Murdie 1969; Salins 1971; Rees 1979; Hamm 1982).

How these dimensions relate to residential energy consumption can be assessed by turning to existing models and research results concerning determinants of home energy usage. Most models of residential energy use show it to be directly influenced by (1) weather conditions; (2) characteristics of the home, including its heating, cooling, ventilating, and other equipment and appliances; and (3) energy-related behavior such as the control of thermostats and the day-to-day operation of appliances. Other factors such as price, attitudes, social norms, income and demographics are seen as indirectly affecting energy use, through their influence upon (2) and (3) (Van Raaij and Verhallen 1983; Brown and Rollinson 1985). Since this chapter focuses upon similarities in the internal variation of energy use within individual urban areas, variations in weather conditions can generally be ignored. Large metropolitan areas with considerable topographical relief may experience significant internal climatic variations including temperature islands and wind tunnel effects (Penz and Yasky 1979). But these are not amenable to generalization across metropolitan areas and are therefore ignored here. Instead, attention focuses upon variations in housing characteristics, energy-related behavior, and the socio-economic and demographic variables which affect them. In this regard, three of the four fundamental dimensions of neighborhood variation appear to have particular significance in the prediction of residential energy consumption--socio-economic status, family status, and residential mobility.

Measures of socio-economic status have been shown to be powerful predictors of residential energy consumption. In particular, energy usage increases dramatically with income (Morrison and Gladhart 1976; Cooper et al. 1983). The increase varies by type of fuel, with electricity consumption increasing faster than that of natural gas (Newman and Day 1975). One study indicated that low-income households use 55 percent less electricity than middle-income households, and 24 percent less natural gas (Grier 1977). A series of studies completed in 1973 and 1974 by the Washington Center for Metropolitan Studies found that the poor use 44 percent less electricity and 20 percent less natural gas than other households (Citizens Energy Project 1980). The implication is that the poor devote a greater portion of their energy budget to the necessity of home heating, and less to appliances and other more "discretionary" uses.

Life-cycle stage, one aspect of family status, is also powerful in its ability to predict residential energy consumption. Consumption follows an inverted U-shaped relationship with life cycle. It increases with each life cycle stage through to the child-rearing years. When the children leave home, consumption declines, but at a slower rate than it increased. As the household passes to the "sole survivor" stage, natural gas and electricity consumption drops rapidly to below that of the "bachelor stage" (Fritzsche 1981; Citizens Energy Project 1980). Warriner (1981) found that the elderly used about one-third less electricity on the average than the non-elderly.

It is critical at this point to note that these findings concerning the role of income and age are based on studies of residential energy consumption per household. They do not take into account the size of dwelling units occupied by different groups. It is likely that elderly and low-income households consume considerably more energy when their usage is standardized for size of house (Warriner 1981; Brown and Rollinson 1985). As a measure of energy use efficiency, BTU's per square footage of heated dwelling space is probably a more appropriate measure.

Several other variables correlate both with life cycle stage and energy consumption. Age of housing unit is one. Older homes generally have lower levels of insulation than newer ones, so that one would expect heat losses to be greater (Rosson and Sweitzer 1981, p. 210). Female labor force participation is also relevant here. Families in which both spouses are employed full-time outside of the home consume less residential energy than families in which one spouse is a full-time homemaker (Fritzsche 1981).

Race and ethnicity have tenuous correlations with residential energy consumption. To the extent that racial minorities are economically disadvantaged, their residential energy consumption will parallel that of low-income households. Otherwise, there is little evidence to suggest that race is an important predictor of energy usage.

There is evidence suggesting that the fourth dimension, residential mobility, may be a significant predictor of residential energy consumption. Expected length of residence is important because of the long payback period of many conservation measures. Thus, owners who expect to move are less likely to install energy conservation measures (Blumstein et al. 1980).

3. RESEARCH DESIGN

3.1 The Study Area

A case study approach is used to study the spatial distribution of residential energy consumption in urban areas. The area chosen for analysis is Decatur, Illinois, a city of approximately 90,000 residents. On the basis of an array of variables pertinent to residential energy usage, Decatur was judged to be representative of a mid-sized Midwestern American city. Specifically, its median family income and occupational structure are similar to that of the state of Illinois. Decatur differs somewhat from state of Illinois averages, however, in terms of its demographic composition (with a higher proportion of elderly), the age of housing (with a higher proportion of homes built before 1940), and housing tenure (with a higher proportion of owner-occupied housing) (U. S. Bureau of the Census 1982a; 1982b).

3.2 The Two Datasets

In order to achieve the objectives of this chapter, two datasets are needed. The first contains census tract data--both the standard set of socio-economic, demographic, racial, and housing variables used in urban ecology studies, as well as energy equipment variables. It allows factor analysis to identify clusters of energy equipment and other characteristics of Decatur's neighborhoods. Data on 24 census variables were compiled for Decatur's 31 census tracts--4 are dwelling unit characteristics (DETACHED, HVALUE, MOBILE, and OLDHOMES); 8 reflect types of heating and cooling equipment (AC--CENTRAL, AC--UNITS, ELECTRIC, FIREPLACES, FURNACE, HTPUMP, RMHEATER, and STEAMHEAT); and 12 are household traits (see Table 1).

The second dataset contains information on individual households living in owner-occupied, detached homes in Decatur, Illinois. The sampling design and survey methods used to collect this information are described in Appendix A. Only households using natural gas as their primary source of home heating were sampled. Since 92 percent of homes in Decatur use natural gas for heat, this restriction excluded only a

TABLE 1. Definitions of the Socio-Economic, Demographic, and Housing Variables

Acronym	Census Tract Data	Survey Data
AC--CENTRAL	Percentage of housing units with central air conditioning	1=housing unit with central air conditioning, 0=other
AC-UNITS	Percentage of housing units with one or more individual air conditioner	1=housing unit with individual air conditioner(s), 0=other
BLACK	Percentage black population	1=household with black respondent 0=other
COLLEGE	Percentage of population with 3 or more years of college education	1=household head with 3 or more years of college education, 0=other
DETACHED	Percentage of housing units which are detached and single-family	No replication possible
ELECTRIC	Percentage of housing units with supplemental built-in electric heat	1=housing unit with supplemental electric heat, 0=other
FEMALEHD	Percentage of families headed by female	1=household headed by female, 0=other
FIREPLACES	Percentage of housing units with fireplaces, stoves or portable room heaters	1=housing unit with supplemental wood heat, 0=other

TABLE 1. Definitions of the Socio-Economic, Demographic, and Housing Variables (contd.)

Acronym	Census Tract Data	Survey Data
FURNACE	Percentage of housing units with central warm-air, floor, wall or pipeless furnace	1=housing unit with furnace, 0=other
HTPUMP	Percentage of housing units with electric heat pump, 0=other	1=housing unit with electric heat pump, 0=other
HVALUE	Percentage of housing units valued at over $80,000.00 0=other	1=housing unit valued at over
MOBILE	Percentage of mobile homes	No replication possible
MORTGAGE	Percentage of households with mortgage	1=household with mortgage, 0=other
OLDHOMES	Percentage of housing structures constructed prior to 1940	1=housing unit constructed prior to 1940, 0=other
PERSONS	Mean persons per housing unit	Persons in housing unit
POP18	Percentage of population under 18	Persons in household under 18 years of age
POP65	Percentage of population over 65	Persons in household over 65 years of age
POVERTY	Percentage of population with incomes below the poverty level	1=households with incomes below $10,000, 0=other

TABLE 1. Definitions of the Socio-Economic, Demographic, and Housing Variables (contd.)

Acronym	Census Tract Data	Survey Data
RENTER	Percentage of housing units renter occupied	No replication possible
RMHEATER	Percentage of housing units with room heaters	No replication possible
SAMEHSE	Percentage of persons residing in in same house in 1975	1=households residing in same house in 1975, 0=other
SINGLE	Percentage of population single	1=single person household, 0=other
STEAMHEAT	Percentage of housing units with steam steam or hot water heating equipment	1=housing unit with steam or hot water heating, 0=other
WEALTHY	Percentage of population with family income over $40,000	1=households with incomes over $40,000, 0=other

small number of sampled households (U.S. Bureau of the Census 1982a). The limitation was imposed because the detailed analysis of utility bill histories (described subsequently) makes assumptions about the end uses of energy based on seasonal variations in natural gas and electricity consumption. The dataset replicates 20 of the census variables compiled in the first dataset. Since the sample does not include renters or occupants of attached or mobile homes, RENTER, DETACHED, and MOBILE are not included. It also does not replicate the variable RMHEATER. In duplicating the remaining 20 variables, many of them were reduced to dichotomous (0-1) measures (see Table 1). The dataset also contains the measures of residential energy consumption shown in Table 2, as determined from utility bill histories accessed through the Illinois Power Company, the gas and electric utility company serving Decatur. The histories document the monthly consumption of electricity and natural gas from May 1981 to April 1982. By assuming that no natural gas is used for home heating between June and September, and that no electricity is used for home cooling between October and May, estimates of energy consumption by end-uses are possible. These energy usage measures enable the role of the census variables to be assessed in terms of their ability to predict home energy usage.

4. RESULTS

4.1 A Factorial Ecology of Census Tract Data

The variables shown in Table 1 were factor analyzed in order to identify clusters of relationships between home energy equipment characteristics and socio-economic, housing, and other traits and to compare these clusters with the standard factors (i.e., socio-economic status, family status, minority status, and residential mobility) of the urban ecology literature. The correlation matrix with diagonals of squared multiple correlation coefficients was submitted to a principal axis factor analysis (Rummel 1972). Two-, three-, and four-factor solutions were rotated (using a varimax rotation); the latter was found to be more interpretable than the two- and three-factor solutions which proved to be too general to clarify relationships. Additional factors were not considered because their explanatory power (i.e., eigenvalues) was considerably less. Altogether, the four rotated factors accounted for 68 percent of the variance contained in the correlation matrix, and were therefore quite successful at characterizing the raw data. Quartimax and oblique rotations were run, in addition to the varimax. These alternative rotations showed fewer clear relationships between the socio-economic and energy characteristics than the varimax rotation.

Table 3 presents the results of the factor analysis, highlighting those variables which are associated with each of the four clusters of variables (i.e., loadings are underlined which have absolute values greater than .5). Each cluster is labelled according to these highlighted variables.

TABLE 2. Definitions of the Residential Energy Consumption Variables

Acronym	Definition	Formula
BTUSUM	Therms of natural gas	$\sum_{i=\text{May } 81}^{\text{April } 82} \text{Therms}_i$
BTUBASE	Therms of natural gas for uses other than space heating	$3\left(\sum_{i=\text{June } 81}^{\text{Sept } 81} \text{Therms}_i\right)$
BTUHEAT	Therms of natural gas for space heating	BTUSUM-BTUBASE
KWHSUM	Kilowatt hours of electricity	$\sum_{i=\text{May } 81}^{\text{April } 82} \text{Kwh's}$
KWHBASE	Kilowatt hours of electricity for uses other than air conditioning	$1.5\left(\text{Kwh's}_{\text{May } 81} + \sum_{i=\text{Oct } 81}^{\text{April } 82} \text{Kwh's}\right)$
KWHCOOL	Kilowatt hours of electricity for air conditioning	KWHSUM-KWHBASE
TOTAL	Total energy consumed in kilowatt hours	29.3 BTUSUM + KWHSUM

TABLE 3. Factor Structure of the Census Tract Data[a]

	Low Economic Status	Rental Housing	Low Family Status	High Residential Stability
WEALTHY	-.87	.01	.07	.18
DETACHED	-.86	-.42	.09	.01
HMVALUE	-.83	-.04	.20	.01
COLLEGE	-.81	.08	.01	.39
POVERTY	.69	.37	-.17	.14
OLDHOMES	.69	.48	.34	.01
FEMALEHD	.61	.39	-.44	.25
RENTER	.49	.76	-.08	.04
SINGLE	.41	-.50	.60	.11
POP18	.16	-.11	-.77	-.03
POP65	.35	.27	.67	.11
SAMEHSE	-.14	-.16	.42	.61
BLACK	.49	.14	-.05	.53
MOBILEHM	.14	-.30	-.10	-.55
AC-CENTRAL	-.87	-.32	-.23	.12
RMHEATER	.78	-.01	-.01	.04
AC--UNITS	.60	-.04	.52	.38
FURNACE	-.26	-.80	.10	.24
STEAMHEAT	.23	.80	.07	-.12
PERSONS	.03	-.05	.15	.46
MORTGAGE	-.16	.15	-.05	-.36
HTPUMP	-.24	-.16	-.21	-.45
ELECTRIC	-.18	.06	-.23	.32
FIREPLACES	-.04	-.12	.03	.15

[a]Entries in the table are factor loadings associated with a varimax rotated solution. Loadings with absolute values greater than 0.5 are underlined.

Factor 1 explains 36 percent of the total variance contained in the correlation matrix. It is labelled "Low Economic Status," loading negatively on WEALTHY, DETACHED, HVALUE, and COLLEGE, and positively on POVERTY, OLDHOMES, and FEMALEHD. This factor is also positively correlated with RMHEATER and AC--UNITS, and negatively with AC--CENTRAL. Thus, economically disadvantaged populations are associated with homes which lack central air conditioning, but which have a large number of individual air conditioning units and supplemental room heaters.

Factor 2 explains 15 percent of the variance and is labelled "Rental Housing." It suggests a close relationship between housing tenure and type of central heating system, loading positively on RENTER and STEAMHEAT, and negatively on SINGLE and FURNACE. Rental units in Decatur tend to have more hot water systems than owner-occupied homes, and fewer forced-air furnaces. This relationship may result from older homes with more outdated and energy-intensive heating systems being subdivided into rental units.

Factor 3 explains 9 percent of the variance and characterizes households of "Low Family Status." It loads negatively on POP18 and positively on POP65 and SINGLE, indicating few young residents but many elderly and single-person households. Like the first two factors, an energy equipment variable is also involved (AC--UNITS). Factor 3 describes areas with a disproportionate number of individual room air conditioners.

Factor 4 explains 8 percent of the variance and is characterized as "High Residential Stability." It loads positively on SAMEHSE and BLACK, and negatively on MOBILEHM. No energy equipment variables are closely related to this factor.

In sum, three of the four standard neighborhood dimensions are identified by the factor analysis of Decatur. The exception is racial and ethnic status, which did not separate out as a dimension, as is the standard finding; nor does it load highly on the economic status factor, as has occurred occasionally in previous urban ecology studies. Presumably it is not a distinct factor partly because race does not relate closely with the 8 measures of heating and cooling equipment. Rental housing, on the other hand, does emerge as an important, distinct factor. Rather than correlating highly with low family status, as in previous research, housing tenure is the core characteristic of a cluster of variables which comprises the second most significant dimension of Decatur's neighborhood ecology. Its strength derives from the close association between rental units', steam heating systems, few furnaces, and few single person households.

4.2 A Regression Analysis of Residential Energy Consumption

This section analyzes the survey data described in Table 1 in terms of their ability to predict the extent to which households consume energy for a variety of residential end uses (see Table 2). The purpose here is to assess the ability of census tract data on socio-economic, demographics, and housing characteristics (including energy-related equipment) to predict home energy usage.

The results of a series of stepwise linear least squares regression analyses are shown in Table 4. While the levels of overall explanatory power are modest (ranging from 2 to 24 percent), significant relationships nonetheless are identified. One of the most striking findings of the regressions is that one measure of socio-economic status and two measures of family status are significant predictors of a majority of the seven components of energy consumption. No measure of racial status or residential mobility enters into a regression equation. While only one of the energy equipment variables (FURNACE) enters into an equation, and none of the housing characteristics is significant, it is important to remember that there are high intercorrelations between the explanatory variables. As the factor analysis showed, many of the energy equipment, housing, and other variables are, in fact, parts of variable clusters. Thus, many of the close correlates of housing and energy equipment traits do enter the equations.

POP65 and SINGLE (measures of family status) and COLLEGE (a measure of socio-economic status) each enter into a majority of the equations, and have consistent signs and comparable beta weights across equations. The prominence of the family status dimension is indicated by the fact that one or more measures of this dimension is in six of the seven equations shown in Table 4. Analysis of each equation is pursued below.

Total residential energy consumption as estimated by the sum of natural gas and electricity usage, on a per square foot basis, increases with elderly household members and single-person households; it decreases with education. These relationships are essentially duplicated for the separate predictions of natural gas and electricity usage, per square foot. Natural gas usage per square foot for purposes other than home heating (included here are water heating and cooking, in particular), increased with household size; holding household size constant, it decreases with the number of household members less than 18 years of age. SINGLE and MORTGAGE are also significant predictors, but are far less powerful. Electricity usage per square foot for purposes other than air conditioning (included here are refrigeration and the use of other appliances, for instance) is not well predicted. It was expected that socio-economic status would be significant. Home heating energy usage, per square foot, increases significantly with the number of elderly persons in a household. Finally, energy consumption per square foot for cooling is best predicted by single-person households, and the relationship is increasing. Recalling the close association of both SINGLE and POP65, on the family status dimension (Table 3), this finding may also reflect the high expenditure of elderly on home cooling, normalized for house size.

5. CONCLUSIONS

This chapter has examined some of the links between the urban ecology and residential energy literatures in an effort to begin to describe the urban georgraphy of home energy consumption. It has been exploratory in

TABLE 4. Regression Analysis of Residential Energy Consumption

Dependent Variable	Standardized Regression Equation	Coefficient of Determination
BTUSUM	.24 POP65*** −.15 COLLEGE*	.09
BTUBASE	.89 PERSONS*** −.50 POP18*** +.21 SINGLE* +.14 MORTGAGE**	.24
BTUHEAT	.31 POP65**	.09
KWHSUM	.26 SINGLE*** −.20 COLLEGE** +.15 POP65*	.17
KWHBASE	.14 FURNACE*	.02
KWHCOOL	.26 SINGLE*** −.17 MORTGAGE* −.15 COLLEGE*	.17
TOTAL	.22 POP65*** −.20 COLLEGE** +.18 SINGLE**	.15

*, **, and *** indicate that independent variables are significant at the 0.05, 0.01, or 0.001 levels, respectively. Significance is based on t-statistics associated with the beta weights shown in the above equations.

nature, due to the reliance on a single case study area--Decatur, Illinois--and other data limitations, most notably the inclusion of only owner-occupied housing in the survey data. Nevertheless, it has provided some insights, due to other aspects in which the data were highly appropriate. The survey-based variables almost exactly replicated the census tract measures in terms of operational definitions, thereby enabling the explanatory power of the census data to be tested. Further, energy consumption data were collected from monthly meter readings, enabling estimates of energy consumption by end-use, and allowing more precise measurements than self-reports or engineering estimates.

The results of the factor analysis of census tract data on socio-economic, demographic, and housing characteristics (including heating and cooling equipment), uncovered the standard socio-economic status and family status dimensions, as well as rental housing and residential stability dimensions. Measures of energy equipment were central to three of the four dimensions. Different residential areas have different types of heating and cooling equipment, and the coorelations are quite close. In particular, "low economic status" is associated with homes which lack central air conditioning, but which have a disproportionate number of room air conditioning units and supplemental room heaters. "Rental housing" is associated with fewer than average forced-air furnaces and many hot water systems. Finally, areas of "low family status" (characterized by elderly and single households) have a disproportionate number of individual room air conditioners. Tracts with "high residential stability" do not closely align with any of the energy equipment variables studied.

The results of the regression analysis of household-level data were not as successful. The seven components of residential energy consumption were not well predicted by the same set of socio-economic, demographic, and housing (including energy equipment) variables as was studied in the factor analysis. The best predicted component was natural gas consumption for uses other than space heating; 24 percent of the variance in this component could be predicted. The most important explanatory variables were the number of elderly household members, whether or not the household had a single member, and whether or not a household head had three or more years of education--consumption per square foot of dwelling space increasing with the first two predictors and decreasing with the last.

These preliminary findings suggest that extreme caution should be used when employing neighborhoods as market segments in efforts to implement energy conservation programs. In certain instances, it may be highly appropriate. An example is when community groups are used to advertise and deliver weatherization services, since locally-based groups have been shown to be highly credible and effective in such roles. In other instances, neighborhoods may be inappropriate. This may be the case for programs which limit eligibility to particular neighborhoods, as does the low-interest loan program in Cleveland which is restricted to five neighborhoods (Krumholz and McDermott 1984) and the information/audit program limited to one area in Memphis (Morgan and Katz 1984).

The results of this chapter suggest that energy inefficiency is not highly concentrated within particular urban subareas. Thus, public programs to conserve energy must be widespread geographically. At the same time they should be tailored to the characteristics of the various types of areas; programs which are implemented without reflecting geographical variations are suboptimal. For instance, requirements that a minimum number of households from each of the various subareas of a city participate in a program [a stipulation which exists, for instance, as part of a weatherization program administered by the City of Clevelend (Mihaly, 1984)] might not adequately reflect variations in household needs. More research is needed before the marketing of energy conservation programs can fully capitalize on intra-urban variations in home energy consumption and associated residential traits. The potential returns on such research would appear to be great.

6. APPENDIX A--THE SAMPLE DESIGN AND SURVEY

The survey data analyzed here were collected through a personally administered survey conducted in 1980 and 1981. The sample of 319 homeowners is composed of two subsamples. For the first subsample, telephone numbers were systematically drawn from the 1980 Decatur Telephone Directory. One digit was then added to each number, thereby reducing the potential bias introduced by voluntarily and involuntarily unlisted numbers (Landon and Banks, 1978). Households were contacted by telephone and screened for the population of interest, after which an in-home interview was arranged. This sampling design resulted in 159 completed interviews, a 22 percent cooperation rate among eligible homeowners, and a slight under-representation of low-income households. The second subsample of 160 homeowners was drawn from the 1981 Decatur City Directory. A stratified sampling design was used to ensure that the total sample of homeowners, by census tract, was proportional to the distribution of the population of interest, thereby obtaining a more representative sample. Households were sent a postcard introducing them to the project, followed by a telephone call or home visit to screen for eligibility and arrange a subsequent interview. A cooperation rate of 36 percent was achieved. Permission to access utility bill histories was obtained from 239 of the 319 homeowners.

7. ACKNOWLEDGEMENTS

This research was supported by the Geography and Regional Science Program of the National Science Foundation (NSF SES-8112007). Assistance with the data analysis was provided by Paul A. Rollinson and is appreciated.

8. REFERENCES

Blumstein, C., Krieg, B., Schipper, L. and York, C. 1980. Overcoming social and institutional bsarriers to energy conservation. Energy 5: 355-71.

Brown, M.A. and Rollinson, P.A. 1985. The residential energy consumption of low-income and elderly households: How "non-descretionary" is it? Energy Systems and Policy forthcoming.

Citizens Energy Project. 1980. Energy and the poor. Washington, D.C.: Citizens Energy Project.

Cooper, M.S., Sullivan, T.L., Punnett, S. and Berman, E. 1983. Equity and energy: Rising energy prices and the living standards of lower income Americans. Boulder, CO: Westview Press.

Fels, M.F. and Woteki, T.H. 1980. Indices of energy consumption: An explorative analysis of a utility's monthly billing data. Energy 5: 1117-30.

Fowler, G.L. and Solzman, D.M. 1984. The ecology of urban residential energy consumption. Unpublished manuscript.

Fritzche, D.J. 1981. An analysis of energy consumption patterns by stage of family life cycle. Journal of Marketing Research 18: 227-32.

Grier, E.S. 1977. Colder...darker: The energy crisis and low income Americans. Washington, D.C.: U.S. Government Printing Office.

______. 1978. Energy consumption in American households, In Energy and the community, eds. R.J. Burby and A.F. Bell, pp. 3-13. Cambridge, MA: Ballinger Publishing Co.

Hamm, B. 1982. Social area analysis and factorial ecology: A review of substantive findings. In Urban patterns: Studies in human ecology, ed. G.A. Thoedorson, pp. 316-37. University Park, PA: The Pennsylvania State University Press.

Kendall, J.W. 1978. Consumer behavior and energy conservation. Position paper prepared for the Consumer Research Branch, Consumer and Corporate Affairs, Ottawa, March 1.

Krumholz, N. and McDermott, M. 1984. Poor people, neighborhood groups and energy conservation: A Cleveland study. In Doing Better: Setting an agenda for the second decade, proceedings of the ACEEE 1984 summer study on energy efficiency in buildings, Vol. H: pp. 63-75.

Landon, E.L. and Banks, S.K. 1978. An evaluation of telephone sampling design. Advances in consumer research 5: 103-08.

Meyers, S. and Schipper, L. 1984. In Families and energy: Coping with uncertainty. Proceedings of a working conference, eds. B.M. Morrison and W. Kempton, pp. 81-102. October 9-11, Michigan State University.

Mihaly, M.E. 1984. No cost/low cost: A community energy-saving effort, In Doing better: Setting an agenda for the second decade, proceedings of the ACEEE 1984 summer study on energy efficiency in bulding, Vol. H: pp. 76-84.

Moore, J.L., Berger, D.A., Griggs, H.M. and Rubin, C.B. 1980. Community Energy Auditing: Experiences with the comprehensive community energy management program. AML/CBSU-TM-43. Argonne National Laboratory.

_____. 1981. The comprehensive community energy management program: An evaluation. ANL/CNSU-TM-89. Argonne, IL: Argonne National Laboratory.

Morgan, S. and Katz, E. 1984. The financing of energy conservation services to low income households: Alternative to grants. In Doing better: Setting an agenda for the second decade, proceedings of the ACEEE 1984 summer study on energy efficiency in buildings, Vol. H: pp. 85-103.

Morrison, B.M. and Gladhart, P.M. 1976. Energy and families: The crisis and the response. Journal of Home Economics 68: 15-18.

Murdie, B.A. 1969. The factorial ecology of metropolitan Toronto, 1951-61. Chicago, IL: University of Chicago, Department of Geography Research paper No. 116.

Newman, D.K. and Day, D. 1975. The American Consumer. Cambridge, MA: Ballinger Publishing Company.

Penz, A.J. and Yasky, Y. 1979. Uncertainties in predicting energy consumption in houses. Energy Systems and Policy 3: 243-69.

Rees, P.H. 1979. Residential patterns in American cities: 1960. Chicago: University of Chicago Department of Geography Research Paper No. 189.

Rosson, P.J. and Sweitzer, R.W. 1981. Home heating oil consumption: Profiling "efficient" and "inefficient" households. Energy Policy 9: 218-25.

Rummel, R.J. 1972. Applied factor analysis. Evanston, IL: Northwestern University Press.

Salins, P.H. 1971. Household location patterns in American metropolitan areas. Economic Geography 47: 234-48.

Savitz, M.L. 1984. Ten years out--a problem not a crisis, In Families and energy: Coping with uncertainty, proceedings of a working conference. eds. B.M. Morrison and W. Kempton, pp. 7-9. October 9-11, Michigan State University.

Seligman, C., Darley, J.M. and Becker, L.J. 1978. Behavioral approaches to residential energy conservation. In Saving energy in the home: Princeton's experiments at Twin Rivers, ed. R.H. Socolow, pp. 231-53. Cambridge, MA: Ballinger Publishing Company.

Thompson, W. 1984. The residential energy consumption survey: A look back and a look forward. In Families and energy: Coping with uncertainty, proceedings of a working conference, eds. B.J. Morrison and W. Kempton, pp. 49-64. Proceedings of a working conference, October 9-11, Michigan State University.

U.S. Bureau of the Census. 1982a. Census of Housing: 1980. Washington, D.C.: U.S. Department of Commerce.

______. 1982b. Census of Population: 1980. Washington, D.C.: U.S. Department of Commerce.

Ban Raaij, W.F. and Verhallen, T.M. 1983. A behavioral model of residential energy use. Journal of Economic Psychology 3: 39-63.

Warriner, G.K. 1981. Electricity consumption by the elderly: Policy implications. Journal of Consumer Research 8: 258-64.

CHAPTER 16

SPATIALLY-ORIENTED ENERGY CONSUMPTION SCENARIOS: METHOD AND APPLICATION TO HOLLAND

Peter Nijkamp
Department of Economics
Free University, Amsterdam
The Netherlands

Douwe Tiemersma
Economic and Social Institute
Free University, Amsterdam
The Netherlands

ABSTRACT. This chapter aims at providing insight into the contents and use of spatially-oriented energy scenarios. First, a brief review of types of energy scenarios will be given, with a specific emphasis on geographical aspects. Then a general class of energy models (the so-called MEDEE-models) will be discussed which have been designed in Europe in order to provide a systematic and comprehensive computer-based picture of all relevant behavioral modules in energy consumption. Next, an adjustment of this general model for specific Dutch conditions (e.g., transportation, housing, agriculture) will be made. Particular attention will be paid to spatial dimensions of this model (urbanized areas versus rural areas, etc.). Finally, a series of scenarios for economic development and energy use in the Netherlands will be presented. The consequences of these scenarios will be assessed by employing the above mentioned adjusted MEDEE-model. Various empirical results will be presented in order to illustrate the mechanism of this scenario analysis and energy modeling.

1. INTRODUCTION

Energy analysis has become a rapidly growing research field at the crossroads of technical and social sciences. The fourfold increase in oil prices in late 1973 and the subsequent economic crisis which has affected the industrialized world since then have stressed the relevance of designing a coherent methodology for energy analysis. Despite the impressive body of knowledge regarding physical and chemical aspects of energy use, the overall field of energy analysis is still poorly developed (see Johansson and Lakshmanan 1985). Three major fields of energy

F. J. Calzonetti and B. D. Solomon (eds.), Geographical Dimensions of Energy, 299–323.
© 1985 by D. Reidel Publishing Company.

analysis should be mentioned in this respect:

- the exploitation and supply of energy resources. Past experiences have clearly demonstrated the inability of many research efforts to provide reliable predictions of the stock of energy sources, the depletion rates and the impacts of international politics.
- the consumption of energy. The past decade has, due to the oil crisis, shown a remarkable shift in energy consumption due to the increased energy efficiency of equipment in industrial installations, agriculture, transportation and households (sometimes up to thirty or forty percent). Demand effects (e.g., substitution mechanisms) are still hard to assess and forecast with satisfactory precision (see, among others, Lakshmanan and Nijkamp 1983).
- the energy technology. In recent years, energy technology has been marked by a wide spectrum of adjustments and innovations (e.g., integrated heating systems, wind and solar energy). Despite the short-term inertia of energy technology, its medium-term flexibility for the design and introduction of technological innovations is remarkably high. Consequently, it is difficult to make satisfactory projections of future developments in energy technology.

Thus, it can be concluded that energy analysis is marked by a high degree of uncertainty. Several attempts have been made to develop analytical tools that are more adequate for energy analysis, both short-term and long-term.

This present chapter will provide a coherent analytical framework for energy analysis, with special attention to energy technology and energy demand. First, a brief survey of some analytical models for regional energy analysis will be given (Section 2). Next, a modular design for complex energy systems will be proposed (Section 3). The main feature of this framework is a coherent representation based on a systems approach for relevant components of an energy system. This approach will mainly be based on the so-called MEDEE model for medium- and long-term energy demand analysis. In a subsequent section, attention will be focused on the use of such models for integrated national-regional energy planning, <u>inter alia</u>, by using various scenarios. Next, in Section 5 an empirical application to The Netherlands will be discussed, followed by a presentation of numerical results. Finally, Section 6 discusses the potential of the MEDEE approach for regional energy planning.

2. METHODS AND MODELS FOR REGIONAL ENERGY ANALYSIS

2.1. Regional Input-Output Models

In many impact studies, regional aspects of energy use are included in a broader framework of production and final demand by means of (inter) regional input-output (I-O) models (see, among others, Lakshmanan and Nijkamp 1980; 1983; Muller 1979). In this way, a consistent configuration of intersectoral, interregional, and intraregional production flows and related energy flows can be obtained. I-O analysis is also important, since it is able to distinguish between energy use for final demand and for intermediate use. Various energy sources can be taken into account by means of an I-O framework. The same applies to environmental impacts. The dynamic I-O model at the core of the strategic environmental assessment system (SEAS) model used in the US is an example of this potential (see Ratick and Lakshmanan 1983). The model MORSE used in the Swedish case of energy planning is also a good illustration of a multiregional I-O model that provides regional-national linkages.

Drawbacks of regional I-O analysis, however, are its rigidity (lack of substitution due to fixed coefficients) and the large data base required, although an obvious advantage is its completeness and consistency.

2.2. Energy Models with Substitution

The choice of inputs (including capital, labor, energy and materials) to produce a certain amount of output, based on input substitution and technological changes, can be analyzed by means of trans-log substitution models (see Hudson and Jorgenson 1976; Ratick and Lakshmanan 1983; Lesuis _et al._ 1980). These models allow for substitution by relating changes in technical coefficients to price effects, based on a duality relationship between the production function and the cost function of a firm. In this way, energy input coefficients can be included as endogenous variables in models of producer behavior, so that a system of relative factor-demand relationships can be derived allowing for the computation of prices, substitution elasticities, and changes in technical coefficients. The trans-log approach can be linked to an I-O model in a straightforward manner. The substitution analysis can be carried out at two levels: one at the so-called KLEM level (capital, labor, energy and materials), and the other at the so-called intra-KLEM level (allowing for substitution between different energy components). A variety of these models depicting economy-energy relationships are part of the battery of integrated models used in the assessment of a variety of energy technologies in the US.

It should be noted, however, that both I-O and trans-log models are strongly demand-oriented, and thus have limited relevance for energy supply planning unless they are used within a more comprehensive economy-energy systems framework.

A further application of the above mentioned class of models can be found in comprehensive (or at least) integrated econometric models (see, for instance, Jorgenson 1976). However, two major drawbacks of econometric approaches to energy analysis are:

- the need for detailed time series data which are in most countries only available from the 1970's onwards;
- the fact that most econometric models are based on a fixed structure and stable patterns, so that bifurcations or singularities can not be taken into account very well. If lack of data precludes a valid

estimation of such econometric models, simulation models may be used (see, for instance, Fromm and Taubman 1968). Such models may be extremely helpful in case of a lack of reliable data, but have also several weaknesses:

- the statistical validation of such models is normally from a methodological viewpoint very limited;
- the data input for such models is often based on unjustified behavioral assumptions.

2.3. Comprehensive Regional Energy System Models

Comprehensive regional energy system models aim at portraying regional economic and energy flows in a coherent way (see Chatterji 1980). They may be decomposed into modules, e.g.:

- an activity sub-module (including _inter alia_ sectoral production and employment levels);
- an energy sub-module (including _inter alia_ various regional energy supply and demand categories);
- a socioeconomic sub-module (including _inter alia_ housing market and demographic patterns);
- a financial sub-module (including _inter alia_ public tax rates and revenue systems);
- an environmental sub-module (including _inter alia_ pollution levels, damage costs and quality-of-life indicators);
- a spatial sub-module (including _inter alia_ transportation and land use variables).

Another good example of a comprehensive demand-oriented approach is the MEDEE model (see Chateau and Lapillonne 1979), which was a cornerstone of the International Institute for Applied Systems Analysis (IIASA's) long-term energy study (Hafele 1982) and which will also be employed in the present study on Dutch energy demand. Such comprehensive models reflect potential information requirements, and may be based on econometric modeling techniques, I-O analysis, mathematical programming methods, or simulation methods. Energy demand models, like MEDEE, are also often used in combination with energy supply models (e.g., MESSAGE or EFOM).

2.4. Energy Circuit Models

Energy circuit models have been developed by ecologists who aim to bridge the gap with economics. The main philosophy is that energy is an important common denominator in all ecosystems (Odum 1976), and is then regarded as the joint currency for ecological "housekeeping" (accounting). This approach assumes that the source and quantity of available energy largely determines the pattern of functional and developmental processes in a region. Therefore, insights into the energetic structure and processes of regional ecosystems are of major importance in understanding the properties and implications of regional ecosystems.

Ecosystem components can then be linked together by means of energy flows. Such energy circuit models are extremely helpful for judging the energy efficiency of a region, as well as the growth in energy consumption. Many of these are simulation models based on a systems approach. Applications of such models to energy-based regional planning and land use problems can be found in Costanza (1975), Nijkamp (1978), and Odum (1976).

2.5. Cartographic Energy Analysis

The spatial dimensions of energy supply and demand can be taken into account by means of so-called cartographic energy methods or heat maps (see Allaert et al. 1977; 1980). This approach is based on a computerized system that allows the user to produce detailed maps (even at a scale of 1:500) for supply and demand of various energy categories. By means of such refined grid systems, one can obtain detailed insights into the energy supply and demand of residential and industrial activities. The spatial orientation of cartographic analysis is very relevant for analyzing the geographical aspects of energy systems (e.g., agglomeration advantages from district heating, energy efficiencies from locally integrated energy supply systems).

Cartographic energy analysis can also describe the energy intensity of supply and demand at any desired spatial scale and in each time period. Consequently, it allows for a permanent monitoring and comparison of energy supply and demand, so that bottlenecks can be identified and overcome by planners.

2.6. Regional Energy Scenario Analysis

Given the complexity and future uncertainty of energy systems, a wide variety of scenarios have been constructed for the design of a feasible and consistent picture of energy supply and demand patterns. Such scenarios do not predict the future, but rather provide a meaningful image of it that can serve as a frame of reference for policy choices. A good illustration of the use of energy scenarios can be found in WAES (1976). In general, scenario analysis aims at designing a limited set of consistent future developments of a complex system, with a particular emphasis on changing trends in public policies or external (global) developments (see also Hafkamp 1984). Such analyses usually have a strong policy relevance, as they are able to demarcate the range of feasible policies and their impacts on the long-term dynamics of main components of a complex system. Clearly, scenario analysis also has a major disadvantage, as it is only based on plausible or feasible future trajectories, without providing a statistically sound behavioral justification. It should be added that scenario analysis can also be combined with econometric and simulation models.

Many energy scenarios, however, do not include a clear regional dimension. Regional energy scenarios may be based on three kind of changes in the energy system:

- Shifts in exogenous circumstances not controlled by either the national or the regional government (e.g., sharp reductions in oil supply by OPEC countries).
- Policy measures taken by the national government, but not controlled by the regional government (e.g., a uniform national energy tax system).
- Policy measures taken by the regional (or sub-regional) government (such as district heating system design). A regionally decentralized energy policy may increase the efficiency of an energy system because more detailed aspects of regional land use, transportation flows, and residential and industrial locations can be taken into account.

A good example of a regional energy scenario analysis can be found in Foell (1980), who designed a large-scale regional scenario including, *inter alia*, four scenario properties (socioeconomic structure, life-styles, technology and environment), and five modules (socioeconomic activity, end-use energy demand, energy conversion, environmental impacts, and preference and decision structures), which have been transformed into a set of operational models. On the basis of alternative assumptions regarding the scenario properties the effects of alternative input data for the energy study could then be assessed. For the preference and decision module, the use of multi-attribute utility theory was suggested.

2.7. Multidimensional Energy Policy Models

As implied above, policy analyses for energy systems may be either *discrete* or *continuous* in nature. An example of a discrete energy evaluation system can be found in Blair (1979), who analyzed various alternative regional supply systems using conflicting criteria such as energy costs, environmental quality, and conservation of natural resources. The preference analysis was based on Saaty's prioritization method (by means of an eigenvalue analysis). In addition, various compound scenarios were created to resolve policy uncertainties regarding energy issues.

Examples of continuous multidimensional energy policy models can be found in Lesuis *et al*. (1980) and Hafkamp and Nijkamp (1983). The latter study is based on a comprehensive, large-scale, multiregional, multisector I-O model including energy and pollution. Three conflicting objective functions are considered: employment, environmental quality, and energy security. The compromise strategies in this model were based on an "ideal-point" approach.

Having provided a brief overview of some relevant methods and models for regional energy analysis and planning (with the main emphasis on the demand side), we will now examine a specific class of *comprehensive* energy models in more detail, that is, the so-called MEDEE models. This will be the subject of the next section.

3. MODULAR DESIGN FOR COMPLEX ENERGY SYSTEMS

In the present section a conceptual model for medium- and long-term energy analysis will be presented based on the so-called MEDEE framework (see Chateau and Lapillone 1979). This model is one of the major tools used by the Commission of the European Community (E.C.) to systematically explore the uncertainties regarding technology and demand for energy in the member states. The model has also been used in other countries (e.g., the USA), while it has also played a role in the international energy study carried out under the auspices of the IIASA (see Hafele 1982).

The MEDEE model is based on a detailed and integrated analysis of the energy demand determinants in a country, and a sophisticated scenario description of its economic, spatial and social developments. It aims at identifying various end-use categories, the socioeconomic needs and the (spatial) economic activities creating energy demand for various modules, as well as the relationships between sub-modules that describe their evolution in the medium- and long-run. The medium- and long-run time horizon has been chosen in order to forecast energy consumption over a time period that is long enough to take into account technological adjustments (for instance, construction time for new energy production facilities or equipment), and to prevent demand forecasts from being affected by ad hoc influences. The design of the MEDEE rests upon an integration of systems analysis and scenario analysis, complemented with either econometric or simulation models. A simplified, conceptual version of this model is represented in Figure 1.

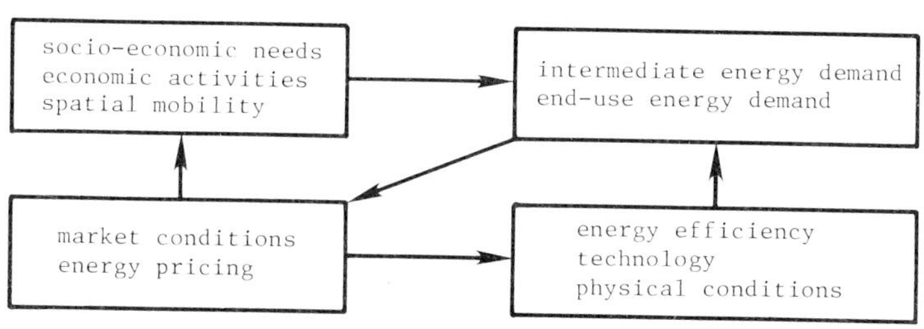

Figure 1. Conceptual version of the energy demand model.

The MEDEE model is next operationalized by using the following steps:
- a disaggregation of energy demand (right upper block of Figure 1) into a set of coherent modules (industry, households, etc.).
- a closer examination of energy demand determinants in each module (i.e., the impacts exerted on the right upper block by the remaining

three blocks) and of relationships between these modules.

- a development of a simulation model by formalizing the relationships within and between modules, and by linking the sub-modules on the basis of the first step.

Altogether the MEDEE model may be regarded as both an engineering and an economic model, thus requiring both detailed technical spatial and socioeconomic data. It has a hierarchical modular structure made up by ordered sub-modules and recursive equations. The output of the model is given in terms of final energy demand distinguished according to users and end-use category. A quasi-operational structure of the MEDEE model is given in Figure 2.

The operational version of the MEDEE model consists of estimated or calibrated relationships describing the evolution of energy demand as a result of changes in socioeconomic conditions, spatial activity patterns, economic activities, behavioral adjustments and technological progress. In case of lack of data, simulations may be carried out, while wild fluctuations are prevented by imposing plausible upper and lower limits on the functioning of the system. In the operational version, three types of variables are distinguished:

- endogenous quantitative variables (determined by the internal mechanism of the model itself).
- exogenous quantitative variables (assumed to be known prior to the prediction of annual valves of the endogenous variables).
- qualitative and quantitative scenario variables (describing the main trends of technical, political, social and economic determinants of the system at hand).

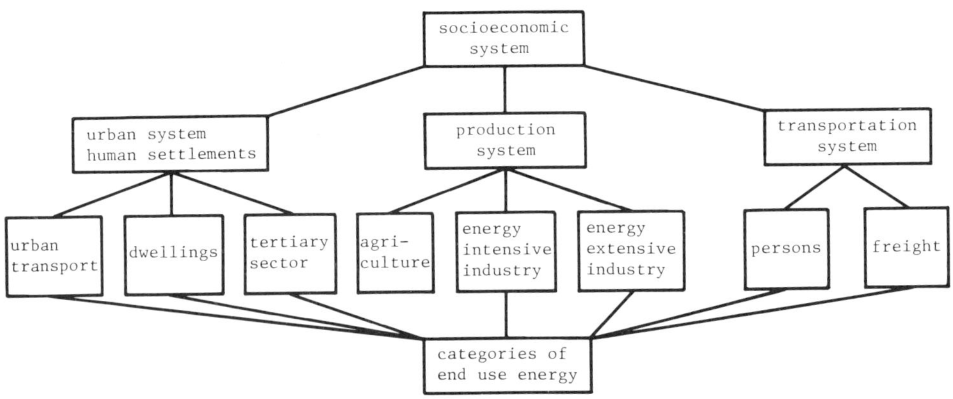

Figure 2. The main structure of the MEDEE model.

In the case of qualitative variables, so-called tabular functions are used in order to transform qualitative scenario indicators into quantitative values in a consistent manner by means of ordering statistics. Each of the three modules (urban system/human settlements, production system, transportation system) will now be briefly represented in Figures 3 - 5. It should be noted that this figurative

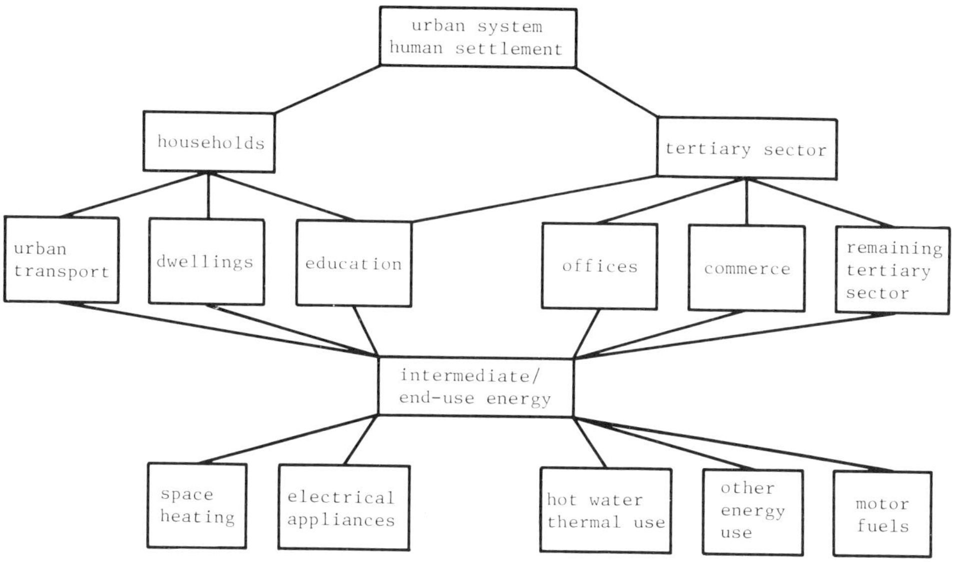

Figure 3. The urban system/human settlement module.

representation of the modules is still a drastic simplification of the actual model structure which is based on numerous sub-modules.

The advantage of the MEDEE model is its flexibility, since it can easily be adjusted to specific economic, spatial or technical conditions in a certain region or country. The computer package which is now available allows the user to select or to suppress certain (sub-) modules according to his or her own needs.

In addition to the modular design of MEDEE, it is important to pay more specific attention to the scenario approach implied by this model. A scenario aims to represent in a plausible and consistent way a trajectory of the main future tendencies of a system. Examples are trend and contrast scenarios. The scenario approach in MEDEE is governed by a triple-layer structure:

- world scenarios: characterize the international environment of the system and the economic/political relationships between the system and this environment (growth of the world economy, international labor division, etc.).
- socioeconomic scenarios: characterize the main economic, social,

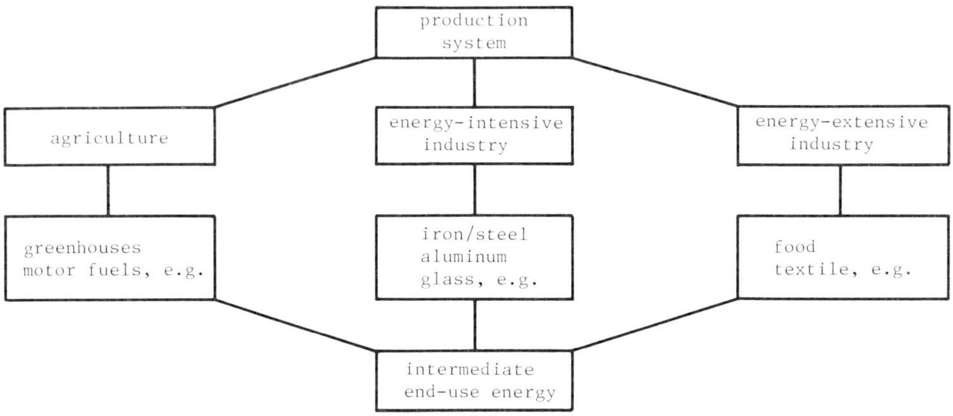

Figure 4. The production system module.

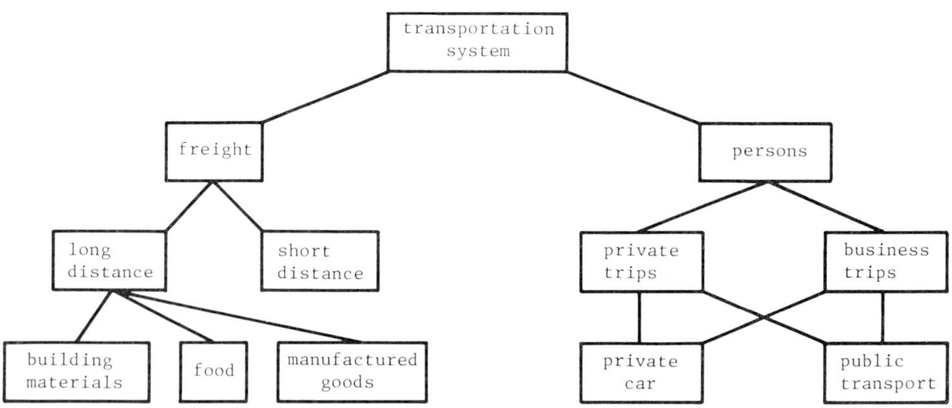

Figure 5. The transportation system module.

political and technological features of the development of the (national of regional) system at hand (urbanization, technology, social policy, etc.).
- modular scenarios: characterize exogenous evolutions of the various individual modules (e.g., sectoral innovation patterns).

The information (qualitative and quantitative) obtained from these scenarios is used in each relevant layer of the modular structure of the model. The information may be obtained directly from policymakers or generated by experts (for instance, by means of Delphi techniques). Given the modular pattern of MEDEE, the formalization of all relevant relationships (either by means of econometric or simulation methods) and the scenario information, one may generate - on the basis of a set of input data on socioeconomic and energy variables - a wide variety of energy demand results for the medium - and long-term horizon of the system under consideration. Clearly, the MEDEE approach also has some limitations, such as a strong demand for consistent scenarios containing very detailed economic, technical and behavioral information.

4. INTEGRATED NATIONAL-REGIONAL ENERGY PLANNING

4.1. Introduction

The above mentioned approach to energy demand analysis has been applied to the Netherlands. Various experiments have first been carried out in order to judge the validity of the model and its outcomes.

A first tentative exercise was to use 1975 data in order to assess energy demand for 1980, 1990 and 2000. The 1980 results were compared with realized outcomes in the same year, while the 1990 and 2000 results were compared with predictions provided by the Dutch Central Planning Bureau. In this way, errors and weak points in the input data and the scenario assumptions could be identified. The first experiments demonstrated that the aggregate final energy demands calculated by the model were only slightly lower than actual energy consumption in 1980. Electricity demand and total demand for fossil fuels (excluding motor fuels) were forecast especially precisely by the model. Discrepancies, however, took place with regard to the use of motor fuels, energy use for heating, and final energy demand for non-energy purposes.

Following these tentative experiments, a more rigorous attempt has been made to build a satisfactory variant of the MEDEE model for the Netherlands. Two important steps had to be undertaken:
- the design of a set of scenarios and of a specific model that would be in agreement with the structure and specific character of energy demand in the Netherlands.
- the design of a more reliable information system on socioeconomic, spatial and energy variables in the Netherlands.

In order to undertake this endeavor, more specific information on the structure of energy demand in the Netherlands is required. This will be the topic of the next sub-section.

4.2 Adjustments of the MEDEE structure

In various respects, the structure of energy demand in the Netherlands is different from that in most other Western European countries. Some important factors determining these features are:

- The presence of a large natural gas reserve
  From the beginning of the 1960's onward, the role of natural gas in the domestic use of energy has increased. Natural gas is still rather cheap compared with competing oil products. Dwellings, factories, farms etc. are well connected to the national gas distribution system. One of the consequences has been that many energy-using installations nowadays are only suited for natural gas. Hence, a shift to oil or coal would require large investments for new installations, much more solidly constructed chimneys, and the creation of storage space for fuels (which is often very difficult in densely built areas).
- The open character of the Dutch economy
  Many production sectors export more than fifty percent of their products, mainly to West Germany and other E.C. countries. Hence, the development of a large part of the domestic energy demand for production as well as transport is directly influenced by the economic developments in these countries and the competitiveness of the Dutch producers. This, in turn, is partly a function of the relative prices of energy.
- The dominance of a few energy intensive production sectors in the total energy demand
  The presence of a big, accessible and modern harbor and a large gas reserve, in combination with a large hinterland (West Germany), has stimulated the growth of the basic chemical industry, which is very energy-intensive. A large amount of oil products and gas is used as a feedstock in this sector (more than 15% net of the total final use of fossil fuels in 1980 - see Central Bureau of Statistics 1981).

In our version of the MEDEE-model, known as the MEDEE-3 model, the energy demand for eight energy-intensive industrial products is modeled separately from the energy demand in the rest of the industrial sectors (the light industries). For the latter products, the demand for fuels and electricity is related to the value added of the sectors concerned, which is determined exogenously (e.g., with the aid of an econometric model). The energy demand for the above mentioned eight products (the large energy consuming industries) is related to the physical production volume and the production processes used. Compared to the standard MEDEE model, the following specific adjustments for the Netherlands had to be carried out. In the first place, the number of large energy consuming industries had to be extended for the Netherlands. To cover the whole basic metals industries, besides steel and aluminum, zinc has to be incorporated. For the same reason, besides glass and cement, bricks and other building materials have to be taken into account. Finally, in order to cover the whole demand for feedstocks in the chemical industry, besides ammonia, chlorine and ethylene (and related cracking products), methanol has to be incorporated. On the basis of a recently published report, much information could be obtained about the processes used and their specific energy consumption for those products (including pulp and paper, the last product being modeled separately-Social Economic Council 1983). The volume of the physical production of these sectoral products is determined mainly endogenously by a linkage to the value added of the main consuming sectors of the products concerned.

The net import or export of these products of the large energy consuming industries is determined in the scenario part of the model by means of a so-called tabular function, related to the scenario indicators 'international division of labor' and 'industrial policy' (for more details see Section 5). Because of the openness of the Dutch economy, this is not very satisfactory for the above mentioned sectors. The (few) firms operating on this market are as much competing on the European market as on the Dutch market. For instance, almost all aluminum produced in the Netherlands is exported, while nearly all aluminum consumed is imported. Hence, at least for the basic metals and chemical products, the relation with domestic consumption is not very relevant and the total production is to be related to the relevant scenario indicators. In the scenario concerned, the development of the total demand for these products in Western Europe, the shift of the production to developing countries (international division of labor) as well as the competitive position of the Dutch manufacturers (industrial policy, but also the relative prices of energy in the Netherlands) should be taken into account.

Thus, the simple, nationally-oriented modeling framework, as existed in the original MEDEE model for France, is not very satisfactory for the Netherlands, because of the large share of the large energy consuming industries in the total demand for energy. At the moment, the production volume is exogenous, but fortunately due to its hierarchical, modular structure, the MEDEE model can easily be extended and adjusted.

The openness of the Dutch economy also makes it rather difficult to model the transport of goods properly. In the original MEDEE model, its development is linked to the growth of the relevant production sectors. This is, however, only valid for domestic commodity transport. The quantity of international transport (even apart from seaship transport, expressed in ton-kilometers) is actually larger than the domestic transport, with a big share of transit (e.g., the transport of iron ores via Rotterdam to West Germany). In order to model the energy demand for commodity transport reasonably well, the external developments which influence this very open system have to be taken into account in the scenarios.

A very energy-intensive production sector in the Netherlands is the horticulture in greenhouses. Almost ten percent of all final use of natural gas is being used for this purpose. In the initial model, this energy demand is included in the total energy demand of the agricultural sector. For the Netherlands, a separate module has to be constructed which takes into account the potential for production growth and for energy conservation in this sector (see Section 6).

Finally, also the module describing the use of different <u>heating systems</u> in dwellings has to be adjusted. Other combined heating systems have to be taken into account, mainly because of the availability of relatively cheap gas in nearly all dwellings (e.g., solar energy in combination with natural gas instead of electric resistance heating).

4.3. The Urban System

In the part of the MEDEE framework describing the urban system, several classes of urban areas are distinguished, varying from very large metropolitan areas (such as Paris or London) to small towns and rural areas. The number of city classes is five or six. This distinction is supposed to be relevant for the development in the dwellings and urban transport sectors, mainly with respect to:

- the share of apartments in all dwellings
- the possibilities for district heating
- the number of motorized urban trips
- the mean length of the above trips
- the share of public transport in urban trips.

Of course, the five or six distinguished classes of urban areas should be rather homogeneous in these respects.

The application of such classification to the Netherlands appeared to be rather difficult, since no similar typology is commonly used in Dutch statistics. In these data bases, one usually finds figures for the provinces, or for classes of municipalities (e.g., bigger and smaller towns, suburbs, rural towns and rural villages).

Another problem is that in the Netherlands a process of deurbanization is taking place, especially in the Western part of the country. This leads to an increasing functional integration and overlapping of the urban agglomerations in the so called Rim-city (ring of larger urban areas in the Western provinces), as well as to an extension to the urban regions outside the Rim-city. Hence, the assumption in the MEDEE framework that urban public transport is mainly used for urban trips (home to work, shopping, etc.) and that inter-urban buses and trains are mainly used for other trip purposes (social visits, recreation, holidays, professional visits) is not valid for the Dutch case. In fact, the inter-urban public transport modes, as well as highways for private cars, are also commonly used for trips within the urban agglomeration (urban trip purposes).

Consequently, an adjustment is made to represent a more satisfactory model structure for passenger transport. Also, the classification of urban areas is adjusted, in order to obtain classes which are homogeneous with respect to the dwelling stock, urban passenger transport and the feasible or expected developments in these activities.

4.4. The Data Base

Thus far, much emphasis has been put on the construction of a reliable data base (see Tiemersma 1984). A completely new data base has been constructed for the industrial sectors concerning their value added, physical production processes used, and specific energy demand for the base year 1975 (as well as for 1980, which may be used as the base year for future experiments with the model). Given this information, appropriate definitions of industrial sectors, products and processes, as mentioned above, are introduced. In addition, a completely new data base for the passenger and commodity transportation system is constructed, although it may be adjusted to new circumstances, when the structure of some modules is improved. It is difficult to obtain reliable data about the tertiary sector (schools, offices, retail business and others).

The very detailed information concerning dwellings needed for the model is hard to obtain (insulation coefficients and energy use for different age classes, and for single family dwellings as well as apartments). Despite limitations in the data base, the scenario analysis may still be fruitful, as it is mainly focusing on relative growth (instead of absolute figures), while the uncertainties are also covered by the tabular function approach. In some cases, where lack of reliable statistical data is accompanied by a very detailed model structure, modules should be made more compact. In other cases, however, the structure of the model is to be maintained and attempts have to be made to collect data in ways other than via the usual statistics published by the statistical agencies. Especially for the industrial sectors, this appears to be a necessary (and also feasible) strategy.

5. APPLICATION OF E.C. SCENARIOS TO THE NETHERLANDS

5.1. Introduction

The MEDEE model has been used by the European Commission in order to obtain projections of energy demand of the E.C. member states, on the basis of three different scenarios (see Commission of the European Community 1984). In this section these scenarios will briefly be described, while next the results of this exercise for the Netherlands will be presented, as an illustration of the functioning and possibilities of the MEDEE model. The improvements of the structure and the data base, mentioned in the previous section, have not yet entirely been completed, so our results are only indicative. Therefore, the present quantitative results should thus be treated with caution, but in the present context they may well serve as an illustration of the ways in which the model may be used.

5.2. The Three E.C. Scenarios

The three scenarios used by the European Commission differ mainly from each other with respect to the following key assumptions:

- the world economic growth
- the position of the E.C. with respect to the world economy
- the change in the world price of crude oil.

Other differences in these three scenarios have been derived from these key variables.

The first scenario, Ouverture (free competition), is based on the development of worldwide free competition, without any intervention by the E.C. countries.

The price of crude oil, and similarly the prices of other forms of energy, is assumed to be very unstable: falling first, but rising very sharply around 1990. Economic growth is assumed to be rather weak in Europe, with a transfer of basic industries (steel, petrochemicals) to (developing) countries with energy resources. There are only a few incentives for energy saving.

The second scenario, Concertation (cooperation), is founded on increased international cooperation and a revival of the world economy through an international program of development for developing countries. The price of crude oil increases only gradually and is predictable. Hence, it is conceivable that substantial energy savings will be realized, especially in industry. Economic growth is rather high, after a period of recovery. Within the E.C. countries, investments will be implemented in new industrial sectors as well as in new production processes.

In the third scenario, Europe (European Community), it is supposed that world-wide cooperation will fail and Europe seeks its own solutions. Economic growth will still be rather high, partly because of a revival of the internal E.C. market. The import of oil and other energy forms will be reduced and internal energy prices stabilized at a rather high level. Thus, energy savings are stimulated in all sectors (also by government policy and subsidies).

The economic growth in the Netherlands and the growth of the value added of different sectors is supposed to evolve according to the trajectory shown in Table 1. As stated earlier this development is exogenous to the MEDEE model. Other exogenous variables concern the total population, the number of school children, and the size of households, which are the same for all scenarios.

In the scenario part of the model, so-called tabular functions are used. In these tabular functions, the states of two generally qualitative scenario desciptors are used as arguments for the values of an endogenous variable.

Derived variables are inter alia: the number of demolished dwellings (a function of the housing policy), the share of households with a car (a function of the economic growth indicator), the amount of urban transport offered (a function of transport policy and land use policy), the rate of energy conservation in industry (a function of energy policy and the speed of the development of oil prices), and the use of new processes in large energy consuming industries (which is also determined by energy policy and prices).

Table 2 shows the values of some intermediate variables used in the

Table 1. Growth of Gross National Product and of Value Added of Economic Sectors. (mean annual growth: %)

		Ouverture	Concertation	Europe
GNP	1970-1980	2.8	2.8	2.8
	1980-1990	1.5	2.1	2.1
	1990-2000	2.4	3.4	3.0
Agriculture	1970-1980	4.1	4.1	4.1
	1980-1990	1.7	1.7	1.7
	1990-2000	1.5	1.5	1.5
Energy	1970-1980	5.7	5.7	5.7
	1980-1990	2.0	2.5	2.5
	1990-2000	2.0	2.5	3.0
Industry	1970-1980	2.2	2.2	2.2
	1980-1990	1.4	1.7	1.7
	1990-2000	2.3	2.9	2.8
Building and Construction	1970-1980	1.1	1.1	1.1
	1980-1990	-1.6	0.0	0.0
	1990-2000	1.0	3.0	2.5
Tertiary Sector	1970-1980	3.4	3.4	3.4
	1980-1990	1.8	2.5	2.4
	1990-2000	2.6	3.9	3.3

Table 2. Some intermediate results for the three scenarios.

Variable	Unit	Year	Ouverture	Concertation	Europe
Old Dwellings (<1975)	10^6	1980	4.1	4.1	4.1
		2000	3.3	3.2	3.2
New Dwellings (>1975)	10^6	1980	0.6	0.6	0.6
		2000	2.7	2.8	2.8
Share of Single Family Dwellings	%	1980	0.69	0.69	0.69
		2000	0.72	0.73	0.73
Insulation Coefficient New Single Family Dwellings	Kcal/h^{o}Cm^3	1980	1.4	1.4	1.4
		1990	1.2	1.0	1.9
		2000	1.1	0.9	0.8
Share of Central Heating by Gas in Single Family Dwellings	%	1980	0.56	0.50	0.56
		1990	0.73	0.68	0.73
		2000	0.81	0.78	0.83
Domestic Commodity Transport by Trucks	10^9 ton/km	1980	23.2	23.2	23.2
		2000	52.9	36.1	36.1
Domestic Commodity Transport by Barges	10^9 ton/km	1980	5.9	5.9	5.9
		2000	5.9	6.2	6.2
Number of Private Cars	10^6	1980	3.62	3.62	3.62
		2000	6.41	6.59	6.55
Production of Steel	10^6 ton	1980	4.34	4.34	4.34
		2000	3.22	4.90	6.41
Production of Ammonia	10^6 ton	2000	1.75	2.34	2.72

MEDEE model. The difference between the scenarios is very clear with respect to the production of intermediate goods such as steel or ammonia. Table 3 shows the final energy demand for the three scenarios, resulting from the MEDEE model. This demand is derived from the demand of different sectors, by use, as is shown ad 2-5 in the table.

In the Europe scenario, energy conservation is stimulated the most, which results in a rather low energy demand for dwellings and the tertiary sector, as is shown in Table 3, ad 5. In this scenario, however, the heavy industries show further growth, resulting in increasing energy demand for industry. Hence, the total final energy demand is almost the same as in the scenario Ouverture. The demand of fuels for non-energy uses increases very much in this scenario (see ad 3), as does the demand for electricity (see ad 1), which will lead to a much larger need for primary energy.

The MEDEE model calculates final energy demand, without including the production of electricity or the (central) combined production of electricity and heat. This should be treated by an energy supply model (like EFOM), which is also used by the European Community. Furthermore, the distribution of the fuel demand in industry for heating purposes over solid fuels, gas and oil products is exogenous to MEDEE-3. The outcomes in terms of fuel demand can also be used as an input for the EFOM model, which optimizes the above mentioned distribution over the three kinds of fuels.

Finally, it should be noted that the projected differences between the scenarios in terms of final demand for energy after 20 years are fairly small. Thus, the three respective scenarios are not extremely discriminating between these results.

6. REGIONAL DIMENSIONS OF THE MEDEE APPROACH

The MEDEE approach is very well suited for distinguishing between homogeneous energy demand sectors, because of the hierarchical, modular structure of the model. As far as spatial differences in the development of sectors influence energy demand, spatial distinctions are to be made in order to obtain homogeneous sub-sectors - as is the case in the MEDEE model with respect to the urban system. If necessary or desired, this may be extended to other parts of the model.

A complete regionalization is also possible by specifying the variables for the regions taken into account. The development of the regional variables may be derived from the national structure by an extension of the regional variables. For a national model like MEDEE, however, several hundreds of values are already required for the data base. This number will be much higher if many of these data have to be specified for, say, four or five regions. Many of these data may not be available and consequently have to be synthethized. Furthermore, for the construction and pre-recording of the tabular functions, many reports and experts have to be consulted.

The MEDEE model as such may also be applied to a region instead of a

Table 3. Results of the MEDEE model for the Netherlands, for the three E.C. Scenarios: final energy demand (million ton of oil equivalents).

	Ouverture			Concertation		Europe	
	1980	1990	2000	1990	2000	1990	2000
1. Total final demand[1]	43.6	46.3	52.0	45.8	52.2	45.1	51.3
Petroleum products	13.9	13.1	14.6	13.8	15.1	12.6	13.9
Gas	23.2	25.7	28.1	24.4	26.6	24.3	25.4
Solid fuels	1.3	1.7	1.8	1.6	1.8	1.8	1.9
District heat	0.2	0.2	0.3	0.3	0.5	0.5	0.9
Electricity *	5.0	5.6	7.2	5.7	8.2	5.9	9.1
New energy forms **	0.0	0.0	0.0	0.0	0.0	0.0	0.1
2. Final demand industry	13.9	14.1	15.5	14.8	16.7	14.9	17.0
Low temperature	7.3	8.3	9.9	8.4	9.8	8.4	3.7
High temperature	4.4	3.4	2.8	3.8	3.8	3.8	3.4
Specific electricity *	2.2	2.4	2.8	2.6	3.1	2.7	6.0
3. Demand for non-energy uses[2]	4.2	3.8	3.4	4.2	4.2	5.2	6.0
4. Final demand transport[3]	8.6	9.4	11.1	9.5	11.3	9.5	11.2
Road	6.7	7.1	8.3	7.2	8.7	7.2	8.1
Railways	0.2	0.2	0.3	0.2	0.2	0.2	0.2
Waterways	0.7	0.8	0.9	0.8	0.9	0.8	0.9
Air	1.0	1.3	1.6	1.3	1.4	1.3	1.6
5. Dwelling and others[4]	21.0	22.7	25.1	21.6	24.3	20.7	23.1
Heating	16.0	17.0	17.9	15.9	16.0	15.0	14.6
Hot water	1.2	1.6	1.7	1.4	1.7	1.4	1.6
Cooking	0.4	0.4	0.5	0.4	0.5	0.4	0.4
Other thermal uses	0.9	0.8	1.0	0.9	1.2	0.8	1.0
Specific electricity *	2.3	2.7	3.8	2.8	4.7	2.9	5.3
Motor fuels	0.2	0.2	0.2	0.2	0.2	0.2	0.2

* consumption equivalence (1 kWH = 860 kcal).
** for example, solar energy.

[1] excluding non-energy uses.
[2] feedstocks for ammonia and ethylene, cokes for anodes in aluminum production, bitumen for building sector.
[3] excluding motor fuels for agriculture.
[4] dwellings, service sector and agriculture.

country. In such a case, problems with data availability may also exist. On the other hand, modules not necessary for the region concerned can easily be suppressed, and others may be extended. However, in such a case, it seems to be wise to create a new, region-specific model, using the MEDEE approach, and if desired the SAMSS software, which is used for the MEDEE model. In this case, the model can be specified in relation to the specific energy demand structure of the region, the relative weight of the energy demand sectors in the region, as well as the data available. Such a region-specific MEDEE model will now briefly be described.

This approach is presently used for the Westland, a region South of the Hague, with about 30 km^2 of greenhouses. These are responsible for about 85% of the entire energy demand in the region. In the region-specific model, this horticulture sector is evidently described in much detail, while other sectors are treated more superficially. The usually large energy consuming industries can in this case be left out of consideration.

Table 4 shows the sectors and sub-sectors distinguished in the Westland model. The structure of the greenhouses module is shown in Figure 6. Both physical production and the heat and electricity demand per m^2 are influenced by the development of breeds, cultivation techniques, and insulation of the greenhouses. The horticulture categories consist of three kinds of vegetables, five kinds of flowers, potted plants and breedings. For the development of insulation and the stock of heating facilities in dwellings, a separate model is used, the output of which is input for the dwelling module in the Westland model.

Thus far, two reports have been published concerning the Westland project (see Samenwerkende Bureaus voor het Energieplan Westland 1984a and 1984b). The first report presents the developments expected until 1995 in the form of a reference scenario. The second report presents the possibilities for new heat-power systems, use of waste heat from the Europoort industrial area near Rotterdam, *inter alia*. The final report is due out in 1985.

It appears that also in the Westland project the MEDEE approach offers a very useful framework for the integration of all kinds of statistical data, reports and expert knowledge for the construction of a suitable energy demand model and of coherent scenarios.

Table 4. Sectors and sub-sectors in the energy demand model for the Westland region.

Sector	Sub-sectors
I. greenhouses	10 cultivation categories
II. other agriculture	not applicable
III. industry	metal processing, chemical industry, other industries
IV. other business sectors	auction buildings, wholesale trade, retail-sector, services, public sector, etc.
V. public utilities	pumping stations for water sewage and draining, drinking water pumping stations, other public utilities (e.g., streetlights)

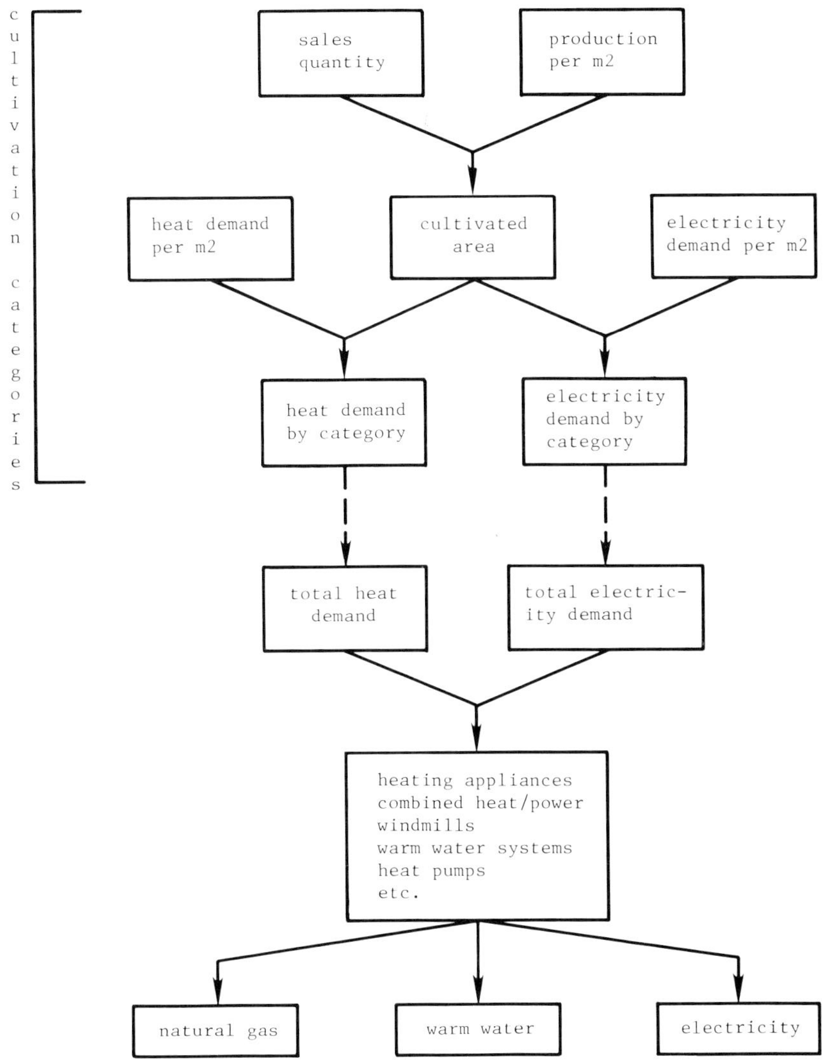

Figure 6. The greenhouse module.

7. REFERENCES

Allaert, G., Bergmans, P.P., and Rul C. 1977. De Warmetekaart voor Belgie. Stero 11: 8-33.

__________, Rul C., and Rosier M. 1980. Warmtekaart van Belgie, Final Report. Ghent: Dept. of Physical Planning, State University of Ghent (mimeographed).

Blair, P. 1979. Multiobjective regional energy planning. Boston/The Hague: Martinus Nijhoff.

Central Bureau of Statistics. 1981. De Nederlandse Energiehuishouding, 1980 4, Staatsuitgeverij, The Hague.

Chateau, B., and Lapillonne B. 1979. Long-term energy demand simulations. In Energy Models for the EC, ed. A. Strub, pp. 120-28. Brussels: European Community Directorate-General XVII.

Chatterji, M. 1980. Energy modeling with particular reference to spatial systems. Regional Science and Urban Economics 10: 325-42.

Commission of the European Community, Directorate-General for Science. 1984. Scenarios a l'Horizon 2000, Pays Bas. Brussels: Research and Development.

Costanza, R. 1975. The spatial distribution of land use subsystems, incoming energy and energy use. M.S. Thesis, Department of Ecology, University of Florida, Gainesville, Florida.

Foell, W.K. 1980. A systems approach to regional energy/environment management. Regional Science and Urban Economics 10: 303-24.

Fromm, G., and Taubman P. 1968. Policy simulations with an econometric model. Washington, D.C: The Brookings Institution.

Hafele, W. 1982. Energy in a finite world: paths to a sustainable future. Cambridge, Mass.: Ballinger Publishing Company.

Hafkamp, W. 1984. A triple layer model. Amsterdam: North-Holland Publishing Co.

__________, and Nijkamp P. 1983. Conflict analysis and compromise strategies in integrated spatial systems. Regional Science and Urban Economics 13: 115-40.

Hudson, E.A., and Jorgenson D.W. 1976. Tax policy and energy conservation. In Econometric studies of U.S energy policy, ed. D.W. Jorgenson, pp. 9-94. Amsterdam: North-Holland Publishing Co.

Johansson, B., and Lakshmanan T.R., eds. 1985. Regional impacts of large-scale energy development. Amsterdam: North-Holland Publishing Co.

Jorgenson, D.W., ed. 1976. Econometric studies of U.S. energy policy. Amsterdam: North-Holland Publishing Co.

Lakshmanan, T.R., and Nijkamp P., eds. 1980. Economic-environmental-energy interactions. Boston/The Hague: Martinus Nijhoff.

_________, 1983. Systems and models for energy and environmental analysis. Aldershot: Gower Publishing Co.

Lesuis, P., Muller F., and Nijkamp P. 1980. An interregional policy model for energy-economic-environmental interactions. Regional Science and Urban Economics 10: 343-70.

Muller, F. 1979. Energy and environment in interregional input-output models. Boston/The Hague: Martinus Nijhoff.

Nijkamp, P. 1978. Theory and application of environmental economics. Amsterdam: North-Holland Publishing Co.

Odum, E.P. 1976. Ecology. London: Holt, Rinehart and Winston.

Ratick S., and Lakshmanan, T.R. 1983. An overview of the strategic environmental assessment system. In Systems and models for energy and environmental analysis, eds. T.R. Lakshmanan and P. Nijkamp, pp. 126-52. Aldershot: Gower Publishing Co.

Samenwerkende Bureaus voor het Energieplan Westland. 1984a. Eerste deelrapport van het integraal energieplan voor het Westland. Amsterdam: Bosboom en Hegener.

_________. 1984b. Tweede deelrapport van het integraal energieplan voor het Westland-energie-opties. Nijmegen: Hakoning.

Social Economic Council (Commissie voor Ontwikkelingsproblematiek van Bedrijven). 1983. Energie en Economie. The Hague: SER.

Tiemersma, D.N. 1984. An improved data base for the MEDEE-3 model for the Netherlands. Amsterdam: Economic and Social Institute, Free University.

WAES. 1976. Energy demand studies. Workshop on Alternative Energy Strategies, Cambridge, Mass: MIT Press.

CHAPTER 17

HOUSE PRICES AND HOUSE BUYERS: DOES ENERGY MATTER?

Allen L. White
Boston Planning Group, Inc.
Ten Post Office Square
Boston, MA 02109

ABSTRACT. Increasing energy costs have stimulated billions of dollars of investments in upgrading the energy performance of single family dwellings. These investments should translate into higher prices for energy-efficient houses if home-buyers are energy-conscious and a local housing market is tending toward equilibrium. An analysis of this relationship is performed on a sample of 1981 transactions in Newton, Massachusetts, a community whose socioeconomic characteristics are likely to produce measurable price effects of energy improvements. Using both hedonic price equations and a behavioral analysis of house-buyers, the results suggest a substantial increment in prices for houses with a basic set of energy improvements. At the same time, individual buyers generally describe energy as a low priority in selecting and pricing a new residence. This discrepancy implies that the energy proxy is capturing the effects of the buyer's broader appraisal of the physical condition of the unit at the time of purchase.

1. INTRODUCTION

Escalating energy costs during the last ten years have introduced a new variable into the dynamics of residential housing markets. While the consumer price index (CPI) for housing overall rose to 297.1 in May 1983 (1967 = 100), fuel (coal, oil, bottled gas) and gas/electricity prices reached indices of 621.0 and 429.1, respectively, the highest of any component of the CPI for housing (U.S. Bureau of the Census 1983). In 1978-1979, $55 billion was spent on all residential energy consumption, $34.7 billion in single family, owner-occupied dwellings (U.S. Energy Information Administration 1980). Expenditures by residential property-owners for energy-related improvements, maintenance and repairs totaled $3.4 billion, $2.3 billion in single family, owner-occupied dwellings. In terms of both operating and total housing costs, energy claims an increasing percentage of consumer budgets.

F. J. Calzonetti and B. D. Solomon (eds.), Geographical Dimensions of Energy, 325–352.
© 1985 by D. Reidel Publishing Company.

If energy costs represent an increasing burden on household budgets and housing markets are reasonably sensitive to these trends, it is reasonable to expect home-buyers to demand, and be willing to pay for, additional efficiency in purchased units. Such efficiency may assume the form of specific durable goods investments (e.g., heating systems, storm windows, insulation) which, when other attributes are held constant, should be reflected in house prices. An energy-conscious homebuyer might be expected to seek out and evaluate the recent utility bills of a house he might purchase and that these bills, like the specific improvements themselves, will enter the decision calculus. Alternatively, one might anticipate somewhat less rational economic decision-making based on the mere presence of certain durable goods. In the absence of a uniform standard of energy-efficiency universally available to home-buyers, the capitalization process is likely to remain imperfect and subject to a host of institutional, geographical and temporal variations across residential housing markets.

This study examines the influence of energy considerations in the purchase price of houses in Newton, Massachusetts during 1981. The principal objective is to extend the limited research to date by investigating home-buyer attitudes toward energy efficiency and the extent to which these attitudes affect house prices. The approach combines (1) conventional hedonic price equations, and (2) a behavioral assessment of individual homebuyers involved in searching for and purchasing a residential property. Thus, we shall attempt to determine both if energy-efficiency is a significant determinant of house prices, and to explain the process which underlies such a determination.

The chapter is organized into seven sections. In the next section, a review of selected literature draws upon both earlier house price studies in the appraisal and economics literatures, as well as a number of surveys of consumer attitudes. This is followed by presentation of a conceptual framework for understanding how energy considerations might affect the purchase price of a home, with particular emphasis on information flows between actors in the purchase process. This conceptual framework is translated into a set of hypotheses and related models upon which subsequent analyses are based. Next, a description is given of the data sources, questionnaire and administration of the survey instrument, as well as the empirical findings. Qualitative evidence compiled in the survey is then used to provide insights into the motives and information acquisition of homebuyers. Finally, the chapter closes with a summary of the findings and suggested areas for future research.

2. CONCEPTUAL FRAMEWORK

Antecedents to the current study are found in the economics, real estate and city planning literatures dating back to the late 1960s. Researchers have drawn primarily from the economic theory of consumer behavior (Lancaster 1971) to develop multivariate models aimed at quantifying the relationships between locational, environmental and structural characteristics of a housing unit and the price paid for the unit (Butler 1980; Lucas 1975; Thrall 1983). Underlying this approach is the premise that a consumer's housing "services" attempt to maximize utility by his or her selection of a bundle of goods within a fixed budget, which collectively satisfies personal tastes. Proximity to workplace, an ethnically homogeneous neighborhood, and a four-bedroom unit are examples, respectively, of locational, environmental and structural attributes of a house which a buyer may seek.

Extending this logic to the case of energy is straightforward (Johnson 1981). Input in the form of durable goods investments e.g., a boiler, insulation, weather-stripping produce heating and cooling, the output. Depending on one's tastes and budget, choices among the type and quantity of technological inputs will vary. In times of surplus energy supply and low prices, housebuyers are little concerned with such durables since heating, cooling and appliance operation can be achieved by consuming inexpensive fuels. As energy prices increase, a trend toward substituting energy-saving durable goods for fuels occurs. In a reasonably competitive local housing market, while holding other attributes constant, houses with such features should attract higher prices than units without such features. How much higher depends on the energy-sensitivity of the prospective buyer and the dollar value of fuel savings generated by their installation. The latter provides an estimate of the stream of benefits of a set of improvements, as long as appropriate adjustments are made for other household characteristics (e.g. numbers, age and income of occupants) and time-specific weather characteristics which are known to affect energy consumption.

The translation of this theoretical construct into empirical studies typically takes the form of a hedonic price equation of the form:

$$P = f(L, E, S)$$

where P = selling price of the house
L = a set of locational attributes
E = a set of environmental attributes
S = a set of structural attributes

Within this framework, energy investments fall within the S-type attributes and may assume the form of either (1) a dummy variable which defines the presence of a specific energy feature; (2) an estimate of the dollar cost of energy investments; or (3) a surrogate for overall energy-efficiency, usually the annual fuel bill for each house in the sample. Use of any one implies a set of assumptions which are not met entirely in the real world but should at least be considered in model specification and data compilation (Freeman 1979; Maclennan 1977; Mark 1977). The major assumptions and related analytical implications are as follows:

The housing market is tending toward equilibrium; that is, prices are at levels which are sending reasonably efficient signals of the value consumers are placing on individual house attributes. This implies informed decisions on the part of housebuyers and minimal transaction costs or other impediments to changing residence. In reality, of course, purchase decisions are made with incomplete information with respect to the availability of housing attributes. This is certainly true for energy-efficiency, the determinants of which are neither well understood by nor always available to homebuyers.

The sample units under study comprise a definable, segmented market in which one price structure prevails. When segmentation and different price structures exist, cross-sectional studies of spatially dispersed samples aggregate different markets and yield imprecise price estimates. In contrast, a study covering a short time period and confined to a spatially compact and socioeconomically homogeneous area yields more reliable estimates of implicit prices.

Individuals and households are homogeneous with respect to the preferences for specific housing attributes. Individuals, in fact, may place widely different values, for example, on commuting time, accessibility to workplace or to the central business district. For reasons of age, size and racial character, households weigh attributes differently.

Demand-side variables are the major determinants of house prices. House price studies are dominated by demand, or consumption-oriented, variables with little attention devoted to supply-side and financial variables. Yet it is reasonable to assume that comparable dwellings in submarkets with substantially different vacancy rates will command different prices. The vacancy rate, itself, may depend on certain environmental attributes contained in the price equation. Alternatively, vacancy rates may have less to do with environmental conditions than regulatory constraints (e.g., building permits, zoning) imposed by local governments. Mortgage rates and financing instruments also may affect house prices, particularly in recent years when the ability of first-time home-buyers to enter the market has been severely constrained by high interest rates (Schelhardt 1983). These factors may complicate the interpretation of the hedonic price coefficients, particularly in samples which are highly aggregated, drawn from disparate submarkets, and which include transactions over several years. While these problems cannot always be eliminated, they can be mitigated by the selection of a spatially and temporally limited sample of house purchases.

3. EMPIRICAL STUDIES

Notwithstanding these limitations, hedonic price techniques have been applied in a wide range of empirical studies, although notably few have focused on energy variables (Freeman 1979; Johnson 1981). The appraisal and assessment community has demonstrated substantial interest in multivariate techniques generally (Anderson 1979; Cho and Reichert 1980) and in their application to the valuation of energy features in residential properties (Brown 1980; Isakson 1977; Isakson 1981; Ragas and Wyatt 1978; Society of Real Estate Appraisers 1980; Webb 1980).

Empirical analyses also have been conducted through numerous attitudinal surveys of homeowners and homebuyers. In a 1978 study of 600 households in North Carolina, Kaiser et al. (1980) found that 87 percent of prospective homebuyers would consider purchasing more energy saving features in their next house. Ninety percent indicated a willingness to spend $200 more initially to save $50 per year, 85 percent would spend $600 to save $100 a year, and 80 percent would spend $1,200 more for a heat pump which would reduce heating bills by one-third.

In a national survey of 1,400 households which purchased new homes during 1977-1978, 60.3 percent indicated that energy features were "important" in their purchase decision (Anonymous 1981). Energy conservation equipment was the fifth most frequently mentioned feature on a list of ten. When asked what features would be sought in a next home, energy efficiency was mentioned by 78.6 percent of the respondents, ranking fourth behind workmanship, neighborhood quality and builder reputation. In questions comparable to those posed by Kaiser et al., 90 percent said they would be willing to pay $500 for extra insulation, and 84.7 percent were either very or somewhat willing to spend $1,200 - $1,500 for thermal panes or storm windows. In general, a strong preference for energy-efficient housing and a willingness to pay extra held "for every type of buyer from every region of country and across all price ranges." (Anonymous 1981, 196).

Corroborative findings were gathered by Dow Chemical (1980) in a national survey of 3,000 home-buyers. For the entire sample, 95 percent considered energy-saving features to be very or extremely important in a new home, and 83 percent believe that a well-insulated home is worth more at resale than a home without insulation. Furthermore, respondents indicated a willingness to pay an average of $1,362 up front for an insulation package which annually would save 25 percent on home heating bills. This approximates the 4-6 year simple payback found in other studies and which seems to govern residential energy conservation decisions.

These findings are consistent with three multivariate analyses of house prices in which a proxy for energy efficiency is entered into the hedonic price equation. Halvorsen and Pollakowski (1981) examined the effect of price changes in space heating fuels on house prices between 1970-1975 in a well-defined Seattle neighborhood. Rapid oil price hikes in the post-1973 period provided an opportunity to gauge consumer sensitivity to fuel type, a proxy for energy efficiency. In times when oil prices substantially exceeded gas prices, one would expect homebuyers to bid up the price of gas-heated homes after allowing for a time lag between price shifts and market response. As the oil/gas ratio decreases toward 1.0, prices for oil and gas-heated houses should converge.

To derive the necessary coefficients, Halvorsen and Pollakowski constructed a hedonic price equation (n = 269) using the form:

$$H_{jt} = f(Z_j, t, tZ_j)$$

where H_{jt} = the price of house j (in natural log form) in period t

Z_j = is a set of attributes, including space heating fuel, for house j
t = time of purchase
tZ_j = a set of interactive terms, e.g., 1974 sale of an oil heated house

The latter variables capture the effects, ceteris paribus, of changes in relative fuel prices on house prices.

The semi-log form of the equation yields an R^2 of .745 and signs in the anticipated direction. Fuel type appears to influence house prices, as the coefficient of the oil/time interaction term tracks the oil/gas price ratio during the study period. For three time periods--1974, 1/75-6/75, and 7/75-12/75--oil-heated homes sold for 3.55, 18.5 and 0.8 percent less than gas-heated homes. For a typical gas- versus oil-heated house during the 1974-1975 period, the price differential for the three time intervals is \$761 (1974), \$4,597 (1/75-6/75) and \$233 (7/75-12/75). The second figure, it should be noted, well exceeds the price of fuel conversion or the value of short-term losses. The author interprets this as an overreaction on the part of homebuyers to rapidly escalating oil prices during the study period.

Congel et al. (1982) investigated the implications of energy efficiency--defined as the relative amount of heat loss detected in an aerial infrared photograph--for appraising comparable units. The objective was to assist appraisers in assigning value to energy efficient housing using the comparable units approach. In a sample of single family units in Lubbock, Texas during 1978-1979, conventional structural characteristics were specified in either continuous (e.g., square feet of living area, number of baths) or dummy (presence of a two-car garage, air conditioning) form. Notably absent are both environmental and locational variables, a curious omission for a city-wide sample. All homes used in the analysis sold for less than \$40,000 and were smaller than 1,500 square feet to minimize biases due to cost and size. Energy efficiency was entered as a dummy variable, specifically, 1 if the structure "appears dark" on the infrared photograph and 0 if it shows a relative amount of "lightness."

The study explored two models using house price in its linear form as the dependent variable. The second model is simply a reduced form of the first, using only those variables with coefficients significantly different from zero at the 95 percent confidence level. R^2 for both models is .72 and the energy dummy variable is \$3,415 and \$3,247 for models A and B, respectively. This indicates that, ceteris paribus, homebuyers are willing to pay at least \$3,000 more for a relatively thermally efficient dwelling. While the measure of energy-efficiency is admittedly crude, it is also true that the homebuyer's perception may be equally inexact. A rough indicator of the type utilized in this study may be preferable in quantifying consumer behavior, an issue to be explored in detail later.

A third and, we believe, the most exhaustive study to date was conducted by Johnson (1981) using a sample of 1,317 single-family detached units sold in the Knoxville area during 1978. Thirty variables were included in a standard form:

$$P_i = f(U_i, S_i, N_i, L_i)$$

where
P_i = sale price of house i
U_i = the adjusted fuel bill for house i
N_i = a set of neighborhood attributes for house i
S_i = a set of structural attributes for house i
L_i = a set of locational attributes for house i

For the entire sample, the estimated linear equation yields an R^2 of .80 with 29 coefficients attaining a significance level of at least .05 and all 18 predicted signs consistent with expectations. For the key energy variable, the equation yields a .02 level of significance and a coefficient of 20.73. This suggests that a one dollar reduction in annual fuel bills will increase the market value of a unit, _ceteris paribus_, by $20.73. The standardized parameter estimates, which allow for a comparison of relative of importance of each variable, place the energy variable fifth in significance behind size, lot, age and air-conditioning variables. Interestingly, substituting a simple storm window dummy variable as the energy proxy increases the R^2 to .85, and yields a higher significance level of .01. The coefficient is 2,021.39, implying a willingness to pay an additional $2,021 for a house equipped with storm windows. Together, these findings suggest that buyers are willing to pay more for energy-efficiency housing, but the identification of such housing is more directly tied to the presence of a familiar feature which may not necessarily accurately measure the energy performance in the purchased unit.

4. THE DECISION PROCESS

To what extent and from what sources do home-buyers obtain energy information which ultimately creates the price effects implied by these studies? Figure 1 depicts a simple model of the residential purchase process and the potential effect or role of energy considerations at each decision point. Initially, a decision to relocate is prompted by an event internal (e.g., change in family size, desire to purchase instead of rent a home) or external (e.g., job change, deteriorating school system) to the household. One such impetus may be escalating energy costs in the form of the home-owner's appraisal that the investment required to achieve affordable energy costs in the present residence is not cost-effective. This may be particularly true in the case of an elderly household, wherein surplus space and relatively fixed incomes make energy an especially burdensome expense. In other instances, energy cost pressures may be lumped together with other maintenance, repair and operating expenses to trigger a relocation decision.

Confronted with dissatisfaction in the current dwelling, opportunities to adjust _in situ_ are sometimes possible. Space constraints may be alleviated by adding a bedroom, neighborhood decline may be reversed by collective action, or investment needs may be met by creating a rental unit. In a similar fashion, energy costs may be managed by altering living habits (e.g., lowering thermostats, closing

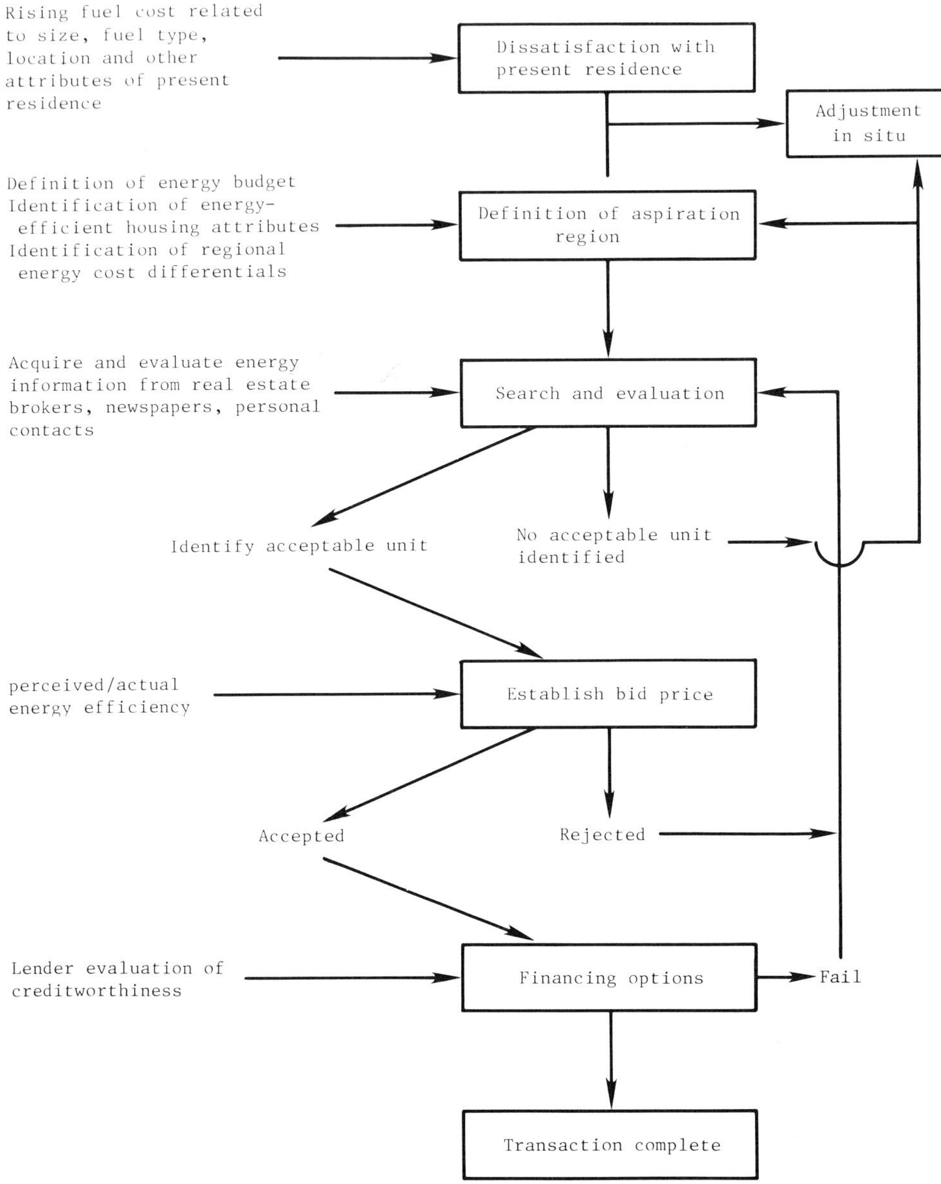

Figure 1. The role of energy in the residential relocation process.

off rooms) and/or by investments in durable goods to upgrade energy efficiency.

If adjustments in situ are not feasible, definition of an aspiration region ensues (Maclennan 1977). The aspiration region sets locational, financial, structural and other boundaries on the search process, in effect, melding the desirable with the feasible. Energy considerations may surface at this point through their role in the household budget, the relationship of structural or site characteristics to anticipated energy costs, or regional differences in household energy demand (Cowan 1980). The latter, for example, is clearly an influence in population shifts to the Sun Belt, especially among the fixed-income elderly.

Once an aspiration region has been defined, search and evaluation proceeds. Information flows play a crucial role in narrowing the aspiration region to a specific dwelling unit. The real estate broker, newspapers and friends/relatives are the primary actors during this stage. The broker typically acquires some knowledge of the energy conditions of the unit, ranging from the identification of storm windows or insulation to monthly utility bills or a detailed energy audit. The availability and use of this information is critically important to understanding the consistent statistical relationship between energy-efficiency and house prices. If such information is routinely sought and obtained by home-buyers, then this relationship makes sense. If not, then a statistical association may be indicative of another more complex determinant of house prices.

The search and evaluation process either produces an acceptable dwelling or compels the home seeker to (a) redefine his aspiration region or (b) reconsider adjustment in situ. If the former occurs, a bid price is eventually offered based on a valuation of structural, environmental and locational characteristics. Energy attributes might affect this price by virtue of their impact on monthly carrying costs, anticipated comfort, or simply lifestyle preferences unrelated to potential energy savings. In the absence of a standardized measure of energy performance, the value assigned by the home-buyer will depend on a variable mix of perceived and real benefits.

The bid price on a unit is either accepted, rejected or negotiated between buyer and seller. If accepted, and assuming financing is required, a lending institution enters the process. In assessing the risk level associated with a loan, the institution considers both the credit-worthiness of the buyer and the characteristics of the dwelling unit. In periods of high interest rates, and mortgage delinquencies (New York Times 1982), energy costs should be a central concern. Not surprisingly, Olin (1981) reports that 50 percent of all savings and loan associations do consider energy in their appraisals and 25 percent do so in their underwriting. Energy-efficiency, according to Olin, should be considered because it directly affects the borrowers' ability to meet monthly mortgage obligations, enhances resale value, and increases the probabilities of refinancing. In dollar terms, all this should translate into a higher allowable loan-to-value ratio, thereby placing a marginally affordable home within the reach of the buyer.

5. PRICE ANALYSIS

The house price model conforms to the standard hedonic price equation structure:

$$P_i = f(L_i, N_i, S_i, F_i, E_i)$$

where P_i = selling price of house i
L_i = a set of locational attributes of house i
N_i = a set of neighborhood attributes of house i
S_i = a set of structural attributes of house i
F_i = a set of financial attributes related to the sale of house i
E_i = a proxy for energy-efficiency of house i

The specific variables, the expected signs and data sources are listed in Table 1. The choice and number of independent variables was influenced by several considerations. First, the study area, described in greater detail below, is Newton, Massachusetts, an old, built-up suburb contiguous to the city of Boston. The city, as part of its tax reassessment effort, developed its own multiple regression model of house price determinants, the coefficients of which were provided to the author by the Chief Assessor (City of Newton 1982a). Although the data did not include standardized coefficients nor tests of significance, the tabulations did provide useful guidance in selecting variables from among the many typically included in house price studies. Variables which were included most frequently in the assessor's models--LOT, STYLE, and TRAFFIC--were also included in the present analysis. To these were added other variables included in earlier studies which appear, though somewhat less frequently, in Newton's reassessment calculations: BATH, FIRE, KITCHEN, SPACE AND GARAGE.

An accessibility variable, TRANSIT, is included to account for accessibility to rapid transit. Since the sample is drawn from within a single municipality, this variable is a more accurate measure of accessibility than linear distance to the city's multiple shopping districts or to downtown Boston.

In addition to these locational, neighborhood and structural variables, financial indicators also are incorporated in the model. RENTAL is aimed at capturing the value associated with the widespread (usually illegal) practice in Newton of creating an accessory apartment within an existing single family dwelling (Clay 1982; Siegel 1982; White 1982). It is hypothesized that the existence, or potential, of a rental unit increases the house's sale price, owing to its contribution to mortgage and operating costs.

A second financial variable is the mortgage interest rate. Since interest rates topped 11 percent beginning in late 1979, the rate itself began to affect the bid price for a given unit (Harney 1982). In some areas, house price stablilization and decline between 1979-1982 has been attributed to the depressive effects of high interest rates (*New York Times* 1982).

Table 1. Variable Names, Descriptions and Expected Signs in Hedonic Price Equation for Newton Sample, 1981

Independent variable: Sale price of house, P_1

Variable	Description	Expected Sign
LOT	lot size (ft^2)	(+)
BATH	number of baths	(+)
SPACE	square feet of living space	(+)
FIRE	number of working fireplaces	(+)
GARAGE	number of garage spaces	(+)
KITCHEN	kitchen remodelled in last 5 years, D=1[a]	(+)
STYLE	Victorian style, D=1[a]	(+)
TRANSIT	within 10 minute walk of train, subway or express bus, D=1[a]	(+)
RENTAL	contain or plan to have rental units, D=1	(+)
INT	interest rate paid on the mortgage	(-)
ENSAVE	approximate percent reduction in energy consumption attributable to conservation measures in place	(+)

[a]D=1 signifies dummy variable

Sources: All variables except P_i compiled from author's survey of homebuyers, Newton, Massachusetts, 1981; P_i derived from monthly summary of transactions, Middlesex County Transfer Directory, 1981.

A number of options designed to measure the effect of energy-efficiency on house prices were considered. These may be grouped into two broad categories, each of which implies substantially different assumptions about consumer behavior. First, one may assume that the buyer will seek, obtain and evaluate information which measures energy performance and then incorporate that evaluation into the bid price for the unit. This is implied in the Johnson (1981) study which uses an annual adjusted utility cost as the independent variable.

Other such indices may be constructed, for example, energy use per square foot, energy cost per square foot, or predicted energy use in a home equipped with certain energy features versus a base case (Rosenfield and Wagner 1982; Schuck and Millhone 1982). If standardized performance indices were readily available to prospective homebuyers, houses could be compared and their performance could be readily integrated into the purchase decision. In the absence of such information, however, one must assume that a significant relationship between house prices and energy efficiency derives from the buyer's ability to compile and evaluate such indicators.

The second category is more impressionist in nature, based on the premise that even an energy-conscious homebuyer is unlikely to seek and evaluate specific quantitative evidence of efficiency. Instead of utility costs or energy consumption per square foot, the mere presence of familiar energy features such as storm windows, insulation or weather-stripping build a perception of energy-efficiency and thereby increase the bid price, *ceteris paribus*. Because the buyer cannot be expected to calculate the energy savings associated with each feature, the incremental valuation may bear little relationship to the dollar savings which will be generated over the life of the equipment. Pursuing this line of reasoning, a package of energy features collectively may present a picture of energy-efficiency which exceeds the sum of each separate feature. Both Congel et al. (1982) and Halvorsen and Pollakowski (1981) use a variant of this second approach. The former defines energy-efficiency by employing a simple binary measure of heat loss, whereas the second study correlates house prices with fuel type using a dummy variable specification.

For ease of interpretation and consistency with the earlier assumptions underlying hedonic price analyses, this study utilizes a ratio-scale variable which approximates the total percent energy savings likely to result from the installation of one or more common conservation measures. The selection of the first option as opposed to the simple summation of total measures (the second option) reflects the non-linear or non-log linear increments in savings associated with each package of measures. While appealing from a behavioral standpoint, the total measures approach would violate a basic assumption necessary to estimate a buyer's willingness to pay. ENSAVE, the energy variable used in this analysis, ranges from 0 to .50 depending upon the estimated total percent reduction in energy consumption attributable to features identified by the surveyed homebuyers as present at the time of purchase. Two options are specified, the first a linear functional form, a log form, the second an adjustment for the price of primary fuel--oil, gas or electricity--per million Btu's during 1981. This refinement brings the energy variable a step closer to approximating the real economic benefits likely to accrue to the home-buyer.

5.1 Data Compilation and Sample Area

Most data were collected through a telephone survey of households who purchased homes in Newton between January and December, 1981 and appear in the Middlesex County Transfer Directory. Since approximately one month transpires between a transaction and its appearance on the realtor's listing from which the actual names were drawn, most closings actually occurred between December 1980 and November 1981.

An initial round of 131 buyers were selected at random. Purchases by trusts, corporations, or identifiable foreign-born buyers were excluded from the sample, the latter to avoid communications difficulties for telephone interviewers. Condominiums also were excluded to avoid complications of shared heating and cooling systems. After these exclusions, 93 households remained. A second round of 40 houses using the same criteria then was added. Ultimately, 83 usable interview forms were completed, representing 10.4 percent of the 803 single family house transactions recorded during the referenced 12 months, and 63.8 percent of all households contacted.

Newton is a community of 84,000 located immediately west of Boston, and contains some 29,081 housing units of which 20,132, or 69.2 percent, are owner-occupied (City of Newton 1982b). The City is an old, built-up inner suburb of predominantely single family homes. The housing stock includes a cross-section of large 19th century wood frame structures, colonial, tudor, ranch and Cape Cod style homes. The mean value of non-condominium housing units in 1980 was $81,200, or 10.3 percent above the 1980 national average of $72,800 for existing homes (Case 1982).

Newton consists of ten villages, each with its own business district and distinctive socioeconomic profile (Figure 2). Although village boundaries differ according to census tract, assessor and zip code definitions, the community may be roughly divided into: (1) North Newton--ethnic, working class neighborhoods of small single family and multi-family homes located north of the Massachusetts Turnpike; (2) Central--affluent, professional households with large Victorian colonial and brick units, located between the Massachusetts Turnpike and Route 9; and (3) South Newton--middle and working class, relatively new housing stock, located south of Route 9.

A reputation for superior schools, public services and environmental amenities keeps Newton's overall vacancy rates at about two percent, and even less for single family homes. The city's accessibility to Boston further reinforces high prices and low vacancy rates. The median asking price or mean value of owner-occupied non-condominium units reported in 1980 was $90,600 (City of Newton 1982a). At the same time, the city's housing turnover rate is high, reflecting its attractiveness to Metropolitan Boston's mobile academic and professional population. These conditions combine to create a

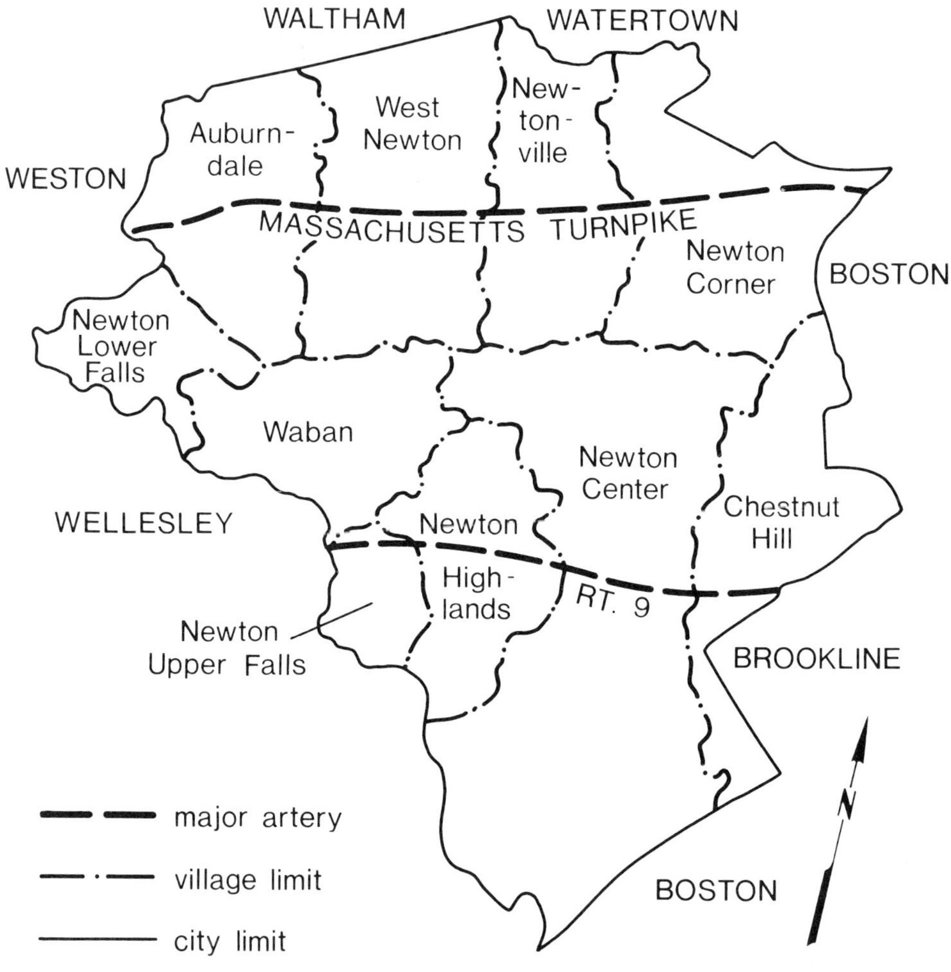

Figure 2. Village of Newton, Massachussetts(defined by postal codes).

favorable setting for the present study in that homebuyers are sophisticated and informed, selling prices are competitive, the study area is geographically compact and key environmental and locational attributes are fairly uniformly distributed across neighborhoods.

Table 2 summarizes the characteristics of the sample transactions. Housing prices averaged $124,053. In addition to neighborhood amenities, the high average prices are explained by particularly spacious houses and moderate to large lots. Houses in the sample contain an average of 8.7 rooms, 2.1 baths, over 2,200 square feet of living space and a quarter acre of land. Prices also reflect some income bias among respondents. The affluent, educated home-buyer is more apt to participate in a survey than the population as a whole.

Table 2. Selected Characteristics of Sample Houses in Newton (n = 83)

Variable	Mean	Maximum	Minimum
Price ($)	124,053.0	262,000.0	66,000.0
Lot Size (ft^2)	11,289.4	44,000.0	1,500.0
Number of Rooms	8.7	18.0	4.0
Number of Baths	2.1	4.0	1.0
House Size (ft^2)	2,211.9	6,000.0	800.0
Garage Spaces	1.3	4.0	0.0
Working Fireplaces	1.3	7.0	0.0
Age of Home (years)	56.1	158.0	11.0
Mortgage Value (dollars)	69,087.0	130,000.0	0.0
Interest Rate on Mortgage (%)	17.8[a]	19.5	12.0[b]

[a] including five cases of zero mortgage
[b] excluding five cases of zero mortgage

Source: Author's 1982 survey of Newton homebuyers.

Household characteristics reflect the housing attributes (Table 3). The average household income of $47,000 is well above Boston metropolitan and national figures, as are educational levels (93.9 percent college graduate) and professional status (70.7 percent). A composite picture of the homebuyer in the survey sample suggests a mid-thirties couple with one or no children who moved to Newton from within the Boston metropolitan area. Thus, our findings are most applicable to this type of home-buyer, a limitation which constrains their reliability in evaluating other socioeconomic groups.

The energy characteristics of the sample houses are summarized in Table 4. Responses were recorded only as yes or no, with occasional subjective judgments left to the interviewer when a "some" or "in fair condition" response was offered. The most common energy feature is storm windows, a popular, cost-effective and relatively expensive improvement found in 81.9 percent of the houses sampled. Attic/ceiling insulation (45.8 percent) and weather-stripping (43.4) are the next two most frequently identified features. To place these figures in perspective, a sample of national percentages in 1978 were: storm windows--62.8; attic insulation--77.1; and wall insulation--53.8. (U.S. Energy Information Administration 1980). These differences are attributable in large part to the old age of Newton's housing stock versus the nation as a whole.

Table 3. Selected Characteristics of Average Homebuyers in Newton, 1982 (n = 83)

Household income ($)[a]	47,444
College graduates (%)[a]	93.3
Age of household head (years)[a]	35.2
Professional occupation (%)[a]	70.5
Households three persons or less (%)	57.8
Owned previous residence (%)	43.4
Previously lived in	
Newton (%)	28.9
Metropolitan Boston (%)	54.2

[a]estimate based on grouped data for head of household

Source: Author's 1982 survey of Newton homebuyers.

Table 4. Energy Characteristcs of Sample Houses (N = 83)

Feature	% of Homes With Feature
FURNACE	12.0
HOTWTR	24.1
WTHRST	43.4
CURT	4.8
FSWITCH	8.4
ATTC	45.8
WALL	26.5
STORM	81.9
PIPE	24.1
SHOW	10.8
CLOCK	24.1
WOOD	3.6
SOLAR	1.2

Total Number of Features	% of Homes
0	4.8
1	13.3
2	24.1
3	25.3
4	7.2
5	10.8
6	10.8
7	3.6

Source: Author's 1982 survey of Newton homebuyers.

Table 5. Parameter Estimates for Model Using Percent Energy Savings[1]
Dependent variable: Selling Price

	Expected Sign	Coefficient	Standard Error
ENSAVE	+	94,271.57 (2.33)*	40,416.71
LOT	+	0.87 (0.86)	1.03
BATH	+	20,392.03 (2.70)**	7,558.78
SPACE	+	15.33 (2.40)*	7.48
FIRE	+	-540.14 (0.08)	7,249.51
GARAGE	+	8,337.55 (1.13)	7,387.05
KITCHEN	+	1,667.40 (0.16)	10,573.98
STYLE	+	-12,069.13 (0.85)	14,116.18
TRANSIT	+	15,248.30 (1.16)	13,163.61
TRAFFIC	+	16,327.56 (1.37)	11,915.56
RENTAL	+	23,572.05 (1.19)	19,819.50
INT	-	-352.28 (1.02)	345.70
CONSTANT		-11,142.09	
R^2	.74		

t-statistic in parentheses.
Significance levels for one-tail test: **.01, *.05
[a]Savings estimates based on Rothchild (1981).

Table 4 also reveals a wide variation in the number of features identified in individual houses. Very few (4.8 percent) had no features, while 62.7 percent had between one and three. Four or more features were identified in 32.4 percent, or nearly one-third, of the units.

5.2 Model Estimations

Table 5 presents the parameter estimates for the linear model. With the exception of FIRE and STYLE, coefficients are in the expected direction and their magnitudes generally are reasonable. Deferring for a moment a discussion of ENSAVE, it is evident that three size-related variables--BATH, SPACE and LOT--offer most explanation to differences in house prices. The coefficients imply that holding other variables constant, any additional bath adds $20,392 to house price, an additional square foot adds $15.33, and an additional square foot of lot space adds $0.87. Only BATH and SPACE, however, are significant at the .05 or .01 level.

The two variables whose signs contradict expectations are STYLE and FIRE, both of which are associated with large, older homes prevalent in several of Newton's villages. Although neither achieve an acceptable significance level, their negative signs do suggest either our misjudgement of the buyer's valuation of historically interesting architecture or problems of multicollinearity among the independent variables, a situation typical of many house price studies (Anderson 1979). In light of the strong showing of ENSAVE, it appears that the latter is the case.

The ENSAVE coefficient of 94,271 implies a strong effect of energy performance on house prices. Consider a house containing storm windows, attic insulation and weather-stripping/caulking. Together, these three common features reduce energy consumption by an estimated 33 percent. Applying the parameter value to this figure, an additional $31,100 is added to a house price, holding other variables constant. The explanatory strength of BATH, SPACE and ENSAVE together is demonstrated by an R^2 = .65 (versus .74 for all variables) after the three are entered in a stepwise regression.

In rethinking the results of the fist model, three changes may enhance its performance. First, ENSAVE may be capturing the effects of house condition in view of its unexpectedly large coefficient and the omission of any explicit measure of physical condition in the model. The negative signs of STYLE and FIRE lead further credence to this idea, insofar as they are acting as a surrogate for condition. In other words, the effects of energy measures may be entangled with the buyer's appraisal of overall condition, an understandable situation since many energy improvements are integrally related to the proper maintenance of both structural and cosmetic house features. For example, the addition of insulation and weather-stripping/caulking greatly reduces temperature fluctuations in ceilings and walls, and thereby prevents damage to plaster, concrete, brick and other building materials. Under these circumstances, the correlation between energy performance and overall condition is likely to be strong, especially in large, older Victorian homes which are well represented in our sample. In the next model, we attempt to rectify this situation by adding AGE, a variable intended as a surrogate for overall house condition.

Second, this relationship may be clarified if ENSAVE itself is refined to reflect anticipated savings based on prices for the specific primary heating fuel used in each house at the time of purchase. Price differentials in Newton in 1981 were substantial: gas cost residential consumers \$6.07 per MMBtu, oil \$8.69 per MMBtu and electricity \$28.42 per MMBtu. In the second model, ENSAVE accounts for these differences with the objective of isolating the effects of energy performance from other condition-related variables.

Third, the shortcomings of linear hedonic models have been examined by many authors, although the linear functional form together with the semi-log form remain the most common in house price studies (Halvorsen and Pollakowski 1981). Both forms have their drawbacks. The first model, for example, implies that the marginal valuation of each percent energy savings is equally valued by home-buyers across all savings levels, in this case, 0-50 percent. A log form, on the other hand, assumes that increased savings has the same percentage impact on prices. The most appropriate functional form cannot be specified *a priori*, which makes experimentation a worthwhile exercise in hedonic price analysis. Thus in our second model, the house price is incorporated in its log form and parameters re-estimated with the other changes described above.

Table 6 presents the results of the second model. Note that the anticipated sign for STYLE and FIRE has been changed to negative in concert with earlier conclusions. Other coefficients except that for AGE show the expected signs. The unsatisfactory performance of AGE may be attributable to its unreliability in measuring the actual physical condition of a property. In the absence of a more direct measure based on on-site inspection, no other age variable was available for inclusion in this analysis. The strongest explanatory power is again represented by size-related variables--BATH, SPACE, LOT and ENSAVE--while the overall performance of model (R^2 = .71) approximates its predecessor. Significance levels of .05, however, are achieved only by BATH and ENSAVE. A comparison of the beta weights for ENSAVE in both equations suggests a very modest dilution of the explanatory power of ENSAVE after inclusion of AGE. This result, of course, may be attributable in whole or in part to other revisions in the model specification.

In sum, the hedonic models produce convincing, though somewhat ambiguous, findings with respect to market responsiveness to energy investments. Regardless of model specification, the proxy for energy efficiency performs well in a limited sample of 1981 transactions in an affluent suburban housing market. The uncertainties remaining stem from the recognition that energy performance is closely tied to the overall physical condition of a housing unit and the purchaser's evaluation of that condition.

The behavioral underpinnings of our quantitative analysis suggest

Table 6. Parameter Estimates for Logarithmic Model of Selling Price, with a Revised ENSAVE and Age of House (AGE)

	Expected Sign	Coefficient	Standard Error
ENSAVE	+	0.069 (1.69)*	0.041
LOT	+	0.000009 (1.27)	0.0001
BATH	+	0.174 (3.16)**	0.055
SPACE	+	0.00007 (1.31)	0.00005
FIRE	-	-0.028 (.42)	0.067
GARAGE	+	0.640 (1.03)	0.062
KITCHEN	+	0.040 (.53)	0.075
STYLE	-	-0.063 (.04)	0.144
TRANSIT	+	0.106 (1.19)	0.102
TRAFFIC	+	0.091 (1.35)	0.085
RENTAL	+	0.128 (.89)	0.145
INT	-	-0.003 (1.28)	0.003
AGE	-	0.0007 (.28)	0.003
CONSTANT	10.762		
R^2	.71		

t-statistic in parentheses.
Significance levels for one tail test: **.01, *.05.

two scenarios, both of which would lead to the same findings. The first depicts a home-buyer who seeks, acquires and places a value on energy information obtained prior to purchase. The second depicts a home-buyer sensitive to the overall physical condition of a house but unaware, or vaguely aware, of the energy performance of the unit. The association between the two may not explicitly enter the decision calculus but nonetheless manifests itself in the price relationships examined thus far. The following section investigates buyer behavior in an effort to identify which of these scenarios best describes the typical house purchase decision.

6. BEHVIORAL EVIDENCE

Figure 1 depicts numerous opportunities for the home-buyer to incorporate energy criteria into the relocation, search and price bid decision. Energy costs initially may be an impetus to changing residences and the type of house sought. The information compiled on a particular dwelling and the price offered may be subject to the influences of anticipated energy performance. And mortgage lending practices are increasingly sensitive to the effect of energy costs on the credit-worthiness of borrowers. The quantitative analysis suggests that energy criteria are in fact operating in the local housing market. We now examine the congruity between the analysis of price determinants and what buyers themselves say about their energy consciousness.

6.1. Reasons for Moving

Two reasons dominated the decision to seek a new residence. For 30.1 percent of the sample, space needs prompted the move, a finding consistent with the importance of size-related variables in the regression analyses. For another 28.9 percent, the purchase decision was prompted by a desire to own instead of rent a home. Only 2.4 percent mentioned energy as a primary reason for moving. Among "other" responses, commuting distance and investment considerations were occasionally mentioned. Thus, no substantial evidence suggests that energy considerations were important at this early stage in the relocation process.

6.2. Features Sought

Once the decision to relocate was made, two housing characterstics most commonly guided the search process. For 37.8 percent, size was the first priority, the corollary of the motive identified in the decision to move. Another 29.3 percent identified "neighborhood quality" -- shopping, transit, recreation--as an important criterion in their search for a new residence. Among the anticipated responses, affordable energy costs was identified by only 2.4 percent of the respondents. The "other" category is revealing by both the number and substance of responses. Yard space, another size variable, was mentioned by 14.5 percent; location, mostly to workplace, by 13.3 percent: condition of the house by 9.5 percent; and investment potential by 7.2 percent. Again, no evidence was found to support the strong role of energy implied in the earlier analysis.

6.3. Awareness of Energy Features

A rough indicator of energy awareness may be derived by summing the "don't know" responses for all respondents when asked if a particular energy feature was present in their new homes at the time of purchase. Allowing for the probability that a fraction of "no" responses were, in fact, "don't know" (that is, people are reluctant to confess ignorance), 79.5 percent of the respondents recorded zero and 14.5 percent recorded one "don't know." This suggests that the large majority of buyers are both familiar with energy features and have taken the time to ascertain their presence. It cannot be automatically concluded, however, that the necessary inspection occurred prior to the time of purchase or that the sale price was affected by its outcome.

6.4. Energy Improvements After Purchase

Some 8.4 percent of the respondents volunteered information concerning their conservation investments _since_ the purchase date. While the number is small, some investments were substantial, ranging up to $3,000 within one year. The inclination to offer such information prompts two lines of thinking. First, an additional number of respondents would have provided comparable information with encouragement from the interviewer. Second, the buyer may incorporate energy-efficiency by deducting all, or a portion of, anticipated energy and related home improvements from the bid price. This second possibility reinforces an emerging sense that energy-efficiency is melded with a broader appraisal of the house's physical condition and that this condition, rather than energy _per se_, underlies the strong performance of the energy variable in the hedonic price equations.

6.5. Sources of Information

In searching for a new house, real estate agents were the most commonly used information source, mentioned by 56.6 percent of the respondents. Newspapers were mentioned by 39.8 percent and family/friends by 14.5 percent. In contrast to the process of identifying available units, the most common source of energy information was the seller, mentioned by 57.8 percent of the respondents. This is followed by the building inspector (30.1 percent), personal inspection (21.7 percent), and the real estate agent (13.3 percent). The building inspector as a surprisingly prominent source of energy information was not anticipated. Traditionally, the inspector's primary role is to investigate the structural integrity of the dwelling. Advice regarding needed energy investments, especially heating systems, evidently has become part of his function. Because the house inspection is mandatory in most communities, enhancing the inspector's energy audit skills would appear to be an effective vehicle for infusing energy sensitivity into the housing marketplace.

6.6. Types of Information

Research reviewed earlier relied on various measures of energy-efficiency; for example, utility bills, storm windows, level of heat loss, and fuel type. In our sample, only two specific types of information were commonly obtained: utility bills (38.6 percent) and the presence of insulation (24.1 percent). Utilty bills, of course, are in large part determined by insulation levels, storm windows and fuel type. Thus, it may be argued that buyers with a knowledge of utility costs would not be inclined to seek additional information. However, less than one half of the sample obtained estimates of utility costs and only a fraction of these obtained the information in a written, reliable form. Even for those who did have access to utility bills, a probable response to the information is a simple expensive versus inexpensive dichotomy. In other words, a house is judged to be either expensive or inexpensive within the context of the buyer's previous residence and current budget constraint. Johnson (1981) notes a similar discrepency between the statistical strength of her utility variable and the 43 percent of the sample transactions where utility costs were not reported. Some proportion probably obtained verbal utility cost data directly from the seller, as in Newton, where 28.9 percent of the respondents fit this category. Nevertheless, only a minority of buyers obtain utility information from any source, certainly in numbers fewer than the earlier price equations imply.

6.7. Anecdotal Evidence

Several respondents offered additional insights, both solicited and unsolicited, into the role of energy considerations in their purchase decisions. In one instance, a buyer required insulation and exterior repair work to be completed prior to closing the sale, conditions which were formally written into the purchase and sale agreement. Another respondent reported that structural and site characteristics suitable to energy improvements, specifically solar equipment, influenced his choice of house, though not decisively.

Numerous buyers expressed a concern for the cost of energy and other home improvements after purchase. A few expected to make major improvements (such as installing a new furnace) after closing on the houses. Previous price studies of energy investments have employed both types of variables with consistently significant results, indicating buyer responsiveness to both specific energy features or single variable proxies of energy-efficiency.

7. CONCLUSIONS

This study examined house price variations and related buyer behavior in a sample of transactions in Newton, Massachusetts during 1981. Based on real estate records and survey data, information on conventional house price determinants was compiled together with a detailed listing of energy features present at the time of purchase. In addition, information was gathered on the motives for moving, criteria in selecting a new residence, and sources and types of energy information acquired.

The results of the price equations in both linear and logarithmic functional form indicate a strong positive effect of energy improvements on prices, second only to various size characteristics of the dwelling and its lot. The magnitude of the parameter estimates, however, are substantially larger than expected or justifiable on a life cycle benefit basis, raising the possibility that energy performance in the eyes of the homebuyer is subordinate to a broader appraisal of the physical condition of a property. An attempt to control for condition using age as a surrogate yielded unsatisfactory results.

An analysis of individual home-buyer decision-making provides some insights into this situation. The home-buyers interviewed placed little importance on energy considerations in their decision to relocate and in choosing a new house. Space needs, home ownership and neighborhood quality were the dominant concerns. In general, respondents were knowledgeable about specific energy features present at the time of purchase. Sellers and building inspectors were the most common information sources and utility costs (especially verbal quotations) and insulation levels were the most frequently obtained types of information, although less than half the sample obtained either.

The picture that emerges from this sample of affluent, educated homebuyers is one of only modest energy awareness, certainly at levels below those suggested by the parameter estimates derived in the price equations. The findings point to a process in which energy-efficiency in the form of durable goods investments, rather than ultimate costs or thermal performance, contributes to the buyer's broader appraisal of the physical condition of the dwelling unit. It is the price effects of this appraisal which we believe are reflected in the parameter estimates.

Three issues raised in this study merit further research. First, an application of the basic model to other housing markets and socio-economic groups would be a valuable next step in validating and refining the findings reported herein. Alternative functional forms and variable specification need to be tested.

Second, experimentation with various indices of housing condition, a variable conspicuously lacking in most house price studies, is necessary to pursue the line of inquiry which evolved from this study. Census and real estate data are limited in this area. Appraisal and the assessment data may be a richer source, especially in areas where recent property revaluation has generated a current data base.

Finally, as a matter of public policy, it is clear that the development and application of standardized measures of residential energy-efficiency would facilitate the responsiveness of housing markets to energy investments. Such indicators, if placed in the hands of realtors, building inspectors, and tax assessment offices, would provide prospective buyers with an understandable and comparable measure of economic returns with which bid prices could be suitably adjusted.

8. ACKNOWLEDGEMENTS

This research was supported by a grant from the Lincoln Institute of Land Policy. The author gratefully acknowledges Lewis Branzburg, Kitty Cox, Tom Mangione and James Paull for reviewing drafts of the survey questionnaire; Richard Goldstein, Tom Kontuly and Bernie Reddy for guidance in the quantitative analysis; and an anonymous reviewer for many valuable comments of both a quantitative and theoretical nature. Any remaining errors, of course, are the sole responsibility of the author.

9. REFERENCES

Anonymous. 1980. What makes today's buyer buy? Builder January 1: 180-99.

Anderson, J.E. 1979. Ridge regression: a new regression technique useful to appraisors and assessors. The Real Estate Appraisor and Analyst November-December: 48-51.

Brown, G.T. 1980. The impact of energy on value. The Real Estate Appraisor and Analyst November-December: 5-12.

Butler, R.V. 1980. Cross-sectoral variation in the hedonic relationship for urban housing markets. Journal of Regional Science 20: 429-54.

Case, K.E. 1982. Land prices, housing prices, and housing production: Does the market work or doesn't it? Paper presented at a conference sponsored by the Joint Center for Urban Studies of MIT and Harvard University, and the Lincoln Institute of Land Policy, Cambridge, MA. Draft.

Cho, C.C., and Reichert, A.K. 1980. An application of multiple regression analysis for appraising single-family housing values. Business Economics 15: 47-52.

City of Newton. 1982a. Assessor's Office. Unpublished property revaluation data.

______. 1982b. Department of Planning and Development. Unpublished 1980 Census data.

Clay, P. 1982. Improving the utilization of the existing housing stock: The case of accessory apartments. Paper presented at a conference sponsored by the Joint Center for Urban Studies of MIT and Harvard University, and the Lincoln Institute of Land Policy, Cambridge, Draft.

Congel, J.B., et al. 1982. Measuring energy efficiency for selection and adjustment of comparable sales. The Appraisal Journal 50: 71-78.

Cowan, E. 1980. Energy costs and home values. The Appraisal Journal 48: 445-48.

Dow Chemical Company. 1980. Perspective on energy: America's homeowners speak out. Summary of a national survey of homeowners.

Freeman, A.M. III. 1979. Hedonic prices, property values and measuring environmental benefits; a survey of the issues. Scandanavian Journal of Economics 81: 154-73.

Halvorsen, R., and Pollakowski, H.O. 1981. The effects of fuel prices on house prices. Urban Studies 18: 205-11.

Harney, K.R. 1982. Financing determines home selling price. Washington Post May 23.

Isakson, H.R. 1977. The real estate appraiser and the energy crisis. The Real Estate Appraiser and Analyst. September-October: 4-13.

______. 1981. Residential energy audits: a new source of appraisal data. The Appraisal Journal 49: 74-84.

Johnson, R.C. 1981. Housing market capitalization of energy saving durable goods investments. ORNL/CON-74. Oak Ridge, TN: Oak Ridge National Laboratory.

Kaiser, E.J., Marsden M.E., and Burby, R.J. 1980. The adoption of energy conservation features in new homes: current practices and proposed policies. In Energy in the cities symposium, ed. Joel T. Werth, Chicago: American Planning Association, PAS NO. 349.

Lancaster, K. 1971. Consumer demand: A new approach. New York: Columbia University Press.

Lucas, R.E.B. 1975. Hedonic price functions. Economic Inquiry. Journal of the Western Economic Association 13: 157-78.

Maclennan, D. 1977. Some thoughts on the nature and purpose of house price studies. Urban Studies 14: 59-71.

Mark, J. 1977. Determinants of urban house prices: A methodological comment. Urban Studies 14: 359-63.

New York Times. 1982. Delinquent mortgages up. March 30.

Olin, H.B. 1981. Paper presented at the 1981 Graduate Seminar for Supervisory Executives, U.S. League of Savings Associations, University of Illinois, Urbana, IL.

Ragas, W.R., and Wyatt II, J.M. 1978. Residential energy capitalization techniques and single family underwriting. The Real Estate Appraiser and Analyst January-February: 13-17.

Rosenfield, A., and Wagner, B.S. 1982. Technical issues for building energy use labels. Paper presented at a conference sponsored by the American Council for an Energy-Efficient Economy, Santa Cruz, CA.

Rothchild, J. 1981. Stop burning your money: The intelligent homeowner's guide to household energy savings. New York: Random House.

Schelhardt, T.D. 1982. First-time homebuyers plunged in 1981 to 13.5% of all purchasers, study says. Wall Street Journal June 15.

Schuck, L., and Millhone J. 1982. Defining energy efficiency in residential lending practices. Paper presented at a conference sponsored by the American Council for an Energy-Efficient Economy, Santa Cruz, CA.

Siegel, E. 1982. When tenants help pay the rent. Boston Globe August 27.

Society of Real Estate Appraisors. 1980. Energy considerations in real estate appraising.

Thrall, G.I. 1983. The proportion of household income devoted to mortgage payments: a model with supporting evidence. Annals of the Association of American Geographers 73: 220-30.

U.S. Bureau of the Census. 1980. Statistical abstract of the United States: 1980 (101st edition) Washington, D.C.: U.S. Government Printing Offices.

CHAPTER 18

RESIDENTIAL ENERGY CONSERVATION AMONG THE ELDERLY

Susan M. Macey
Department of Geography
Rutgers University
New Brunswick
New Jersey 08903 USA

ABSTRACT. This study analyzes the number and type of home heating energy conservation measures adopted by a sample of Lexington, Kentucky elderly to: determine the reasons certain measures are adopted; to analyse the extent of participation in public programs; and to suggest ways to assist the elderly in implementing energy conservation measures. The respondents are conservation conscious, but have special needs and constraints. Expense, concern for comfort, and perceived need act as major constraints in the adoption of conservation measures, while low participation in public programs is linked to lack of awareness of the programs and their potential benefits.

1. INTRODUCTION

In the United States, the proportion of the population aged 65 and over exceeds ten percent, and is projected to double over the next fifty years (Torrey 1979). At the same time the importance of conservation as a means of saving money for consumers, and decreasing the potential for a national energy shortage has been well documented (Olsen 1981; Mathews et al. 1982). While some studies have recognized age as an important factor affecting the ability to conserve (Newman and Day 1975; Rosson and Sweitzer 1981; Macey and Brown 1983), no work has specifically focussed on the elderly population and their problems in conserving energy. Yet this group is particularly vulnerable to increasing energy costs for several reasons.

First, the elderly population generally occupies older dwelling units that have less insulation and are more prone to loss of heat through infiltration (Smiley 1979; Rosson and Sweitzer 1981). Partly as a result of the reduction in the size of their activity space with retirement, they have reduced access to information, and rely more heavily on relatives for needed assistance (Rogers and Shoemaker 1971; Neugarten 1975). This coupled with a dramatic and continuing trend toward older persons maintaining separate households, rather than living

F. J. Calzonetti and B. D. Solomon (eds.), Geographical Dimensions of Energy, 353–371.
© 1985 by D. Reidel Publishing Company.

with relatives or in institutions (Neugarten 1975; Tobin et al. 1977), suggests a growing communication gap. Third, many elderly are on a fixed income (Newman and Day 1975). They comprise about twenty percent of the population with incomes below the poverty level (Wiseman 1978). Therefore they feel the impact of energy price increases more than the rest of the population (McConnel and Deljavan 1983). Concurrently, they may be unable to afford to install or improve energy conservation measures which would reduce their heating cost. Furthermore, several studies identify the poor and elderly as the two groups in society least able to lower fuel bills by reducing consumption, due to their already low energy consumption level (Office of Technology Assessment 1979; Fritzsche 1981; Van Raaij and Verhallen 1983). Finally, many older people have a reduced physical capacity compared with the remainder of the population. Nearly twenty percent of people 65 years old and over have health problems that limit their mobility, and over 75 percent suffer one or more chronic health conditions (Wiseman 1978). Thus little or no cost home energy conservation measures such as caulking or turning down the thermostat may often be ruled out as viable alternatives for many members of this sector of society.

Therefore, it is important to understand the special preferences and contraints which affect the energy conservation behavior of the elderly population. Geographers have contributed much to an understanding of the elderly population (Warnes 1982; Rowles and Ohta 1984). Similarly, within the behavioral geography tradition, much work has been done on preferences, constraints, and the adoption process (Cox and Golledge 1981). This work has provided insights into the variables that affect a wide variety of behaviors including migration, travel, and more recently, energy conservation (Pipkin 1979; Smith et al. 1979; Brown and Macey 1983). The primary goal of this study is to analyze the number and type of home heating energy conservation measures adopted by the elderly population, and their reasons for adoption/nonadoption. A second goal is to investigate the extent of elderly people's participation in energy-related public programs with a view to proposing changes that would assist the elderly in their energy conservation behavior.

2. RESEARCH DESIGN

The population of interest is restricted to 60 years or older residents of Lexington-Fayette Urban County, Kentucky. The sample was drawn from the pool of 900 Older Research Volunteers maintained by the University of Kentucky's Multidisciplinary Center of Gerontology. The pool was screened to include only residents of self-contained units (that is apartments or houses), and to exclude residents of rooming houses, rest homes or hospitals. A sample of 350 (approximately half of those eligible) was ramdomly drawn.

Since the prime objectives of the study are focussed around the decision making of the elderly, a questionnaire forms the main data base. This questionnaire sought information on knowledge and use of a range of home heating conservation measures and public programs; perceived advantages and disadvantages of three selected measures varying in frequency, cost and degree of personal expertise involved; information source usage; and data on dwelling unit characteristics such as size, age and building material.

Many of the questions in the questionnaire were left open-ended to elicit the specific beliefs and attitudes which underly adoption or nonadoption by the elderly population. For each consenting respondent, the questionnaire was supplemented by demographic and socioeconomic data (including length of residence, income, education, and employment status) filed with the University of Kentucky's Center of Gerontology.

Subjects were mailed the questionnaire and a prepaid return envelope in June, 1983. Given the great difference in housing characteristics, and level of responsibility for implementing energy conservation measures, it was hoped the sample would allow for a comparison between owners and renters. However, of the 224 completed questionnaires returned, only 27 were renters. This number was deemed too small for meaningful statistical analysis. Therefore, the results are based only on the 197 questionnaires completed by home owners.

The data thus obtained were computer analysed in two stages. First, descriptive statistics were generated to provide a background on respondents and their energy conservation practices, and to check that assumptions were met for the further statistical procedures to be used. Second, simple correlational analyses tested relationships between adoption of conservation measures and such variables as knowledge, information use, income, education, dwelling unit characteristics, and beliefs concerning adoption. Unless otherwise stated, all significance values are at the .05 level or better.

3. RESULTS

3.1. Demographic Background

Overall, the sample has a lower income, less education, and occupies older housing than the county as a whole (Table 1). This conforms to the general picture of the elderly as a group in society (Wiseman 1978; Greenberg 1982). Looking at the characteristics of the sample in more depth confirms the traditional portrait of elderly people in the 'empty nest' and 'sole survivor' life cycle stages (Table 2); that is, having a large proportion of single household heads (31 percent), while at the same time occupying a relatively large living space for the number of occupants. However, when compared with the results from other studies, the sample is atypical in its relatively high mean self-rating of health (Greenberg 1982), dwelling unit energy efficiency (Rosson and Sweitzer 1981), and the number of conservation measures adopted (Olsen 1978; Jackson 1980).

Table 1. Demographic and Dwelling Unit Characteristics

Characteristic		Sample (N=197)	Fayette County* (N=204,165)
Population	% 65 and over	100.0	8.6
	% male	43.3	51.8
	% female	56.7	48.2
Annual income	% under $15,000	51.2	34.9
	% $15,000 - $49,999	45.9	59.3
	% $50,000 and over	2.9	5.8
Education	% less than 12 years	16.4	2.8
	% high school graduate	59.8	71.6
	% college graduate	23.4	25.6
Race	% white	96.3	85.5
	% other	3.7	14.5
Housing	% built: 1970 - 1983	3.6	31.0
	1960 - 1969	17.2	27.4
	before 1960	79.2	41.6

*Source: U.S. Department of Commerce (1982)

3.2. Conservation Level

All respondents had adopted at least one energy conservation measure, with over 95 percent of the sample reporting the adoption of at least four measures. The most measures adopted was twelve (one respondent), while the mean number adopted was seven. This result contrasts with the observations of other studies (Olsen 1978; Jackson 1980) to the extent that they characterize the elderly as knowing of few conservation measures and conserving little energy. However, it is necessary to weigh both efforts to conserve and amount of energy conserved. Depending on the type of energy conservation measure adopted, the latter may far outweigh the effort required, or vice versa.

As can be seen in Table 3, the type of measures adopted varied greatly in terms of expense, difficulty of installation or adoption, innovativeness, and level or frequency of respondent action required. Low or no cost measures were the most widely adopted, and expensive or innovative measures the least adopted. Using storm/thermal windows, or plastic over windows in winter was the most frequent energy conservation measure (reported by 88.8 percent of respondents). Over 80 percent of dwellings in the sample were fully equipped with storm/thermal windows. For these respondents, adoption of this measure would only involve replacing a window or labor costs if it was necessary to pay someone to install them each season. The second most frequently adopted measure (accounting for 86.3 percent of the sample) was wearing warmer clothing at home in winter. It is also perhaps the easiest measure to adopt, requiring no additional expense or development of a special routine. Having the furnace serviced, and changing the furnace filter at least

Table 2 . Socioeconomic, Demographic and Dwelling Unit Variables

Variable	Mean or Mode	Standard Deviation or Range
Socioeconomic/Demographic		
Respondent's age (in years)	71.2	5.8
" education (in years)	11.9	1.9
" sex (male=1)	.4	.5
Single head (yes=1)	.3	.5
Length of residence (in years)	19.7	11.0
Health[a]	1.8	.7
Income[b]	5.0	1.0-7.0
Dwelling unit		
Main fuel (gas=1)	1.0	.2
Highest utility bill ($)	120.77	56.31
Lowest utility bill ($)	33.32	29.40
Budget billing (yes=1)	.5	.5
Area (square feet)	1749.5	734.9
Age (in years)	34.1	18.2
Building material (brick=1)	.8	.4
Number of energy conservation measures adopted	7.0	2.2
Efficiency rating[c]	.7	1.2

[a]Health is measured on a four-point scale ranging from "excellent" (1), to "poor" (4).

[b]Income is multichotomous, so its average and dispersion are measured by the mode and range.

[c]Efficiency is measured on a five-point scale ranging from "very inefficient" (-2), to "very efficient" (+2).

twice per year were measures adopted by approximately 80 percent of all respondents. Both represent widely-known, easy to adopt behaviors. While the former involves more expense, both options are important in maintaining furnace efficiency and reducing maintenance costs.

Three other measures adopted by over half of all respondents were closing off unused rooms, setting the thermostat to 65 degrees Fahrenheit at night in winter, and using an electric blanket (Table 3). For the majority of the population, these measures should require little effort to adopt. Closing off unused rooms and turning down the thermostat involve no monetary expense. However, depending on the size, design, and space utilization of the dwelling unit, the former may not be a viable alternative for some people. In the case of turning down the thermostat, there is another special consideration for the elderly: their health. While it has been shown that the same criteria for comfort apply to all age levels (Rohles and Johnson 1972), there is a growing concern over accidental hypothermia (that is, a sudden drop in deep body

Table 3. Measure Adoption Rates

Measure	Number of Adopters	Percent of Total
Using storm/thermal windows or plastic on at least 75 percent of windows . .	175	88.8
Wear warmer clothing at home in winter . .	170	86.3
Furnace serviced at least once per year .	163	82.7
Changing furnace filters at least twice per year	156	79.2
Installed/improved caulking	153	77.7
Closed off unused rooms	127	64.5
Winter nighttime thermostat setback to 65°F or below	119	60.4
Used an electric blanket	107	54.3
Added attic/ceiling insulation	98	49.7
Installed/improved weatherstripping . . .	80	40.6
Put up heavy/insulated drapes	64	32.5
Added crawl space/basement insulation . .	23	11.7
Added wall insulation	16	8.1
Installed a wood stove	14	7.1
Caulk any of the exterior of the house at least once per year	13	6.6
Installed an automatic setback thermostat	13	6.6
Installed a heat pump	3	1.5

temperature which may be triggered by even mildly cool temperatures of 60 to 65 degrees Fahrenheit, particularly in those 75 years of age and older). Hypothermia is estimated to be the sixth leading cause of death among the elderly population in the United States. Therefore "the conflict between the need to preserve health and the requirement to conserve energy is more sharply drawn for the elderly than for any other group in our country" (Rohles 1981, p 146).

The importance of adopting measures in combination (for example, wearing warmer clothing, turning down the thermostat, and using an electric blanket) must be stressed. The correlation between the latter two measures was significant at the .05 level, suggesting respondents are using energy conservation measures in combination. However, the moderate adoption rate for these three measures shows room for improvement. Those households not currently using a 65 degree nighttime setting should be made aware of the advantages of adopting complementary conservation measures (Office of Technology Assessment 1979; 1980).

Adding attic/ceiling insulation is a relatively expensive, one-time measure. Forty-five percent of respondents moved into homes with some attic/ceiling insulation in place, and had not added more such insulation. Twenty-five percent had added to the existing insulation, and an equal number had added all attic/ceiling insulation present in the home (Table 3). However, the respondent may have added the insulation a considerable time ago - the average length of residence in the current dwelling unit was 19.7 years. Consequently, the adoption of this measure need not be a reflection of present ability to pay for energy conservation measures. Finally, only seven respondents reported having no insulation.

The adoption of all these measures is in keeping with respondents' environmental orientation (that is, predisposition towards matters affecting the environment). Many expressed concern about the impact of wasting energy, and wanted to do something about it (Table 4). There

Table 4 . Environmental Orientation Measures*

Measure	Mean	Standard Deviation
I am very concerned with the impact of wasting energy (N=178)	1.48	.75
I enjoy doing my own home improvements and repairs (N=164)	.70	1.30
I sometimes spend money now to save money in the future (N=172)	1.28	.85
I believe that I can reduce the amount of energy I need to use (N=170)	.58	1.20
I believe that the energy situation has gotten worse in the past year (N=171)	.42	1.34
Technical gadgets interest me (N=158) . . .	.17	1.30
It's better to stick by what you know than to be trying new things (N=172) . . .	.01	1.35
I am generally concerned about the energy situation and want to do something about it (N=174)	1.15	1.00

*Variables are measured on a five-point scale ranging from "strongly disagree" (-2), to "strongly agree" (+2).

was also a general willingness expressed to spend money now in order to save money in the future.

The adoption rate for weatherstripping was approximately half that of caulking, though the two measures are comparable in the work and expense involved (Table 3). Further questioning about the frequency of caulking revealed that only a small number of respondents (6.6 percent) caulked any of the exterior of the house at a one year interval or less, the frequency commonly recommended. Infiltration is a major source of heat loss, especially in older homes (Steadman 1975; Socolow 1978; Penz and Yasky 1979). With the average age of dwelling units in the sample being 34 years, the low frequency of weatherstripping and caulking would seem to indicate an area for potential energy saving.

Measures requiring a high initial monetary outlay with a relatively long payback period, such as adding wall, crawl space or basement insulation, installing a wood stove, and installing a heat pump, were adopted by relatively few respondents (Table 3). Similarly, the automatic setback (or clock) thermostat, a relatively new device, was adopted by only a small proportion of respondents. While most respondents enjoyed doing their own home repairs, their overall interest in "technical gadgets" was much lower (Table 4). This, in combination with the cost involved, may account for their low adoption rate.

3.3. Adoption Reasons

To establish the underlying reasons for adoption/nonadoption, three conservation measures were chosen for more detailed analysis. They include one major relatively expensive measure (attic or ceiling insulation), one relatively inexpensive measure (caulking the exterior of the house once a year), and a no cost measure (turning the thermostat down to 65 degrees Fahrenheit or below at night in winter). These measures also vary in their required frequency (respectively, once or twice during residence, approximately once a year, to daily in winter), personal action and the degree of technical information involved (highest for installing insulation, to little or no expertise required in turning down the thermostat).

Reasons for adding/not adding insulation are outlined in Table 5. The major reasons for installation were perceived need, and the ensuing savings in energy and money. Those not adopting this measure cited difficulty of installation, expense, or considered it unnecessary because the present level of home energy efficiency was deemed sufficient. Given that only 3.6 percent of the houses in the sample were built after 1970, it is unlikely that the vast majority of respondents started with insulation levels meeting the United States minimum recommended level (U.S. Department of Agriculture 1978) and therefore would not require more. The operative difference between adopters and nonadopters appears to be their different perceptions of energy and monetary saving potential and the benefits of higher insulation levels. While an economic barrier to adoption does exist for some, the main barrier (perceived need) is one that should be amenable to change through the input of information.

The second measure chosen for more detailed study, setting the thermostat to 65 degrees Fahrenheit or below at night in winter is a well-known no cost measure. It also requires daily human action. Saving energy and money, perceived need, and comfort figured prominently in the decision to adopt (Table 5). However, the major difference between adopters and nonadopters is in their desired comfort level. More than half of the nonadopters said this setting was below their comfort level. Twenty respondents were concerned about the time it took to reheat the house in the morning, and eight cited health problems as their reason for using a higher thermostat setting. Thus, while health and health related concerns act as a barrier to adoption for a few people, comfort is by far the major obstacle to adoption of a lower thermostat setting

Table 5. Reasons for Adoption/Nonadoption of Selected Measures

Reasons	Number (Percent of Total)	Percent of Adopters	Percent of Nonadopters	X^2
Attic/ceiling Insulation (N=197, 98, 99)				
Need	69 (35.0)	42.9	27.3	4.6*
Comfort	9 (4.6)	9.2	0.0	--
Save money/expense	37 (18.8)	26.5	11.1	6.7**
Save energy	54 (27.4)	51.0	4.0	52.3***
Investment/payback	6 (3.0)	1.0	5.1	--
Easy/hard to install	21 (10.7)	9.2	12.1	n.s.
Health	2 (1.0)	1.0	1.0	--
Lack knowledge	6 (3.0)	2.0	4.0	--
Other	11 (5.6)	6.1	5.1	n.s.
Nighttime Setback (N=197, 119, 78)				
Need	47 (23.9)	32.8	10.3	11.9***
Comfort	72 (36.5)	24.4	55.1	17.9***
Save money	28 (14.2)	19.3	6.4	5.4**
Save energy	69 (35.0)	37.8	30.8	n.s.
Use electric blanket	19 (9.6)	10.9	7.7	n.s.
Too long to reheat	20 (10.2)	3.4	20.5	13.4***
Health	8 (4.1)	0.0	10.3	--
Other	12 (6.1)	6.7	5.1	--
Caulking (N=197, 13, 174)				
Need	132 (67.0)	23.1	70.1	--
Save money/expense	7 (3.6)	0.0	3.8	--
Save energy	26 (13.2)	38.5	11.4	--
Bother/no time	6 (3.0)	0.0	3.3	--
Difficult to install	13 (6.6)	0.0	7.1	--
Sealed house unhealthy	2 (1.0)	0.0	1.1	--
Part of painting	72 (36.5)	15.4	38.0	--
Other	3 (1.5)	23.1	0.0	--

*, **, and *** represent significance at the .05, .01 and .001 levels respectively.

-- number/cell size too small to allow computation

n.s. not significant

among the elderly population. The dissemination of information showing negative side effects, such as loss of comfort, can be overcome without undue effort or cost may help change negative attitudes towards turning down the thermostat. Further, misinformation which now acts as a barrier, such as the belief held by nearly a third of all nonadopters that turning down the thermostat uses more energy than maintaining a constant temperature, may be dispelled by the spread of accurate information.

Need and savings again figured prominently as reasons for adoption/nonadoption of caulking at a one year interval (Table 5). This common, relatively inexpensive measure requires more physical input than turning down the thermostat, but only at a one year frequency. While the number of adopters is too small to allow statistical comparison, it is clear that the major reasons for adoption are perceived need and energy savings. The length of the questionnaire precluded asking detailed questions on the type of caulking used, the frequency with which the caulking was checked, or if partial repairs were made at a more frequent interval. Any of these factors could explain the low adoption rate for caulking on an annual basis. However, it does appear that many elderly people are conscientious in using this conservation measure. Over a third of all respondents considered caulking a part of the regular maintenance or painting routine for the house. The frequency at which respondents caulk (every two years on average) may also reflect the quality of material used and who carries out the activity. Whereas the person mainly responsible for turning down the thermostat would be the respondent or his/her spouse, under half of the respondents would caulk personally or have it done by a spouse (Table 6). Difficulty or bother

Table 6 . Person Responsible for Carrying Out Selected Measures*

Measure	Percent of Total: Self and/or Spouse	Percent of Total: Other
Adding attic/ceiling insulation (N=159)	37.1	62.9
Nighttime setback (n=175)	97.8	2.2
Caulking the house exterior (N=172)	48.2	51.8

*Response to the question: "Who would most likely carry out the activity for your household?"

to install was mentioned by only ten percent of the sample as a reason for nonadoption, indicating that for the majority of respondents who need it, getting outside help is not an obstacle. However, nearly one-fifth (17 percent of those surveyed) never caulked the exterior of the house, so the potential exists for further savings through the use of this measure.

Of the socioeconomic and dwelling unit characteristics listed in Table 2 , only two were significantly related to adoption behavior. Respondents with lower utility bills had adopted more energy conservation measures, and more storm window coverage. These results contrast with those reported in studies of the general population. While findings on socioeconomic characteristics have not been consistent, some evidence has been found to support links between adoption of energy conservation measures, and such variables as education, income, and dwelling unit characteristics (Newman and Day 1975; Steadman 1975; Rosson and Sweitzer 1981).

Similarly, few significant relationships were found between information source usage and adoption of measures. While on average respondents received more information from utility companies and television (Table 7), these were not significantly related to

Table 7. Information Sources*

Source	Mean	Standard Deviation
Relatives (N=141).	1.72	.90
Neighbors (N=145).	1.65	.89
Other friends or acquaintances (N=143) . .	1.87	.89
Federal government (N=148)	1.97	.98
Kentucky Department of Energy (N=141). . .	1.79	.99
County Extension Service (N=142)	1.69	1.01
Columbia Gas Company (N=168)	2.80	.84
Kentucky Utilities Company (N=166)	2.79	.84
Newspaper/magazine articles (N=166) . . .	2.85	.80
Books or technical papers (N=140).	2.03	1.00
Television/radio advertisements (N=157) .	2.38	.90
Retail stores (N=136).	1.60	.80
Civic or social clubs (N=136).	1.40	.73
Consumer groups (N=138).	1.65	.92
Installers/manufacturers of energy conservation products (N=143)	2.06	1.02

*Response to the question: "For each of the following, please indicate whether you get a lot (4), some (3), a little (2), or no information (1) about ways to conserve energy".

adoption. A larger number of energy conservation measures was adopted by respondents with more interpersonal communications (that is, they received more information from friends and neighbors), as well as those with more information from newspapers, books, federal and consumer reports. The relative significance of personal communications and information from utility companies is the reverse of that found for the general population (Brunner and Bennett 1977; Craig and McCann 1978). Thus, campaigns using public service notices in newspapers and neighborhood programs would appear to be of more use in reaching the elderly population than pamphlets included with utility bills.

3.4. Additional Measures

Respondents were also asked what additional measures they would like to adopt and why, as well as their reasons for not adopting these measures.

Just over one third (63 respondents) wished to add to their conservation efforts. Those measures mentioned by five or more respondents are given in Table 8. Attic/ceiling insulation topped the list, followed

Table 8. Additional Conservation Measures Desired*

Measure	Number of Respondents	Percent of Total
Attic/ceiling insulation	21	10.6
Wall insulation	17	8.6
Storm/thermal windows	11	5.6
Automatic setback thermostat	5	2.5
Better furnace	5	2.5
Wood stove	5	2.5

*Response to the question: "Are there any energy conservation measures that you would like to use or install?"

closely by wall insulation, and improvement in storm or thermal window coverage. Other measures mentioned included various means of decreasing heat loss (that is, improving the thermal integrity of the dwelling), and a few specifically aimed at increasing efficiency of the heating system. Three respondents mentioned an interest in solar heating.

The overwhelming concerns of those wanting to conserve further were saving energy and saving money (Table 9). Thus, in spite of evidence

Table 9. Reasons for Wanting to Adopt Further Energy Conservation Measures and Reasons for Not Adopting

Reason	Want to adopt Number (% of total)	Nonadoption Number (% of total)
Comfort	13 (6.6)	0 (0.0)
Save money/expense	17 (8.6)	36 (18.3)
Save energy	40 (20.3)	1 (0.5)
Easy/difficulty to install	2 (1.0)	6 (3.0)
Need/improve efficiency	13 (6.6)	0 (0.0)
Uncertain of benefit	1 (0.5)	4 (2.0)
Other	10 (5.1)	20 (10.2)

that elderly people have relatively low energy consumption levels (Cunningham and Lopreato 1977; Office of Technology Assessment 1979; Fritzsche 1981), approximately one-third of those in the sample feel that they can increase their energy and monetary savings. Secondary concerns were for improving energy efficiency and comfort. These reasons echo the type of energy conservation measure respondents wish to adopt.

They are primarily geared to reducing home energy consumption through improving the heat holding capability of the structure.

Given the nature of the measures respondents wished to adopt, it is also understandable that expense was the major reason cited for nonadoption. This pattern of responses gives the impression that respondents feel they are doing all they can to conserve without resorting to relatively expensive measures. Concomitantly, elderly people (usually on fixed incomes) are unable to adopt the energy conservation measures they would like to because the high cost of these measures presents a major obstacle. Consequently, targeted public programs may be needed to assist the elderly population in adopting certain conservation technologies.

3.5. Public Programs

Five current public programs are considered here to determine level of participation by the elderly population, and how these programs might be improved to increase participation. Two programs, the Kentucky Cooperative Extension Service Analysis Program, and the federal Residential Conservation Service (RCS) audit provide information on conservation measures. The former is a free service in which the homeowner completes a short questionnaire concerning characteristics of his/her dwelling unit and mails it to the Extension Service where it is analysed, and literature on appropriate conservation measures is returned to the homeowner. The RCS Audit mandated by the 1978 National Energy Conservation Policy Act (U.S. Congress 1978a) requires all major gas and electric utilities to offer on-site home energy audits to residential customers. For a fee of $15, a trained representative of the utility inspects the home and provides an analysis of its energy efficiency along with information on energy conservation measures, their cost, effectiveness, and pay back period, available contractors, and federal tax credits. All customers in this study should have received a pamphlet on the RCS audit program (included with their utility bill) in the two months prior to the start of interviews.

The other three programs (federal tax credits, weatherization grants, and energy bill subsidies), involve financial aid, practical assistance with installation of energy conservation measures, or assistance with payment of heating bills. Federal tax credits, established by the Energy Tax Act of 1978 (U.S. Congress 1978b) are available for investments in specified energy conservation measures up to $2000 made to the taxpayer's principle residence between April, 1977 and January 1, 1986. A minimum expenditure of $67 is required to claim a credit on the individual's federal income tax return. Weatherization grants and energy bill subsidies are available only to low income households. These programs are generally administered by local nonprofit organizations. Weatherization grants may include material and labor assistance, as well as funds to make improvements such as caulking, weatherstripping, and putting plastic over windows in winter. Both programs are subject to the availability of federal funding.

Particularly in abnormally cold or prolonged winters, demand for program participation is often greater than funds permit.

A large majority of respondents (72.7 percent) had no knowledge of the Cooperative Extension Service Audit Program, and only three had used it (Table 10). These figures are only slightly lower than those for the

Table 10. Knowledge and Use of Public Programs

Program	No knowledge No. (%)	Knowledge only No. (%)	Knowledge and used No. (%)
Kentucky Cooperative Extension Service CHEAP[a] (N=187)	136 (72.7)	48 (25.7)	3 (1.6)
Residential Conservation Service audit[b] (N=187)	56 (29.9)	116 (62.0)	15 (8.0)
Federal tax credit (N=190)	25 (13.2)	106 (55.8)	59 (31.0)
Weatherization grant for low income families (N=175)	54 (30.8)	120 (68.6)	1 (0.6)
Energy bill subsidies for low income families (N=175)	48 (27.4)	122 (69.7)	5 (2.8)

[a]CHEAP - Computerized Home Energy Analysis Program

[b]The RCS program had just been initiated in the area at the time of interviewing. Therefore, the measure for this program is intention to use.

state as a whole (Arcury, et al. 1981). The program has not been widely publicized, and while the questionnaire is relatively easy to complete, obtaining it, filling it out, and returning it requires great motivation on the part of the individual.

Although all respondents should have received a mailing on the RCS audit from their utility company shortly prior to this study, nearly 30 percent stated they had no knowledge of the program (Table 10). However, the RCS audit was more widely known than the Extension Service Audit, and its intended use was much higher. Thus, while research indicates that poor and elderly people have low participation in such programs (Brown and Macey 1985), the intended participation rate here is higher than the five percent per annum rate expected for the population as a whole (Office of Technology Assessment 1980). The most frequent reason cited for intending to adopt the audit was to save money, with saving energy second in importance. However, the overwhelming response by nonadopters indicates that a large number see no benefit in having the audit done. A small percentage (13.6 percent) felt constrained by the expense (the $15 fee), and some respondents were distrustful of the information that would be provided.

While a higher percentage of respondents knew about the weatherization grants and energy bill subsidies, participation in these programs was minimal (Table 10). With over half of the respondents in the under $15,000 income bracket, and 30 percent earning under $10,000 per annum, greater participation would be expected. Some explanation for the low adoption of energy bill subsidies may be found in the fact that nearly half of all respondents (49.7 percent) take advantage of "budget billing". This is a system in which monthly utility bills are set equal to the yearly average. Thus the shock of high winter heating bills is absorbed over the course of the year.

Federal energy tax credits had the highest awareness and participation level of all five programs. Over 85 percent of the sample know of the tax credits, while 31 percent had used them (Table 10). As might be expected, with the tendency of this program to favor those in the upper income brackets, and the minimum expenditure ($67) required to make a claim, more expensive measures such as attic/ceiling insulation and storm/thermal windows accounted for nearly three-quarters of the claims. A further 20 percent of the claims were for wall or basement/crawl space insulation, and furnace modifications.

A significant positive correlation exists between income and knowledge of these programs, so awareness amongst those eligible appears to be one obstacle to adoption. The significant relationship between income and the use of tax credits found here supports evidence in other studies (Brown and Macey 1985) that this program is not encouraging of lower income group (and thus elderly population) participation, or the adoption of less expensive energy conservation measures.

Knowledge of tax credits and subsidies is also significantly related to education, further suggesting the importance of access to information as a limiting factor. None of the other socioeconomic or dwelling unit characteristics (Table 2) were significantly related to knowledge of, or participation in the five public programs. Information from federal sources, consumer reports, newspapers, and installers or manufacturers was significantly related to knowledge of tax credits, grants and subsidies. Actual participation was significantly linked to information source usage in only one case, a significant correlation between use of tax credits and information from federal sources. As suggested earlier, while the elderly receive more information from utility companies than from any other source, this information is not used by them. Again, public notices may be an effective means of heightening public program awareness and participation by the elderly population.

4. CONCLUSION

In contrast to the results of other studies (Olsen 1978; Jackson 1980), the findings here indicate elderly people are energy conscious and utilize several energy conservation measures. However, the type of measures they adopt are more likely to be low or no cost measures such as using storm/thermal windows or plastic over windows, wearing warmer clothing indoors in winter, and maintaining furnace efficiency through servicing and changing filters.

While comparable to the general population (Arcury et al. 1981; Macey and Brown 1983), the adoption rates found here suggest room exists for the elderly population to increase energy conservation through the further adoption of such low or no cost measures as weatherstripping, caulking, and turning down the thermostat. Adoption/nonadoption of these measures is linked to perceived need, comfort, and savings, reasons also given by the general population (Cunningham and Lopreato 1977; Morell 1981; Brown and Macey 1983). Therefore, programs aimed at increasing the adoption of these measures for the population at large should be appropriate for elderly people. Specifically, such programs should emphasize the benefits of these measures, especially saving energy and money. While difficulty of installation and negative health effects were mentioned by a few of respondents, they do not appear to present as major an obstacle, as the literature on the elderly population would suggest (Rohles 1981; McConnel and Deljavan 1983).

Expense of adopting the measures is a prominent reason for nonadoption by the elderly. This concern has not been found to be a major one for the general public (Cunningham and Lopreato 1977; Olsen 1981; Brown and Macey 1983). High cost measures such as adding insulation, or installing a wood stove, heat pump or automatic setback thermostat, were adopted by relatively few in the present study. Yet these measures topped the list of energy conservation options respondents would like to adopt. Thus, while a number of respondents felt these measures could save them energy and money, expense of installation clearly prevented more widespread adoption among the sample population.

The overall pattern of the elderly's energy conservation efforts suggests that they feel they are doing all they can to conserve without resorting to relatively expensive measures. Therefore, five strategies are suggested to increase energy conservation and participation in public programs by the elderly population. First, awareness of the benefits of low cost measures such as weatherstripping and caulking needs to be increased, as does awareness of relevant public programs. Second, negative beliefs concerning the energy saving potential and discomfort produced by turning down the thermostat need to be changed through the provision of accurate information on achievable savings and the possibility of using a combination of measures to achieve a desirable comfort level. These two strategies could be effected through government advertising campaigns specifically designed to reach the elderly. One form might be public notices in newspapers. Third, the benefits of participation in public programs should be stressed in any promotional campaign. Fourth, consideration should be given to reducing or eliminating the Residential Conservation Service Audit fee for low income and elderly households. Lastly, consideration should be given to initiating a program aimed at making high cost energy conservation measures affordable for elderly people and other low income sectors of the overall population. This program might involve direct government subsidies, or the provision of free labor similar to that provided for the installation of measures such as weatherstripping and caulking under the federal weatherization grant program.

ACKNOWLEDGEMENTS

Support for this research has been provided by the University of Kentucky Research Committee and the University of Kentucky Multidisciplinary Center of Gerontology. Appreciation is expressed to the editors and an anonymous referee for helpful comments.

REFERENCES

Arcury, T.A., F.M. Cole, D.G. Colliver, and S.S. Holland. 1981. Energy conservation awareness and action: A survey of Kentucky. Lexington, KY: Survey Research Center of the University of Kentucky. Technical Report No. 1.

Brown, M.A., and S.M. Macey. 1983. Understanding residential energy conservation through attitudes and beliefs. Environment and Planning A 15: 405-16.

______. 1985. Evaluating the impact of two energy conservation programmes in a midwestern city. Applied Geography 5: 39-53.

Brunner, J.A., and G.F. Bennett. 1977. Coping with the energy shortage: Perceptions and attitudes of metropolitan consumers. Journal of Environmental Systems 6: 253-68.

Cox, K.R., and G.R. Golledge eds. 1981. Behavioral problems in geography revisited. New York: Methunen.

Craig, C.S., and J.M. McCann. 1978. Assessing communication effects on energy conservation. Journal of Consumer Research 5: 82-88.

Cunningham, W.H., and S. Lopreato. 1977. Energy use and conservation incentives: A study of the Southwestern United States. New York: Praeger.

Fritzsche, D.J. 1981. An analysis of energy consumption patterns by stage of family life cycle. Journal of Marketing Research 18: 227-32.

Greenberg, L. 1982. The implications of an aging population for land-use planning. In Geographical perspectives on the elderly ed. A.M. Warner, pp. 405-425. Chichester: John Wiley and Sons.

Jackson, E.L. 1980. Perceptions of energy problems and the adoption of conservation practices in Edmonton and Calgary. The Canadian Geographer 24: 114-30.

Macey, S.M., and M.A. Brown. 1983. Residential energy conservation through repetitive household behavior: The role of past experience. Environment and Behavior 15: 123-41.

Mathews, S.A., S.B. Fawcett, R.G. Winns, and R.M. Mathews. 1982. An empirical stategy for identifying residential energy conservation techniques. Journal of Environmental Systems 11: 295-306.

McConnel, C.E., and F. Deljavan. 1983. Consumption patterns of the retired household. Journal of Gerontology 38: 480-90.

Morell, D. 1981. Energy conservation and public policy: If it's such a good idea, why don't we do more of it. Journal of Social Issues 37: 8-30.

Neugarten, B.L. 1975. The future and the young-old. Gerontologist 15: 4-9.

Newman, D.K., and D. Day. 1975. The American energy consumer. Cambridge: Ballinger.

Office of Technology Assessment. 1979. Residential energy conservation. Vol. 1. Washington: U.S. Government Printing Office.

______. 1980. Conservation and solar energy programs of the Department of Energy: A critique. Washington: U.S. Government Printing Office.

Olsen, M.E. 1978. Public acceptance of energy conservation. In Energy policy in the United States: Social and behavioral dimensions ed. S Warkov, pp. 91-109. New York: Praeger.

______. 1981. Consumers' attitudes towards energy conservation. Journal of Social Issues 37: 108-31.

Penz, A.J. and Y. Yasky. 1979. Uncertainties in predicting energy consumption in houses. Energy Systems and Policy 3: 243-69.

Pipkin, J.S. 1979. Problems in the psychological modelling of revealed destination choice. In Philosophy in geography eds. S. Gale, and G. Olsen, pp. 309-328. Dordrecht, Holland: Reidel.

Rogers, B.M., and F.F. Shoemaker. 1971. Communication of innovation: A cross cultural approach. New York: Free Press.

Rohles, F.H. 1981. Thermal comfort and strategies for energy conservation. Journal of Social Issues 37: 132-49.

Rohles, F.H., and M.A. Johnson. 1972. Thermal comfort in the elderly. ASHRAE Transactions 78: 131-37.

Rowles, G.D. and R.J. Ohta eds. 1983. Aging and milieu: Environmental perspectives on growing old. New York: Academic Press.

PART V

Multiregional and Environmental Issues

CHAPTER 19

ENERGY-ECONOMIC MEASURES FOR SELECTED ECONOMIES OF THE WORLD, 1960-81

James Zucchetto
University of Pennsylvania
Department of Regional Science
Philadelphia, PA 19104

ABSTRACT. Several analyses relating energy to gross domestic product for selected economies of the world are presented. Not only are individual economies presented but also results for aggregated market, OECD, planned and regional economies. Some estimates are also made for the indirect energy support of economies embodied in trade. The energy to GDP ratio has declined substantially for many economies in the late 1970's. If one considers embodied energy of trade then the energy efficiency of different economies takes on a reordering in comparison to a consideration of only direct consumption within an economy.

1. INTRODUCTION

Few will dispute the importance of energy to the functioning of human societies and economies. The energy disruptions and price increases that occurred in the 1970's stimulated a huge literature on the subject of energy and human activity. Much of this work focussed on quantifying the use of fuels and electricity in various activities in human economies as well as determining the connection between energy consumption and economic value generated. For example, for a given economy, the total output of goods and services in monetary units and total energy consumption can be determined. The ratio of energy consumption to gross economic output gives a quantitative energy basis to production in economic systems. In this view, the economy is a kind of engine with an input of energy and an output of goods and services (Barnett 1950; Cook 1971; Darmstadter et al. 1977). For national economies, either gross national product (GNP) or gross domestic product (GDP) is used as a measure of economic output. Analysis of an aggregated national economy obviously disguises variations within the system. It is also possible to examine energy/economic relationships at the level of sectors or industries within an economy: great variation across activities will be found (Herendeen and Bullard 1974; Zucchetto and Jansson 1979).

F. J. Calzonetti and B. D. Solomon (eds.), Geographical Dimensions of Energy, 375–391.
© 1985 by D. Reidel Publishing Company.

The macromeasure of the ratio of energy consumption (E) to GDP for a national economy can be interpreted in several ways. It can be thought of as a measure of efficiency or the number of units of energy required to produce a unit of economic output: lower ratios imply higher efficiencies. This argument has been used for comparative analysis among national economies (Schipper and Lichtenberg 1976; Darmstadter et al. 1977; Zucchetto and Walker 1981). However, such studies are fraught with ambiguity because of the differences that exist across countries. Climate, geography, spatial organization, technology, prices, culture, human behavior patterns and economic structure are only some of the variables which must be corrected for in order to interpret the ratio, E/GDP, as a measure of efficiency. These difficulties are eliminated to a large extent if only a given economy is considered over time. Monitoring an E/GDP ratio over time for a given economy might indicate trends in energy efficiency. However, changes in this ratio might also reflect changes in the structure of an economy: for example, energy intensive industries may be locating in other countries.

This last point elicits the notion of the openness of economies. It is rather arbitrary to consider the political boundaries of a nation state as suitable for defining the dependence of a given country on energy. If a country imports steel then it is also dependent on energy consumed in the steel exporting country. It then becomes important to measure the embodied energy of goods and services which is the total direct and indirect energy required to produce a unit of economic output (Costanza 1980). From the point of view of energy/economic relationships for national economies one's view would need to become more global in scope. In addition, although most studies have focussed on the use of fuels and electricity in economic activity, attempts have also been made to evaluate the work that the biosphere contributes to human economies (Odum 1971; 1983; Zucchetto 1975).

This paper takes a rather pragmatic and utilitarian approach to the issue of energy/economic relationships by presenting various kinds of analyses, especially over time, for national, regional and global economies. Explanations are not put forth, although suggestions are made, to explain the trends in E/GDP. Rather, the focus is on determining the quantitative relationship between energy consumption and GDP, and any changes which are occurring over time. This information is important for a number of reasons. Projections of E/GDP ratios will determine the extent to which increased levels of economic output will require corresponding increases in energy consumption. A central debate in the energy literature is the extent to which energy efficiency can be increased. Some argue that vast improvements are possible and thus there is no "iron law" between energy and economic production. These arguments have vast implications to energy and economic planning as well as the need for investment in energy production facilities. Such data as presented in the present paper contribute to the assessment of trends in energy/economic measures. Accurate projections of E/GDP measures, as well as energy supply, can help to determine realistically attainable levels of economic growth. Furthermore, such information is important to the assessment of future environmental problems since the consumption of energy generates a set of environmental impacts. The possibility of a carbon dioxide induced atmosphere depend on such variables as economic growth, energy/economic ratios and fuel mix, among other things. Thus, the present paper can provide some contribution to understanding the coupling of energy and economies.

2. DATA CONSIDERATIONS AND UTILIZATION

Several sources of data have been used in conducting these analyses. Earlier studies used GNP data for different economies converted to 1970 prices and dollars at prevailing exchange rates (Zucchetto and Walker 1981). This work has been updated with the availability of a consistent set of real GDP data corrected for the price structure prevailing in different economies, that is, corrected for purchasing power parity (PPP) (Summers _et al_. 1980). These data were used in several analyses of E/GDP relationships and were expressed in dollars at 1970 prices corrected for PPP (Zucchetto _et al_. 1982) and included the years from 1950-77. There were approximately 110 market economies considered for this data set not including the planned economies. Corrections for PPP turn out to be relatively small for industrialized economies but become significant when comparing countries with sizeable differences in levels of development. Selected results using these data are presented. A more recent data set of real GDP has been published with data expressed in dollars at 1975 prices corrected for PPP, including market and planned economies, from 1950-80 (Summers and Heston 1984). There has not been sufficient time to utilize all of these data but selected countries were chosen to extend the E/GDP analysis to 1981.[1] Energy consumption and population estimates were derived from United Nations and OECD statistical yearbooks and expressed in metric tons coal equivalent (1 gm. coal is ca. 7 kcal; 1 metric ton coal is ca. 7 million kcal).

Although it is recognized that there are inherent problems with much of these data, it is the only large scale set of data over time and countries available. The accuracy of the data is probably best for the industrialized and developed countries and deteriorates with the less developed economies. Statistics on the use of energy forms such as dung and firewood is problematic in many countries. In addition, the collection of data in less developed countries is not well supported and incomplete, although recent initiatives are attempting to upgrade energy data for less developed countries (Darmstadter, personal communication). The data on GDP by country is also problematic because corrections for PPP were accomplished for a set of 34 benchmark countries for 1975 and extrapolated to other countries and years. Thus, there are errors in these estimations. Furthermore, there is the uncertainty as to how large the so-called subterranean economy is in both developed and less developed countries, activity which may constitute a high percentage of the total, but it is not reported in the statistics. Quantitative studies of energy and economic activity should always be considered in light of these uncertainties.

3. CROSS-SECTIONAL REGRESSIONS OF GDP VS. ENERGY

The notion that economic production and energy consumption are intimately related was stimulated by graphs of energy consumption vs. gross national product across countries (Cook 1971; Odum 1971). In this type of graph, plotted for a given year, each point represents a different country with coordinates of energy consumption and economic output. It is generally observed that higher economic output corresponds to higher levels of energy consumption and, typically, linear or log-linear regressions are used to quantify this relationship. Such a graph is a snapshot of the world economy for any given year. If the independent variable is energy then representative points lying above the regression line represent greater than expected economic output: these countries could be thought of as more energy efficient. The slope of such a regression line gives some average indication as to increase in economic output as a country moves from lower to higher levels of energy consumption. On the other hand, if economic output is the independent variable, one could estimate the energy required for a given level of economic production. Cross-sectional regressions were conducted for the ca. 110 market economies and the OECD economies from 1960-80 to determine the extent to which this overall indicator, or relationship, has changed over time. The OECD economies were included in the 110 countries. They were also considered separately because of their more reliable data.

For the market economies, linear regressions (total and per capita) with energy consumption as independent variable and GDP as dependent variable were conducted. Regressions for total output generated correlation coefficients of 0.95 or higher; per capita regressions resulted in values of 0.8 or higher (Zucchetto _et al_. 1982). Figure 1A presents the slopes of these linear regressions. It remained relatively constant from 1960 to the late 1960's, declined into the early 1970's and increased through the 1970's. The data for 1978-80 are not strictly comparable to 1960-77 because of different base year prices and number of countries included. The per-capita results are more variable during the 1960's but the trend is upward in the 1970's.

Cross-sectional linear regressions for the OECD countries indicated correlation coefficients of ca. 0.97 for the total regressions and 0.72 for the per-capita. Figure 1B indicates that, for total GDP vs energy, the slope is relatively constant up until 1967, declines from 1967-71, and exhibits an upward trend through the 1970's. The per-capita results are less variable than for the market economies as a whole: some variability up till 1967, decline into the early 1970's, and a generally upward trend through the 1970's. The results are not, in themselves, indicative of changes in the E/GDP ratios among different economies.

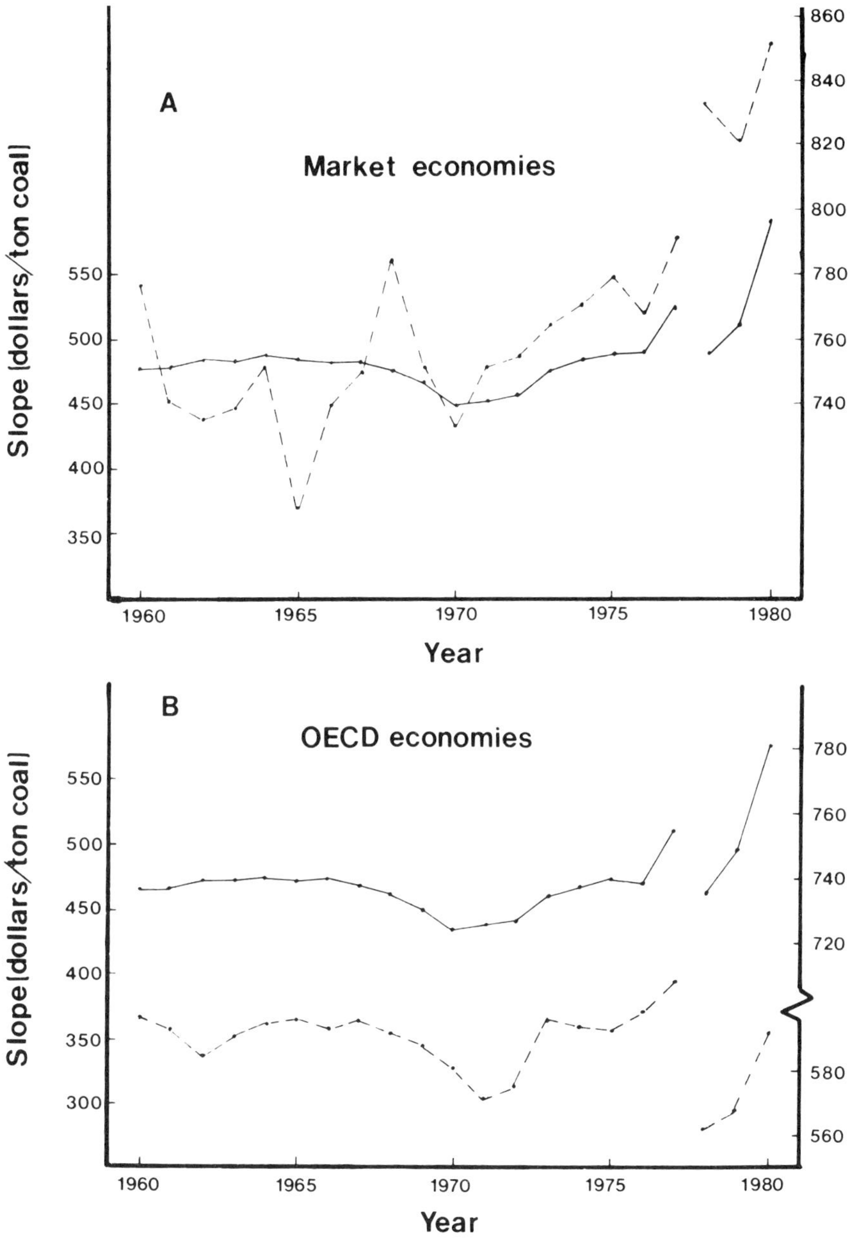

Figure 1. a) market economies; b) OECD countries.

4. ENERGY-GDP MEASURES FOR COUNTRIES AND REGIONS

In this section energy to GDP measures are reported for selected countries, at individual and aggregated levels, over the period 1960-81. With the most recent GDP data at 1975 prices corrected for PPP only the years 1971 to 1981 were chosen and only for the selected 34 benchmark countries. Figure 2 presents E/GDP ratios for the centrally planned economies and, as with other countries, there is great variation across countries ranging from 0.5 to 1.5 kg coal equivalent (C.E.) per dollar. The poorer countries tend to have lower ratios than the more developed countries. In general, there seems to be a downward tendency over this period, although there are fluctuations and somewhat of an upturn for the centrally planned economies as a whole from 1979-80. Figure 3 presents E/GDP ratios for selected countries in Europe with significant declines for the United Kingdom and Belgium but increases for Spain. In general, lower values prevail in comparison to the wealthier of the centrally planned economies, implying more energy efficient economies. Although not presented here, there is great variation in trends of E/GDP ratios over time for different countries: some exhibit rather constant

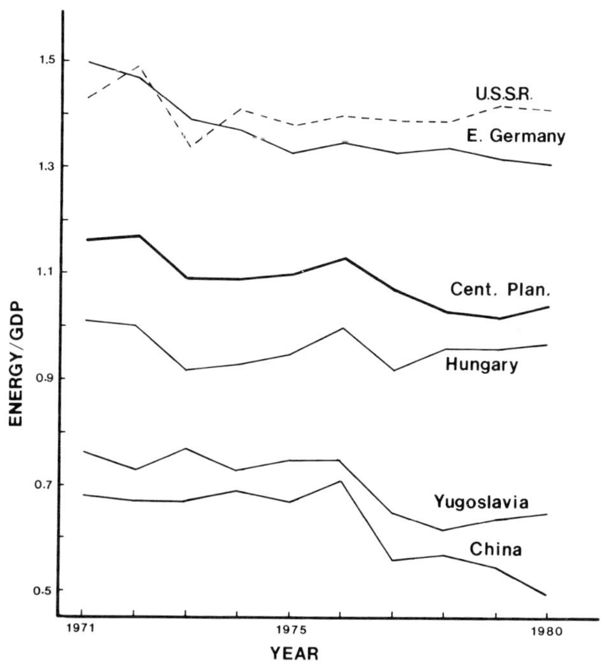

Figure 2. E/GDP ratios for centrally planned economies.

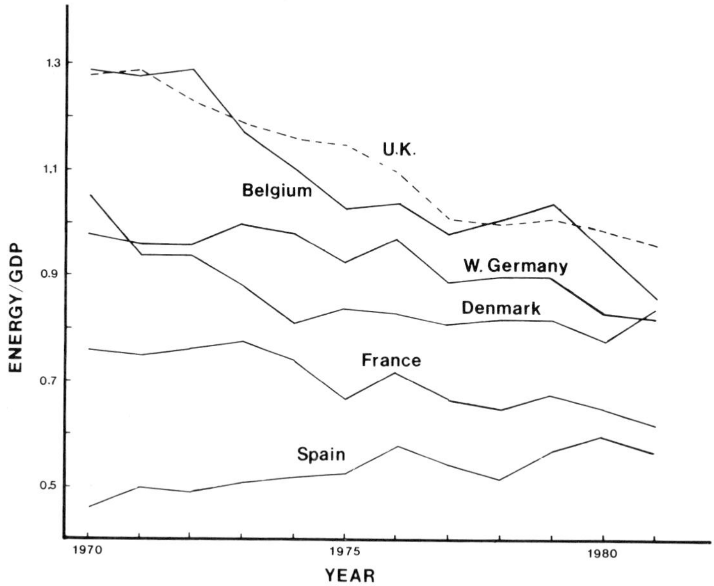

Figure 3. E/GDP ratios for European countries.

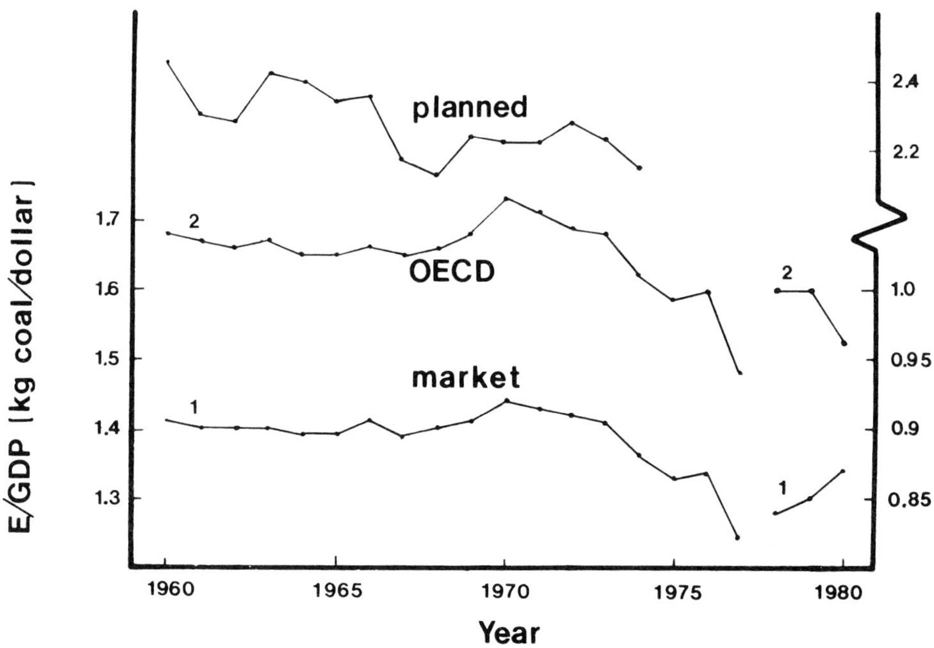

Figure 4. Comparison of E/GDP ratios.

behavior, others are declining while some are increasing. It is difficult to interpret these trends in terms of energy efficiency because of the existence of trade. For example, consider that heavy industry is leaving the U.S. and locating in other countries. The migration of such energy intensive activities would bias the U.S. E/GDP ratio downward and yet those products might still be imported from the country to which the industry moved. Thus, it might be more instructive to look at more aggregated global and regional economies to discern the E/GDP trends. Figure 4 and Table 1 present some results for different regions as well as the market, OECD and planned economies. There was a remarkably constant ratio for the market economies from 1960-69 of ca. 1.4 kg C.E./$ and for the OECD economies of 1.66 kg. C.E./$. These two groups track each other because the OECD economies are such an important component of the market economies. This ratio increases for both in 1969 to 1970 with a decline setting in through the 1970's and a sharp drop from 1976-77. The data for 1978-80 are not strictly comparable because of a smaller set of countries used for the market economy; however, the OECD data indicates downward movement from 1979-80. The data for the planned economies are not strictly comparable to the other groups because exchange rates were used (Yeh 1981); Figure 2 gives a better picture with respect to the planned economies, at least from 1971-80. Table 1 exhibits the differences among regions as well as corresponding trends from 1960-77.

Within country regressions were also performed, for different countries, of GDP (1970 prices corrected for PPP) against energy consumption using data from the years 1960-77. Linear regressions were used and results for selected OECD countries are presented in Table 2. It can be seen that the correlation coefficients are quite high for almost all countries. In addition, the slope could be interpreted as the amount of GDP to be produced for a unit expansion in energy consumption. For example, a 1 million mton expansion would produce 486.5 million dollars in the Netherlands but 1160.7 million dollars in Israel. Thus, this slope could serve as another energy-economic indicator.

5. ENERGY/GDP RATIOS AND LEVEL OF DEVELOPMENT

For both theoretical and practical reasons it is of interest to understand if economies change their energy characteristics during development and, in particular, to ascertain changes that might occur in energy efficiency during development. An exhaustive study of this subject has not been attempted but several results seem to imply that the E/GDP ratio increases with GDP/capita (Zucchetto <u>et al</u>. 1982). First, for each year from 1960 to 1976 for the 110 market economies, regressions of E/GDP vs GDP/capita showed positive correlations. Second, pooling the world data for 1974, 75 and 76 also showed positive correlations; however, the correlation coefficients for these analyses were only in the range of 0.3 to 0.6. Again, this type of macroscopic

TABLE 1

Energy/GDP Ratios for Regions of the World
(kg. Coal Equivalent per Dollar of GDP)+

REGION	1960	1961	1963	1965	1967	1968	1969	1970	1971	1972	1973	1974	1975	1976	1977
WORLD**	1.41	1.4	1.4	1.39	1.39	1.4	1.41	1.44	1.43	1.42	1.41	1.36	1.33	1.34	1.24
AFRICA	0.84	0.7	0.68	0.68	0.71	0.72	0.68	0.68	0.73	0.76	0.75	0.74	0.76	0.79	0.69
ASIA	0.65	0.73	0.72	0.74	0.78	0.79	0.81	0.84	0.82	0.79	0.86	0.88	0.85	0.84	0.74
EUROPE	1.5	1.23	1.28	1.22	1.24	1.25	1.26	1.3	1.29	1.28	1.34	1.24	1.21	1.24	1.15
N. AMERICA & CEN. AMERICA	2.05	2.04	1.97	2.04	2.04	2.07	2.12	2.2	2.19	2.17	2.07	2.04	2.01	2.02	1.89
OCEANIA	----------	?----	------	------	1.46	1.46	1.42	1.42	1.4	1.45	1.47	1.54	1.52	1.59	1.49
S. AMERICA	0.61	0.67	0.64	0.63	0.6	0.65	0.64	0.64	0.63	0.6	0.6	0.6	0.6	0.6	0.58
OECD	1.68	1.67	1.67	1.65	1.65	1.66	1.68	1.73	1.71	1.69	1.68	1.62	1.59	1.6	1.48

\+ GDP expressed at 1970 prices corrected for purchasing power parity.

** World means 110 countries with market economies.

TABLE 2

Selected Results of Within Country Regressions of GDP vs. Energy Consumption from 1960-1977.[a]

Country	Slope(m)	Y-intercept(b)	Correlation Coeff.
Australia	891.45	-22998.51	0.85
Belgium	774.06	-8336.35	0.81
Canada	443.9	1907.96	0.93
Finland	504.52	4576.04	0.96
France	992.04	-7972.91	0.97
Greece	902.17	5577.93	0.98
Ireland	700.53	-13.42	0.88
Israel	1160.69	-101.77	0.91
Italy	575.61	46571.62	0.98
Japan	935.37	11370.04	0.97
Netherlands	486.52	11477.16	0.97
New Zealand	671.58	3513.94	0.96
Norway	727.71	663.4	0.94
Portugal	1343.88	2201.31	0.97
South Africa	529.88	-4728.41	0.95
Spain	887.41	16131.26	0.96
Sweden	591.61	6580.1	0.91
Turkey	1284.5	8816.61	0.99
U.S.A.	435.06	58113.98	0.95
West Germany	754.41	-6759.76	0.97

[a]Equation regressed: GDP = m(E)

where E = units of million metric tons coal equivalent

GDP = units of million of dollars at 1970 prices corrected for PPP.

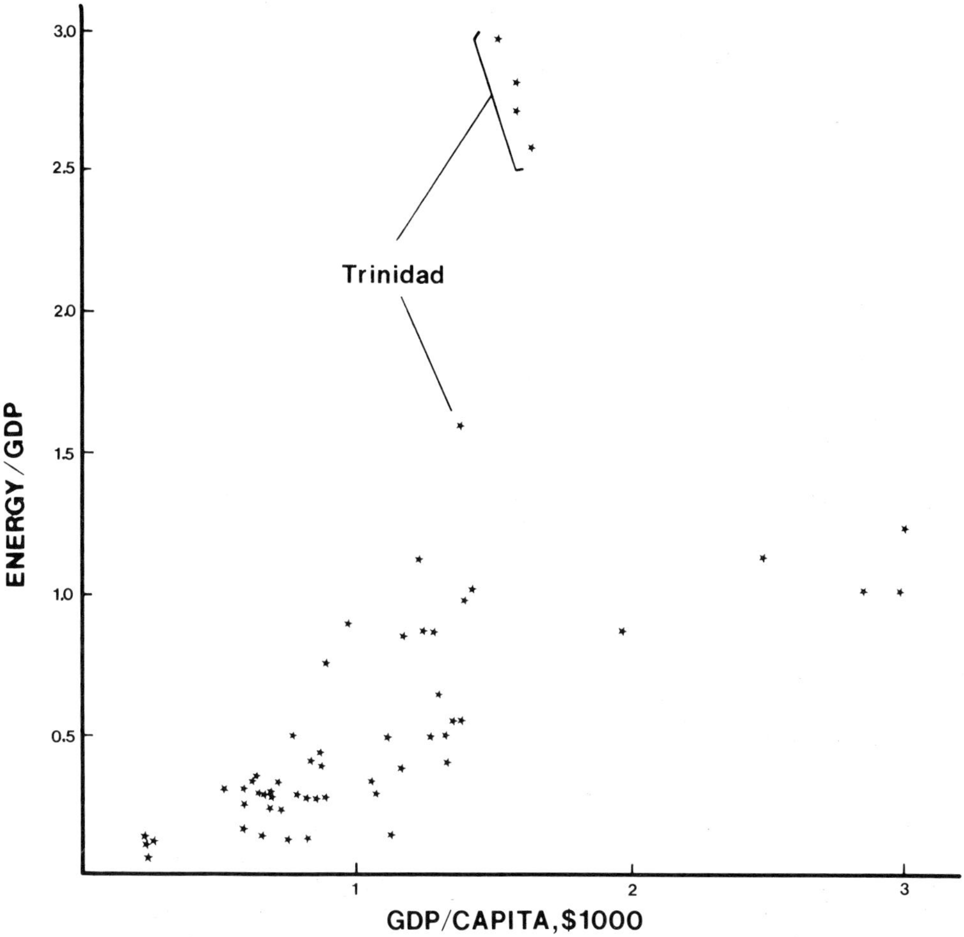

Figure 5. Tropical climate countries.

cross-sectional analysis does not correct for many differences, e.g., in climate. Thus, it could be argued that since many of the undeveloped, low GDP/capita countries are located in warmer climates, their low E/GDP ratios may be a consequence of low heating requirements. In order to correct for climate two groups of countries with two distinct climates were selected (one group tropical climate, the other dry and desert, see Figures 5 and 6) and pooled data for several years plotted. The countries in the tropical climate exhibit a positive correlation between E/GDP and GDP/capita (linear correlation coefficient = 0.3; if Trinidad outliers are left out R^2 = 0.7; Trinidad is an outlier because of much higher than average per capita levels of autos, industrial activity, cement production and nitrogen fertilizer output) as do the dry, desert climate countries (R^2 = 0.55). The points in Figures 5 and 6 are suggestive of logistic or limiting-factor curves with a limiting E/GDP ratio approached at higher levels of development, somewhere on the order of 1.0 kg. C.E. per dollar of GDP.

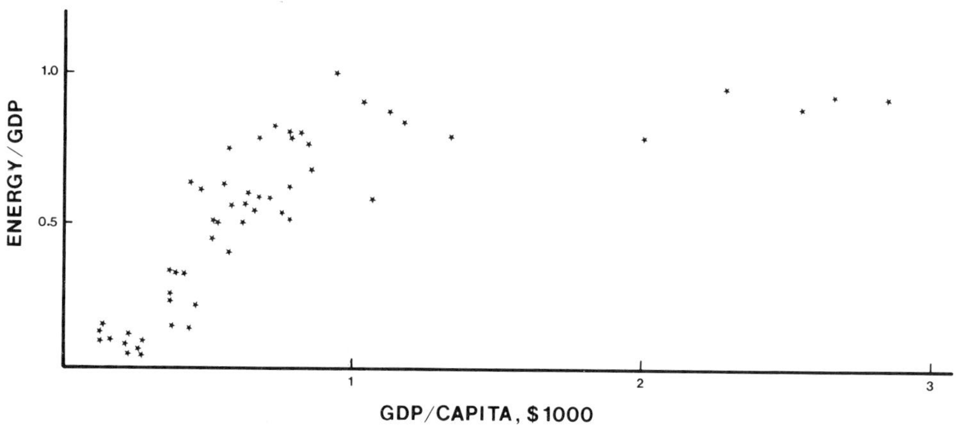

Figure 6. Desert countries.

6. EMBODIED ENERGY AND INTERNATIONAL INTERDEPENDENCE

As in this paper, most energy/economic studies of national economies have not considered the openness of national economies and the indirect energy dependence associated with international trade. A country dependent on imports not only requires energy consumed within its own borders but also within the country from which it obtains its imports.

TABLE 3

Energy Consumption and Energy/Economic Output Coefficients for 1970[a]

REGION	GDP in Million 1970 Dollars (X_i)	Direct Energy Consumption Million Tons Coal	Direct Energy -GDP, KG/$	1970* Embodied Energy Coefficients KG/$	% Change in Energy Coefficients
EUROPE	622900	1350.45	2.17	2.75	26.7%
S. AFRICA	17567	63.19	3.6	4.10	13.9%
CANADA	82890	201.07	2.43	2.78	14.4%
U.S.A	983237	2269.49	2.31	2.40	3.9%
JAPAN	196917	332.37	1.69	1.87	10.66%
AUST. & N.Z.	42900	75.21	1.75	2.04	16.57%
AFRICA	57833	45.73	0.79	1.30	64.56%
AMERICAS +	151300	255.15	1.69	1.98	17.16%
MID-ASIA	47100	85.01	1.8	2.16	20%
ASIA	134100	201.24	1.5	1.81	20.67%
OCEANIA	1300	2.34	1.8	3.08	71.1%
CENT. PLAN.++	858400	1948.21	2.27	2.36	3.97%

a GDP Converted to dollars using exchange rates

* $\underline{e}' = \underline{E}' (\hat{\underline{X}} - \underline{X})^{-1}$

** Not including S. Africa

\+ Not including Cuba

++ From Yea (1981) including Bulgaria, China, Cuba, Czechoslovakia, E. Germany, Hungary, Poland, Romania, U.S.S.R

A detailed energy analysis would involve sectoral or industry breakdown within each economy as well as intersectoral transfers among economies. In lieu of such information, and for illustrative purposes, consider an input-output scheme for a 12 region (sector) world (see Table 3). Although economies are not accurately represented in this way it gives some indication of this embodied energy idea. An energy balance for each region (economy) yields the following set of equations (Herendeen and Bullard 1974).

$$E_j + \sum_{i=1}^{n} e_i X_{ij} = e_j X_j \quad j = 1, 2, \ldots, 12.$$

$$\underline{E}^1 = \underline{e}^1 (\underline{\hat{X}} - \underline{X})$$

$$\text{or} \quad \underline{e}^1 = \underline{E}^1 (\underline{\hat{X}} - \underline{X})^{-1}$$

where $\underline{E}^1$ = row vector of direct energy use in economy "j". $[E_j]$

$\underline{X}$ = matrix of trade flows; X_{ij} = trade from i to j

$\underline{\hat{X}}$ = diagonal matrix of economic production for each economy

e^1 = row vector of embodied energy coefficients (direct plus indirect energy per unit of output). $[e_j]$

The calculation of e_j, the embodied energy coefficients, captures the indirect energy cost, throughout the global economic web, of production in a given economy. If an economy is heavily dependent on imports then its energy support is related to those economies which are connected to it either directly or indirectly. Calculations for 1970 (Table 3) show the difference between the embodied energy coefficients and the direct energy to GDP ratios for a 12 sector world. Note that a different ordering takes place. Some regions, such as Oceania, slip to lower efficiency because of their heavy dependence on imports. During the course of this study it was also hypothesized that as the world economy became more integrated through trade then there would be a convergence of the embodied energy coefficients. This convergence hypothesis suggests that the differences in energy-economic measures will become less important and, in fact, intercountry comparisons would lose some of their significance. To test this notion of convergence, analyses were conducted for 1963, 1970 and 1975 for an 11 sector world economy (centrally planned block excluded) and the dispersion about the mean for the E/GDP and embodied energy coefficients calculated. The results are only suggestive but there seems to be a trend towards less deviation about the mean values and thus a convergence of the embodied energy coefficients (Zucchetto, et al. 1982).

7. WORK CONTRIBUTIONS FROM THE ENVIRONMENT

A complete energy analysis of the work required for different economies should include the contributions from the natural environment. Odum (1983) summarizes many studies which have included such considerations but this field is still in its infancy. Odum and Odum (1983) have conducted a study of 12 countries in an attempt to assess the contributions of the biosphere to their economies. One interesting measure is the ratio of indigenous energies to total energy dependence, a measure of self-sufficiency of an economy (Zucchetto and Jansson 1979). Odum and Odum (1983) note, for example, the small dependence of West Germany on indigenous energy sources and the high self-sufficiency of economies like the Soviet Union or Australia. However, these results are extremely preliminary.

8. DISCUSSION

Better, more extensive and accurate data on economic activity and energy consumption will improve the assessment of trends in energy to economic relationships. As time goes on additional time-series data will resolve the issue of whether efficiency will continue to increase or decrease and how close the approach to thermodynamic limits has become. Recent declines in the E/GDP ratio are partially due to the recent world recession in the post-1979 years since this aggregated measure includes consumer consumption. During hard economic times consumers are forced to be very careful with their use of energy. Perhaps it is necessary for consumers to be somewhat "wasteful" in their use of energy which would stimulate demand leading to an expansionary period characterized by higher E/GDP ratios. Declines in E/GDP ratios have certainly come about because of increases in energy efficiency in industries as well as in the residential and transportation sectors. This improved use of energy has come about because of factors such as price increases, energy awareness and supply disruptions. However, declines in E/GDP ratios may have also income about because of neglect of capital and infrastructure both in the private and public sectors. These investments require activity in energy intensive sectors of an economy; their neglect means living off energy which was expended in an earlier time. The eventual necessity to replace or upgrade deteriorating capital and infrastructure may tend to push E/GDP ratios in upward directions.

The calculation of embodied energy coefficients should include solar energy, as well as fuels, in order to understand how the ecosystems of the various countries of the world contribute to the global economy. In the endeavor to try to develop detailed knowledge for a systems ecology of the globe the interdependence of economies can be extended to materials as well as environmental degradation. In fact, one would prefer to investigate the flows among economies in physical units to avoid the vagaries of exchange rates and monetary fluctuations. The long run goal would be a detailed scientific knowledge of how activity in one economy of the world affects others in terms of resources, environment and economics.

9. FOOTNOTES

[1]Data for 1981 supplied by R. Summers, Dept. of Economics, University of Pennsylvania.

10. REFERENCES

Barnett, H.J. 1950. Energy uses and supplies. U.S. Bureau of Mines Circular 7852 (Oct.).

Cook, E. 1971. The flow of energy in an industrial society. Scientific American 224: 135-144.

Costanza, R. 1980. Embodied energy and economic valuation. Science 210: 1219-1224.

Darmstadter, J., Dunkerley, J. and Alterman, J. 1977. How industrial societies use energy: A comparative analysis. Baltimore: John Hopkins University Press.

Darmstadter, J. Personal Communication, Resources for the Future, Washington, D.C.

Herendeen, R.A. and Bullard, D.W. 1974. Energy costs of goods and services, 1963 and 1967. Center for Advanced Computation Document No. 140, University of Illinois, Urbana-Champaign.

Odum, H.T. 1971. Environment, power and society, New York: John Wiley.

______. 1983. Systems ecology: An introduction. New York: Wiley Interscience.

Odum, H.T. and Odum, E.C. 1983. Energy analysis overview of nations. Working paper. Laxenburg, Austria: International Institute for Applied Systems Analysis.

Schipper, L. and Lichtenberg, A.J. 1976. Efficient energy use and wellbeing: The Swedish example. Science 197: 1001-1012.

Summers, R., Kravis, I.B. and Heston, A. 1980. International com parison of real product and its composition: 1950-1977. The Review of Income and Wealth, Series 26: 19-66.

Summers, R. and Heston, A. 1984. Improved international comparisons of real product and its composition: 1950-1980. Review of Income and Wealth, Forthcoming.

Yeh, B.S. 1981. A Global Economic Model of a Three-Region World, Ph.D. Dissertation, Department of Economics, University of Pennsylvania.

Zucchetto, J. 1975. Energy-economic theory and mathematical models for combining the systems of man and nature, case study: the urban region of Miami, Florida. Ecological Modelling 1: 241-268.

Zucchetto, J. and Jansson, A.M. 1979. Integrated regional energy analysis for the island of Gotland, Sweden. Vol. 11: 919-942.

Zucchetto, J. and Walker, R. 1981. Time-series analysis of international energy-economic relationships. In Energy and ecological modelling, eds., Mitsch, W.J., Bosserman, R.W., Klopatek, J.M. Amsterdam: Elsevier Scient. Publ. Co.

Zucchetto, J., Titus, E., Blanco, S., Graizbord, C. and Walker, R. 1982. Global energy/economic relationships. Working Paper No. 64, Department of Regional Science, University of Pennsylvania.

CHAPTER 20

REGIONAL DEVELOPMENT AS AN ENTROPIC PROCESS: A CANADIAN EXAMPLE

Stephen Lonergan
Department of Geography
McMaster University
Hamilton, Ontario, CANADA

ABSTRACT. The relationship between energy and regional economic growth will be explored by presenting data on regional and national embodied energy flows and relating this information to structural change within spatial systems. Characteristics of the regional energy economy in Canada, such as the energy intensity of various economic sectors and of respective Census Metropolitan Areas, will be highlighted, along with information relating to Canada's energy balance of trade. The contention is made that by using embodied energy flows, one can better understand past fluctuations in the dynamics of spatial systems and, in turn, should design policies that explicitly consider the role of energy in regional development.

1. INTRODUCTION

The spatial implications of the demand and supply of energy have been of considerable interest over the past decade and have stimulated a number of inquiries on the part of geographers and planners into the relationship between energy flow and regional development. Resource rich nations have always had a comparative advantage in terms of development, but accompanying the higher energy prices of the early 1970's was a not so subtle realignment of geopolitical power, or at best the perception of this power by the populous. Developed countries clamoured for "self-sufficiency" of supply irrespective of the higher costs and greater regional inequalities that would accompany such a policy (Wilbanks 1981). The search began for alternative sources of supply, stimulated by heavy investments in energy 'mega-projects', often in frontier areas and with little knowledge of the regional consequences of such investment (Lakshmanan and Johansson 1985). Extensive resource development projects in Canada, including offshore areas (the Arctic and the Atlantic Ocean), tar sands in Alberta and hydroelectric projects in Quebec, have resulted in major flows of capital into remote regions and extensive federal-provincial squabbles over income and ownership questions, but in the relative absence of any

F. J. Calzonetti and B. D. Solomon (eds.), Geographical Dimensions of Energy, 393–409.
© 1985 by D. Reidel Publishing Company.

concern about social or environmental impacts. The *ex post facto* environmental review process that is required in Canada does not have a legal base and has often been used merely to justify development decisions that have already been made. No energy supply project proposal in Canada has ever been denied or even restructured as a result of the environmental review procedure (Lonergan 1984). The goal of self-sufficiency is paramount.

Demand side issues have also been studied for their potential spatial implications. Higher energy prices and reductions in supply, it is generally believed, will eventuate a move towards population concentration (Ostro and Naroff 1980; Wardwell 1980). The dispersed settlement pattern indicative of North American communities can be characterized by an energy inefficient single family home; multi-family dwellings are more efficient because of less exterior wall exposure per unit and the decrease in surface to volume ratio as the size of the structure increases. Suburban living also involves greater distances for personal and work trips, normally by an energy inefficient mode of transportation-the automobile. The implication is that a more concentrated development pattern would facilitate the initiation and utilization of mass transit systems requiring less energy per passenger mile. The savings in space heating and transportation costs would provide ample incentive for the population to become increasingly concentrated.

This argument is strongly refuted by some researchers, based on two propositions. First the perceived benefits resulting from differences in energy consumption will be an insufficient incentive to encourage families to seek higher density living (Small 1980). Individual choices would appear to favor alternative means of adjustment (e.g., smaller cars, more home insulation) requiring less disruption in life-style. Second, those advocating such a shift have not adequately considered the total energy flow in an urban system. The most energy efficient type of human settlement may well be small, polynucleated cities (Odum 1973; Romanos 1978; Til 1979).

The stabilization of world energy prices, present 'excess' supply and government subsidization of conservation measures have, it seems, mitigated the concern over the role of energy in residential (and firm) locational choice behavior. A potentially more important question, however, needs to be addressed. What is the role of energy in the long term development of regions? The following discussion attempts to provide insights into the answer to this question by presenting general design principles for all systems and positing their application in the context of Canadian provincial energy flows.

2. GENERAL DESIGN PRINCIPLES FOR SYSTEM EVOLUTION

The first and second laws of thermodynamics have spawned a number of principles of general system design, including maximization of biomass, maximum stability and maximum efficiency. The following three, however, are of particular interest relative to spatial systems:

1. the theory of dissipative structures (Prigogine 1955; 1978);
2. the maximum power principle (Lotka 1922; Odum 1971; 1983);
3. the theory of minimax dissipation (Kirkaldy 1965; Lonergan 1983; Lonergan and Kirkaldy 1984).

Systems can be viewed as isolated (there is no exchange of matter or energy with the external world), closed (energy is exchanged, but not matter), or open (both matter and energy are exchanged). In isolated systems, entropy increases monotonically until a maximum is reached at thermodynamic equilibrium. Exchange of energy between systems prevents the attainment of this equilibrium, but closed and open systems may exhibit steady-state characteristics when the rate of energy acquisition is equal to the rate of dissipation. These "nonequilibrium" systems that exhibit order in certain circumstances have been termed dissipative structures (Nicolis and Prigogine 1977).

2.1. The Theory of Dissipative Structures

Nonequilibrium, self-organizing systems in the vicinity of equilibrium will tend to assume a steady state characterized by the rate of entropy production of the system reaching a minimum (Prigogine 1978). This state of least dissipation represents an inertial property of the system (Johnson 1981), implying that if a system is perturbed, and entropy production increases, it will tend to return to a state of least dissipation.

Urban and regional systems may also be viewed as dissipative structures, acquiring energy and matter from the outside and using these inputs most efficiently. Allen and Sanglier (1977) and Allen *et al.* (1981) have developed models of urban and regional systems that attempt to describe the self-organization of urban space. Social systems are interpreted as systems far from equilibrium. Their behavior cannot be reduced to deterministic description and is not guided by a general design principle; it is a result of stochastic elements and entirely structure specific. A failure in predisposed macroscopic behavior can result in multiple solutions and bifurcations through the randomness built into the system. The stochastic elements provide an uncertainty in social system evolution that renders future behavior completely unpredictable.

The exploration of urban and regional dynamics by Allen *et al.* (1981) is based on proposing simple nonlinear models and examining the behavior of these models under different conditions and flows. Their aim is "not to establish a realistic model," but merely to investigate possible long term structural change in social systems. Liossatos (1980) has also proposed viewing issues in spatial dynamics from the framework of dissipative structures. Drawing from Schumpeter's argument that capitalism basically involves a process of qualitative change through fluctuation, he presents the concepts of nonequilibrium thermodynamics as presented by Nicolis and Prigogine (1977) as a possible mathematical interpretation. The time evolution of behavior may obey criteria described by a simple differential equation, Liossatos notes, but this represents only a mean value for society. Deviations from this mean that can be assigned to individual actors provide fluctuations and conditions of instability. Qualitative change, therefore, may result from the indeterminate situations arising from a deterministic system.

The deterministic component of system behavior, according to Prigogine, is that systems near equilibrium will tend towards states of minimum entropy production while the behavior of nonequilibrium systems is subject to random processes and is entirely structure specific. Lotka (1922), however, asserted that evolution favored systems that maximize entropy production, described as an imperialistic tendency to which all systems conform, providing the basis for an alternative entropic theory of system evolution. The distinction between near equilibrium or far from equilibrium is, accordingly, unimportant.

2.2. The Maximum Power Principle

Odum (1971; 1983; Ostro and Odum 1976) contends that maximization of power is the basic design principle of self-organizing systems and should be adopted as a fourth law of thermodynamics. Systems that maximize power (which is equivalent to maximizing the rate of entropy production) and use this most effectively will survive in competition with other systems.[1] The principle of maximum power predates even Lotka; Boltzmann noted in 1886 that survival amounts to a struggle for available energy. The implications of the maximum power principle for regional and urban systems, however, have not been well developed. Zucchetto (1975) constructed a simulation model for the City of Miami, Florida, based on embodied energy flows (including infrastructure, trade and direct consumption) and posited that decreasing net energy flows over time resulted in the decline in the importance of the city relative to other regions in the state (prior to 1972). The decreasing ability of the city to capture energy (primarily through tourism), and maximize power, lead to its decline.

Brown (1981) has examined energy flows through the urban hierarchy by similar methods, but no explicit interpretation of these flows within the context of the maximum power principle is entertained. Providing empirical relevance to thermodynamic concepts, such as minimum entropy production and maximum power in the context of social systems, has posed a serious obstacle to further study by Odum's group.

2.3. The Theory of Minimax Dissipation

The maximum power principle and Prigogine's ideas on minimum entropy production seem to be competing theories, but it may be the case that elements of both are appropriate to systems behavior. Johnson (1981) found that ecosystems tend to a state of greatest entropy production while ecodemes (roughly, social groups) approach a state of least entropy production. Prigogine (1978) admits that far from equilibrium, systems with inherent minimum entropy production qualities may behave in the opposite manner. The second aspect of Lotka's principle posited efficiency of use (minimizing entropy production or dissipation), although this seems to have been ignored by Odum in his recent work.

The minimax principle claims that these two forces are present in all systems and tend to compete in promoting a steady state (Kirkaldy 1965). Within a framework of power maximization, systems tend to competitively minimize dissipation. The stable state, accordingly, can be characterized simultaneously as a minimal and a maximal configuration in the rate of loss of available energy. Lonergan and Kirkaldy (1984) have developed a bivariate optimization model of a predator-prey ecology to illustrate the existence of a minimax configuration, and applications of the minimax criterion to social systems have been proposed by Kirkaldy and Black (1972). Post-industrial society is evolving towards an objective of maximum consumption with minimum degradation. Developed societies have very high throughputs of energy, used to pay the real costs of society as well as to improve societal welfare. The minimax principle implies that this throughput should be maximized to ensure survival, but, coincidentally, pollution and wastes, forms of social entropy, should be minimized.

3. ENERGY FLOW MODELS AND THE MINIMAX

The need to assess social and economic activity as an entropic process has been recognized by Georgescu-Roegen (1971) and Miernyk (1982). Spatial economic activities should be viewed in the same light as natural systems where every activity is linked to every other activity. Instabilities and fluctuations that denote qualitative change in systems suggest the existence of stochastic elements in nature's 'grand design'; inertial properties tending systems towards steady state behavior, nevertheless, still exist. Prigogine's conception of systems tending towards an equilibrium state consisting of a minimization of entropy production may only consider part of systems dynamics; systems maximize energy flow and entropy production while competitively minimizing dissipation. Depicting the structural change of urban and regional systems in a similar vein is intuitively reasonable, but pragmatically difficult.

One appealing framework for considering regional dynamics as an entropic process is the 'ecoenergetic' approach (Gilliland 1978; Odum 1983). Energy flows through natural and physical systems have been extensively studied and the analogy to human systems is intuitively appealing. Using energy as a common metric provides not only the entropic framework desired, but a mechanism to link the economy and the environment for evaluation purposes as well. Such an approach is not without critics, particularly due to the use of quality factors and the treatment of labor inputs. Additionally, the inability to adequately define terms such as 'social entropy' or justify why spatial systems necessarily adhere to a minimax criterion underlie the difficulties in justifying such an approach. Studying energy flows through regional systems does however provide and interesting alternative framework for viewing spatial dynamics and, as shown above, may have a strong theoretical base at a general level of abstraction. The problem lies in linking the theory with what one observes; at this stage in their development the general theories are merely suggestive. The following sections present energy flow data for Canada, using the embodied energy measures dictated by the ecoenergetic approach, in an attempt to provide additional insights into regional entropic processes. The methodological framework posed is entirely descriptive. Simulation models that have been developed to mimic systems behavior and illustrate the principles discussed above are prevalent (Odum 1983; Lonergan and Kirkaldy 1984), but have little applicability in a planning context. The use of descriptive information for normative planning, however, is also only appropriate at a general level, but does provide an interesting context for further theoretical refinement.

3.1. Energy and Economic Growth

The link between fuel use and gross national product (GNP) has been well studied in the past two decades (e.g., Schurr and Netscheat et al. 1960). Conservation Foundation (1978) and Cleveland et al. (1984) most recently have shown evidence of a strong correlation between GNP and fuel use as well as between annual rates of change of each. Zucchetto's chapter in this volume discusses trends in energy/GNP ratios worldwide, although Lonergan et al. (1984) noted that international comparisons of specific energy/GNP ratios are of little use due to the variability in initial conditions (e.g., industrial mix, climate and the spatial distribution of the population). Energy/GNP ratios can be useful, however, as indicators of the efficiency of fuel use, in noting the effect of structural change in regions on the energy economy, and as a rough measure of the evolutionary path of a region relative to the theoretical postulates presented previously.

Figure 1 presents energy to gross domestic product (GDP) ratios between 1964 and 1982 in the Canadian economy for Canada, Ontario (the greatest energy consuming province) and Alberta (the greatest energy producing province). Both Ontario and Canada illustrate relatively stable ratios over time with the most notable change occurring subsequent to 1980. In the face of increasing oil prices in the early 1970's, Canada embarked on a policy of federal subsidization of energy prices in an effort to minimize inflationary impacts. Price induced conservation, therefore, did not become a factor until the subsidies were removed with the introduction of The National Energy Program in 1980. Per capita energy consumption in Canada (Figure 2) also rose with minor fluctuations until 1980.

The large difference in energy consumption between Alberta and Ontario, due primarily to structural differences in the manufacturing sector, decreased substantially after 1973, as energy prices increased and the growth in Alberta's GDP far outpaced that of the other provinces. Fuel use per dollar of output in Alberta remains 40 percent higher than in Ontario, however, and with the province experiencing the slowest growth of any region in Canada over the past two years, this difference is expected to increase.

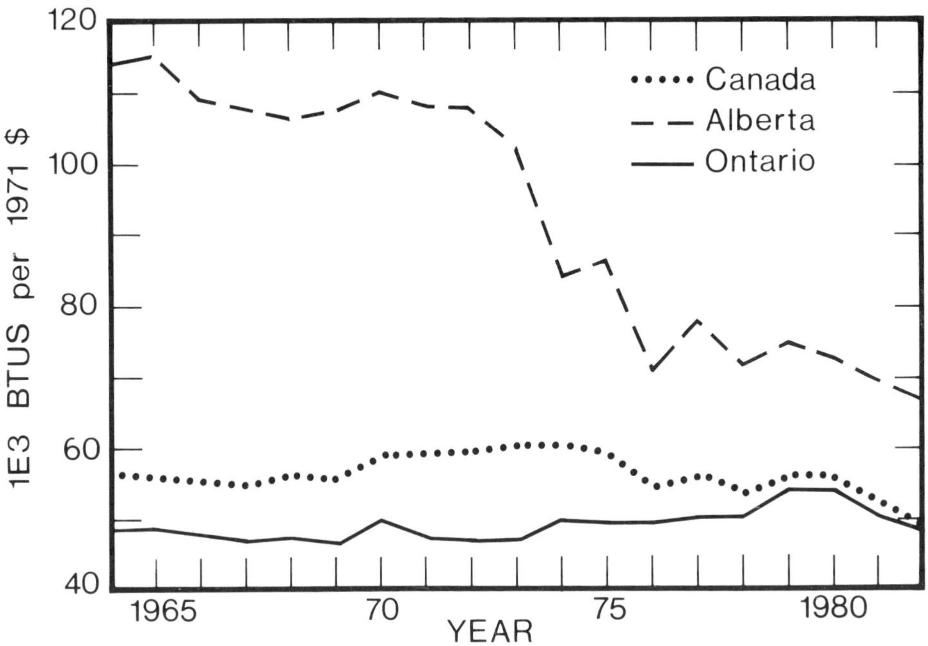

Figure 1. Ratio of energy consumed to gross domestic product, Canada and selected provinces, 1964-1982.

The structural change in the North American economy that has occurred over the past two decades, as regions move from a manufacturing base to a service-oriented economy, is expected to result in lower rates of energy consumption per dollar of economic output (Brumell 1984). No indication of this phenomenon is evidenced in Figure 1. In 1980, energy/GDP ratios were actually higher in Ontario and in all of Canada than in 1964.

Accounting for the relative shift in fuel types over time, as well as spatially, provides slightly different results. Figure 3 illustrates the change in energy/GNP ratios over time for the same regions taking into account fuel quality differences. Fuels differ in their ability to do work, and, accordingly, the quantity of fuel consumed has been weighted by fuel quality, using the transformation ratios from Odum (1983) and Cleveland _et al_. (1984). Standardizing to coal equivalent (CE) Btus substantially decreases the differences in energy/GDP ratios between regions since 1976, and dramatically decreases the efficiency of regions dependent on very high quality energy forms such as electricity. Electricity, which is three to four times higher quality than other fossil fuels, accounted for only three tenths of a percent in Canada as a whole in 1980 (this translates to one percent and thirty percent of the total, respectively, when converted to coal (e units).

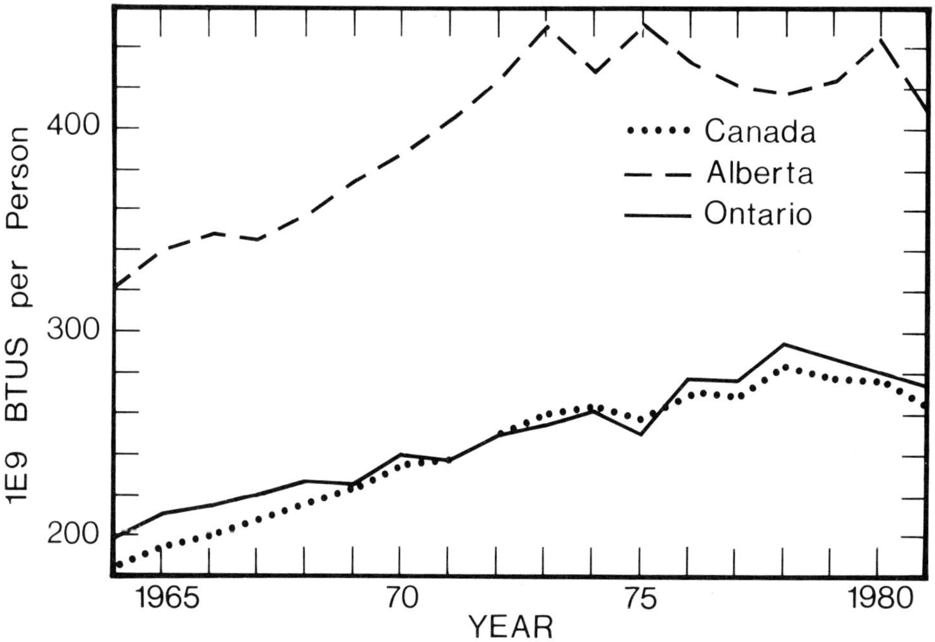

Figure 2. Per capita energy consumption, Canada and selected provinces, 1964-1982.

Correcting for fuel quality provides a further illustration of the increase in energy/GDP ratios in Canada and Ontario between 1964 and 1980. The structural switch to a service-oriented economy in Canada over the past two decades has resulted less in an increase in energy efficiency than simply a switch to higher quality fuels such as electricity. In the U.S., Cleveland et al. (1984) have illustrated that such changes in fuel type used can account for almost 70 percent of the variation in the energy/GNP ratio since 1929. A similar case for Canada is indicated in Figures 1 and 3.

The minimax principle states that systems will evolve by competitively maximizing and minimizing their rates of energy consumption. Changes in fuel efficiency across regions may indicate the attempt at minimizing dissipation, but until 1980 such behavior was not evidenced in Canada. The post-industrial society, with the accompanying demand for more high quality energy, may be equally as energy intensive as its predecessor unless conservation and substantive technological improvements are forthcoming. There seems to be little inherent energy efficiency difference between a manufacturing based economy and a service-oriented one. The decline in energy/GDP ratios since 1980 can be partially explained by the economic recession, but can also be attributed to price induced conservation since energy prices to final demand were deregulated in Canada in 1980.

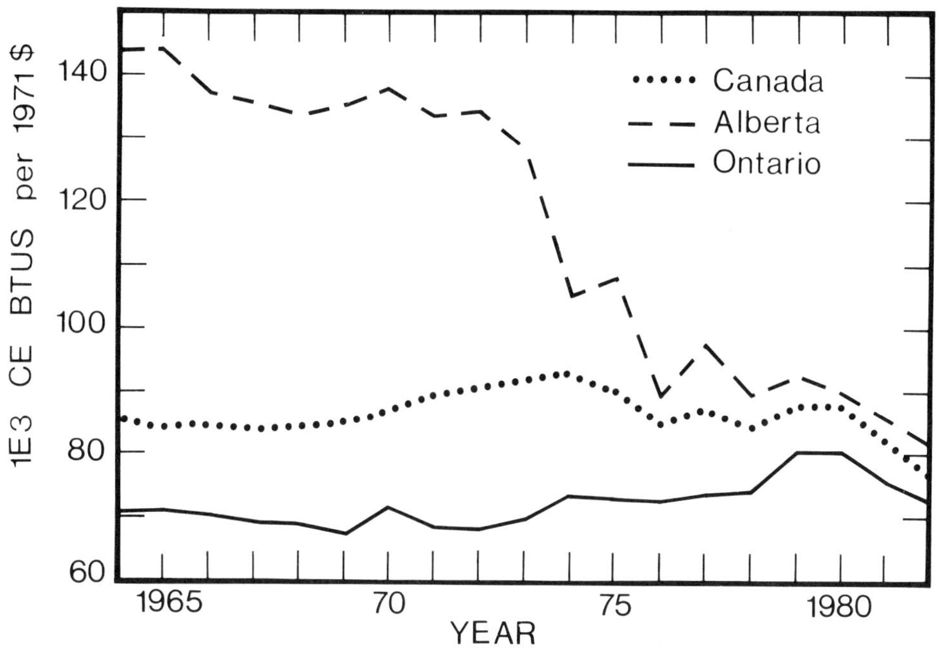

Figure 3. Ratio of total energy consumed (standardized to coal equivalent units) to gross domestic product, Canada and selected provinces, 1964-1982.

3.2. Economic Growth and the Minimax

The ratio of energy used to economic output provides a measure of the energy efficiency of a region and is useful in illustrating the general effects of regional restructuring. One component of this ratio, gross domestic product, can be viewed as the sum of gross consumption, or the real cost of maintaining a social structure, and the amount invested to improve the state of society (Boulding 1970). The objective of advanced regional economies should be to minimize these maintenance costs (payments to labor, land and capital), which can be considered a form of social entropy, while maximizing GDP (or that component of GDP contributing to social welfare). Such a normative planning role is entirely consistent with the minimax framework; systems will try to maximize their energy throughput, to promote growth, and minimize dissipation or maximize efficiency. Overbury (1973) has proposed an input-output framework to estimate the level of 'social entropy' in a regional economy, providing a quantitative framework for promoting the minimax as a planning objective. As a rough approximation of the amount of GDP allocated to improving social welfare, gross fixed capital formation has been used for the national economy and the provinces mentioned above, and the ratio of energy consumed in that component of GDP to gross fixed capital formation over time is presented in Figure 4. Low ratios imply relative system efficiency in the use of available energy. The viability of a region is a function of its investment, or feedback; the amount of flow or addition to society that is stimulated by each Btu invested. Low energy/capital ratios, exhibited by Alberta subsequent to 1974, imply a region that is able to capture large inflows of energy per unit of energy invested. The province also shows much variability in its efficiency of use; declines in the real price of oil may be accompanied by continued fluctuation in energy/capital ratios as maintenance costs increase and more energy is needed to support and create capital stock.

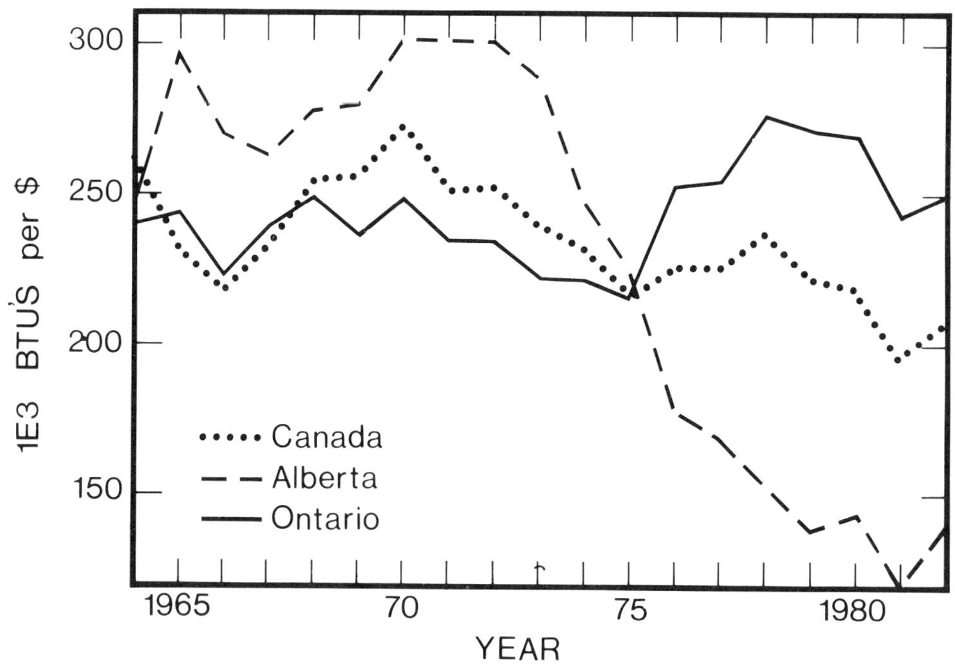

Figure 4. Ratio of energy consumed to gross fixed capital formation, Canada and selected provinces, 1964-1982.

The ratio of energy to gross fixed capital formation represents one simple measure of the energy efficiency of investment. Combining the extension of input-output analysis proposed by Overbury (1973) with the conversion of transactions in the economy to energy units (cf. Herendeen 1974), would provide a more accurate measure of the energy used to maintain societal processes and the amount used to improve the state of society.

3.3. Maximizing Regional Energy Flows

Two general mechanisms exist for regions to acquire energy; investing within the region to capture energy and trading with other regions. Both have an efficiency component attached to them, and both are extremely important to regional economic growth.

A region's energy endowment consists of stocks of fossil fuels and minerals and flows of 'natural' energy (e.g., sun, rain, and wind). Regions that are heavily endowed with natural energy, and are able to capture this energy and use it to stimulate the economy (by capturing more energy), are at a competitive advantage relative to other regions. The energy investment ratio is the ratio of the high quality energy (e.g., fossil fuels) to low quality natural energy used in any process (usually to capture more high quality energy). The ratio is a measure of how effectively a region (or any system) is able to use its natural energy endowment. For example, a municipality that is able to use a swamp or wetlands for tertiary sewage treatment will have a much lower energy investment ratio for that process than a municipality that is required to build a tertiary treatment plant. If the ratio for a region is low, the relative costs of acquiring additional energy are lower, prices are lower, environmental disruption is less and the system expands (it is able to maximize power).

Heavy manufacturing based economies will have higher energy investment ratios, and, as expected, Ontario's energy investment ratio is four times as high as Alberta's. As manufacturing based economies decline, the ratio of fossil fuel to natural energy inputs increases, greater pollution results and prices rise.

Canada is endowed with large natural energy flows, which contribute substantially to the operation of the human economy. Most provinces exhibit natural energy flows far greater than the flows of high quality energy extracted by man (Table 1). Alberta is the only province that has more fossil fuel energy available to it than natural energy (all expressed as CE units). The implications are quite obvious; in a system so dependent on supplies of high quality energy, disruptions in supply or price can have catastrophic effects on the regional economy. Droughts and disease, which affect economies solely based on natural energies, can have equally devastating effects.

A more relevant planning measurement may be the energy return on investment, which is the amount of energy in a useful form yielded per dollar of investment (Cleveland _et al._ 1984). The measure is analogous

Table 1. Ratio of energy produced to natural energy, by province, Canada 1981

Newfoundland	0.164
Prince Edward Island	0.0
Nova Scotia	0.387
Quebec	0.122
Ontario	0.139
Manitoba	0.067
Saskatchewan	0.116
Alberta	1.154
British Columbia	0.153

Source: Statistics Canada 1981; Energy, Mines and Resources 1980.

to the net energy yield in a process. Regions that desire to maximize power and efficiency will invest in processes that yield the best energy return on investment. Economies that exhibit relatively higher energy returns on investment are stronger economically because they are able to do more economic work than regions with a lower energy return on investment.

Interregional trade flows can be viewed in a similar vein. Regions interested in maximizing power must be concerned with their energy balance of trade. Investing in another region, i.e., purchasing their goods, also provides an energy return on investment. Regions that are able to draw on the natural resources of underdeveloped areas are maximizing power, because the energy yield per dollar of raw materials is far greater than the energy yield per dollar of finished goods. Canada, a net energy exporter, exhibits an energy import/export ratio of less than one, compared to 2.2 for the U.S. (Odum 1983). Table 2 gives embodied energy trade flow ratios (imports/exports) for four provinces in Canada, three of which are large raw material exporters. Calculations are based on the energy/dollar ratios of all the provinces (Table 3), as well as the embodied energy in exports from the U.S. and other developed countries found in Odum (1983). Note that in 1971, only one of the four provinces (Ontario) was importing more embodied energy than it was exporting. By 1980, as the world price of oil increased, Alberta exhibited a ratio greater than one. Embodied energy flows between Alberta and the U.S. were roughly equal (using Odum's figure of 2.2 for the U.S.) and the province had the lowest embodied energy consumption/GDP ratio of any in Canada. British Columbia and Quebec, both large exporters of raw materials, export more embodied energy than they import and exhibit negative energy balances of trade. As the price of energy increases, Ontario, which imports over three-fourths of its high quality energy, exhibits a decreasing energy import/export ratio.

Table 2. Ratio of embodied energy trade flows, by province, for 1971 and 1980 (ratio expressed in SEJ of imports/SEJ of exports)

Province	Year 1971	Year 1980
Quebec	0.446	0.618
Ontario	1.385	1.149
Alberta	0.685	1.250
British Columbia	0.370	0.592

Source: Statistics Canada 1981; Table 1 (preceding).

Table 3. Ratio of total embodied energy consumption to gross domestic product, by province, 1980

Province	1E12 SEJ/$
Newfoundland	44.27
Prince Edward Island	3.42
Nova Scotia	3.92
New Brunswick	5.61
Quebec	5.46
Ontario	2.96
Manitoba	6.32
Saskatchewan	5.34
Alberta	2.46
British Columbia	5.67
Canada	4.55

Sources: Energy, Mines and Resources 1980; Statistics Canada 1980; 1981.

4. CONCLUSION

Systems competitively maximize their energy throughput and minimize dissipation. The minimax is an appropriate design criteria for system evolution, as illustrated mathematically by Lonergan and Kirkaldy (1984). The minimax can also be used as a general planning criterion, if one takes a broad perspective on the role of energy in the regional economy. The discussion above has sought to promote an understanding of the minimax by presenting a number of related energy concepts in the context of Canadian regional systems and to illustrate the relationship between energy flows and the entropic frameworks discussed initially.

Any concept of energy must incorporate notions of energy quality and natural energy. Standardizing for energy quality provides a better measure with which to interpret the relationship between energy and economic development, and illustrates that the transition to a service-oriented society (now termed regional restructuring) will not likely be accompanied by greater energy efficiency in the absence of a concerted conservation effort. Price induced conservation was not noticed in Canada prior to 1980, due to federal subsidies, and energy/GDP ratios actually increased during the previous two decades. Taking into consideration natural energy flows is also important, because not only are they the basis of our economy, but they give a more accurate picture of the energy value of certain processes in our society as well.

Regions must also be aware of the energy return on investment; synthetic oil from tar sands has an energy return of less than one-third of the return on Middle East oil per dollar invested. Why are we spending so much on energy mega-projects in Canada when foreign supplies are a much better investment? To attain self sufficiency, we are told. If the minimax is an appropriate framework as an evolutionary theory and as a normative planning principle, the maximization of power relative to these projects may be suspect. The price of self sufficiency may be very high indeed.

5. FOOTNOTES

[1]Energy in the context which it is used by the Odums, and which will be used in the remainder of this chapter, refers to embodied energy, the total amount of energy (direct and indirect) needed for a production process or to provide a good or service. The energy embodied in a table, for example, includes the direct energy used in its construction, as well, as the indirect energy consumed in all inputs (labor, capital) to its production.

6. REFERENCES

Allen, P.M., and Sanglier, M. 1979. A dynamic model of growth in a central place system. Geographical Analysis 11: 256-73.

Allen, P.M. Sanglier, M., Boon F., Deneuborg, J.L., and DePalma, A. 1981. Models of urban settlements and structure as dynamic self-organizing systems. DOT/RSPA/DPB-10/6. Washington, D.C.: U.S. Department of Transportation.

Boulding, K. 1970. *Economics as a life science.* New York: McGraw-Hill.

Brown M.T. 1981. Energy basis for hierarchies in urban and regional systems. In *Energy and ecological modeling*, eds. W.J. Mitsch. R.W. Bossereman and J.M. Klopatek pp. 517-34. New York. Elsevier Scientific.

Brummel, A. 1984. Global scenarios for long-term energy planning. Paper presented at the International Congress of Arts and Sciences Rotterdam, June, 1984.

Cleveland, C.J., Costanza, R., Hall, C.A.S. and Kaufmann, R. 1984. Energy and the U.S. economy. A biophysical perspective. *Science* 225: 1-21.

Conservation Foundation. 1978. The economy can grow with less energy. *The Conservation Foundation Letter*, March 1-8.

Energy, Mines and Resources. 1980. Energy Alternatives. Ottawa: Supply and Services Canada.

Georgescu-Roegen, N. 1971. *The entropy law and the economic process.* Cambridge, MA: Harvard University Press.

Gilliland, W., ed. 1978. *Energy analysis: A new public policy tool.* Boulder, CO: Westview Press.

Herendeen, R. 1974. The use of input-output analysis to determine the energy cost of goods and services. In *Energy*, ed. MaCrakis pp. 141-57. Cambridge, MA: MIT Press.

Johnson, L. 1981. The thermodynamic origin of ecosystems. *Canadian Journal of Fisheries and Aquatic Sciences* 38: 571-90.

Kirkaldy, J.S. 1965. Thermodynamics of terrestrial evolution. *Biophysical Journal* 5: 965-79.

Kirkaldy, J.S., and Black, D.M. 1972. Social reporting and education planning. Toronto: Commission on Post-Secondary Education in Ontario, Ontario Government Bookstore.

Lakshmanan, T.R. and Johansson, B. eds. 1985. *Assessing the regional consequences of large scale energy projects.* Amsterdam:

North-Holland.

Liossatos P. 1980. Spatial dynamics: some conceptual and mathematical issues. Environment and Planning A 12: 1051-71.

Lonergan, S.C. 1983. Regional growth and environmental disruption: a thermodynamic approach. Man, Environment, Space and Time 3: 143-60.

_____. 1984. Evaluating the regional impact of Canadian energy megaprojects: The Syncrude case. In Assessing the regional consequences of large scale energy projects, eds. T.R. Lakshmanan and B. Johansson. Amsterdam North-Holland.

Lonergan, S.C., and Kirkaldy, J.S. 1984. Maximum path probability as an evolutionary criterion. Mathematical Biosciences in press.

Lonergan, S.C., Zucchetto, J. and Nishimura, T. 1984. The dynamics of industrial energy consumption: A U.S./Canada comparison. Energy Policy, in press.

Lotka, A.J. 1922. Contributions to the energetics of evolution. Proceedings of the National Academy of Sciences, 18.

Miernyk, W.H. 1982. The illusions of conventional economics. Morgantown, WV: West Virginia University Press.

Nicolis, G., and Prigogine, J. 1977. Self-organization in non-equilibrium systems. New York. John Wiley.

Odum, H.T. 1971. Environment, power and society. New York: Wiley Interscience.

_____. 1973. Energy, ecology and economics. AMBIO 2: 220-27.

_____. 1983. Systems ecology. New York: Wiley Interscience.

Ostro, B.D., and Odum, E.C. 1976. Energy basis for man and nature. New York: McGraw Hill.

Ostro, B.D., and Naroff, J.L. 1980. Decentralization and the demand for gasoline. Land Economics 56: 169-80.

Overbury, R.E. 1973. Features of a closed-system economy. Nature 242: 561-65.

Prigogine, I. 1955. Introduction to thermodynamics of irreversible processes. Springfield, IL: C. Thomas.

_____. 1978. From being to becoming. San Francisco: W.H. Freeman.

Romanos, M.C. 1978. Energy-price effects on metropolitan spatial structure and form. Environment and Planning A 10: 93-104.

Schurr, S.H. and Netscheat, B., et al. 1960. Energy in the American economy, 1850-1975. Baltimore: Johns Hopkins University Press for Resources for the Future.

Small, K. 1980. Energy scarcity and urban development patterns. International Regional Science Review 5: 97-117.

Statistics Canada. 1980. Quarterly report on energy supply-demand in Canada IV. Ottawa: Supply and Services Canada.

______. 1981. Provincial economics accounts, 1967-1981. Ottawa: Supply and Services Canada.

Til, J.V. 1979. Spatial form and structure in a possible future. Journal of the American Planning Association 45: 318-29.

Wardwell, J.M. 1980. Energy constraints on population dispersal. Paper presented at the Annual Meeting of the Population Association of American, Denver, Colorado.

Vining, D., Zucchetto, J., and Lonergan, S. 1978. An ecological model of the turnaround phenomenon. Paper presented at the Conference on Urban Transportation, Planning and Land Use, Northwestern University, Evanston, IL.

Wilbanks, T. 1981. Energy self-sufficiency vs. interdependency: implications for regional income distribution. Regional Development Dialogue 2: 78-94.

Zucchetto, J. 1975. Energy-economic theory and mathematical models for combining the systems of man and nature, case study: The urban region of Miami, Florida. International Journal of Ecological Modeling 1: 248-68.

CHAPTER 21

CHANGING ENERGY PRICES AND STATE REVENUE

Frank Calzonetti and Robert Hanham
Department of Geology and Geography and
Regional Research Institute
West Virginia University
Morgantown, WV 26505

ABSTRACT. There has been much discussion in recent years about the relationship between changing energy prices and state economic development. The affect of rising energy prices on shifts of income, employment and firms from energy deficit to energy surplus states has been debated in the literature. This chapter focuses upon state revenue changes among energy surplus and deficit states over the period 1967-80. It is found that revenues are sensitive to the type of fuel which comprises a state's energy balance.

1. INTRODUCTION

After many years of relatively inexpensive oil, its real price rose suddenly and substantially on two occasions during the 1970's. This was closely followed by an increase in the real price of other fossil fuels. These events not only raised the importance of energy availability as a component of a country's comparative advantage in the international economy, but they appear to have had much the same effect upon regional comparative advantage within the national economy. This is particularly evident in a country such as the United States, in which there is considerable geographic variation in energy production and consumption. This chapter examines the impact of changing energy prices on state economies. Changing energy prices have been shown to influence state income, employment and revenue. In this chapter we focus upon the relationship between a state's energy position and its revenues.

To some extent, regional energy price differentials have been mitigated by rising industrial process efficiencies in high price regions (Goettle et al. 1977), but the fact remains that exchange ratios between fossil fuels and other commodities have greatly altered since 1974. For example, the exchange ratios between Texas natural gas, Oklahoma crude oil and West Virginia coal on the one hand and New York milk and Illinois corn on the other declined by as much as seventy

F.J. Calzonetti and B.D. Solomon (eds.), Geographical Dimensions of Energy, 411–425.
© 1985 by D. Reidel Publishing Company.

percent between 1970 and 1978 (Miernyk 1982). Similar declines were also registered for dozens of other exchange ratios. However, Garnick (1983, 43) cautions that "...although higher resource prices also entail higher rental incomes to the owners of the resources, these rents do not necessarily accrue to the areas in which production occurs..."

The rise in real energy prices in the United Stated during the 1970s may have had a number of impacts on the structure of the space-economy of the country. Four that have received some attention in the literature are those of employment, unemployment, personal income and government revenue. While data do suggest that states rich in energy have improved their position with respect to these measures, it is difficult to determine if energy is the dominant factor explaining these shifts. Most energy rich states are in the South and West, and the shift in population, income and employment from the "frostbelt" to the "sunbelt" (Landsberg 1982; Sawers and Tabb 1984) or from the "core" to the "periphery" (Norton and Rees 1979) is several decades old and has been intensively investigated.

Efforts to clarify the role energy price increases have had on state economic expansion has been investigated by Miernyk (1977), Hoch (1980), Richardson (1984) and others. The data in Table 1, for example, were used by Miernyk to illustrate the fact that employment

Table 1. Relative Changes in Employment in Standard Federal Regions by Energy Price Responsiveness, 1970-77.

Region	Directly Vulnerable %	Indirectly Vulnerable %	Sheltered %	Total %
New England	- 9.4	- 2.9	8.3	4.0
North Mid-Atlantic	-20.3	-22.5	-1.6	-6.0
South Mid-Atlantic	-10.0	- 1.3	13.3	7.5
Southeast	8.7	12.2	27.5	22.4
Great Lakes	- 5.4	1.3	14.9	9.0
Southwest	26.8	19.4	34.0	32.0
Central Plains	10.8	8.1	17.3	15.7
R. Mtn. - N. Plains	25.0	24.5	45.2	41.7
Far West	10.1	11.9	26.7	23.5
Northwest	16.3	15.8	34.5	30.1
U.S.	- 0.9	2.2	18.2	13.5

Source: Miernyk (1982)

which is either directly or indirectly vulnerable to changes in energy prices was more prone to shift from deficit to surplus regions during the 1970s than was employment in sectors sheltered from such price changes (Miernyk 1984). Employment was classified this way by Miernyk using the results of an input-output analysis that estimated the energy intensity of each sector of the economy. Directly vulnerable activities were defined as those which are energy intensive, whose products are sold in national markets, and which are not bound to their present location by resource or transport considerations. Indirectly vulnerable activities were defined as major suppliers of directly vulnerable activities. Sheltered activities were defined as those which are resource or transport oriented and which serve local markets. These data indicate that vulnerable employment declined in the North Mid-Atlantic, New England, South Mid-Atlantic and Great Lakes regions, predominately energy deficit areas, and increased most in the Rocky Mountain-Northern Plains and Southwest regions, predominately energy surplus areas. Sheltered employment grew in all but one region, but clearly the most in western regions and in the Southeast. Even sheltered employment may be induced to shift to energy surplus regions. Giarratani and Socher (1978) have suggested that since a growing share of economic activity is being drawn toward the market, and since there has been a significant shift in population away from the Northeast to the Southwest and West during the past fifteen years, the latter will attract economic activities by their growing markets. However, at the present time there is little proof that industrial relocation to energy rich states has occurred. Regional shifts in manufacturing that are visible may have deeper causes. There is some evidence that shifts in manufacturing follows a "filtering down" process through a hierarchy of cities (Thompson 1969; Moriarty 1983). This process has been explained in terms of the product life cycle model in which the decentralization of manufacturing is related to the stage of the manufacturing process (Hanson 1979; Rees 1979; Norton and Rees 1979).

Shifts in employment from energy deficit to energy surplus states have clearly benefited the latter at the expense of the former. Such shifts, along with corresponding movements of capital and of white collar workers, have also resulted in a shift in the burden of structural unemployment from surplus to deficit regions. Hanham (1983) has shown, for example, that the average unemployment rate for surplus states exceeded the average rate for deficit states by quite some margin for every year from 1960 to 1973. Since then the situation has been the reverse as rates in deficit states have on average exceeded those in surplus ones.

A further significant impact of rising energy prices on the space-economy has been in terms of per capita income. Miernyk (1982) has argued that the energy price rise of the 1970s substantially altered the terms of trade among regional economies in the United States to favor energy rich areas at the expense of energy poor ones, and that this has brought about a corresponding transfer of real income from the latter to the former. The terms of trade had long favored the industrial regional economies of the Northeast and Midwest (Miernyk 1984), because the price of industrial products had grown at a faster rate than the price of fossil fuels. This situation was, of course, turned around in the 1970s, resulting in a terms of trade shift which favored energy producing regions at the expense of energy consuming ones. Miernyk (1984) has shown, for example, that there was a significant inverse correlation between changes in state per capita income as a percentage of the United States average on the one hand and changes in both the price of electric energy and the price of gas on the other hand between 1973 and 1979. Miernyk (1984) also provides data to illustrate the relationship between the change in regional relative income per capita over the period 1959-79 and a region's energy balance. These data are reproduced in Table 2. They clearly

Table 2. Relation Between Energy Balance in 1978-79 and Changes in Relative Per Capita Income, 1959-1979, in Standard Federal Regions.

Region	Energy production minus consumption (% of U.S. total)	Percent change in relative per cap. income
New England	- 4.0	- 6.6
North Mid-Atlantic	- 7.2	-13.7
South Mid-Atlantic	0.5	0.8
Southeast	- 6.1	18.9
Great Lakes	-15.0	- 2.9
Southwest	29.5	13.3
Central Plains	- 3.4	3.8
R. Mtn - N. Plains	6.9	5.6
Far West	- 4.8	- 6.1
Northwest	3.6	1.2

Source: Miernyk (1984)

indicate the fact that in energy surplus regions relative income tended to grow more than in deficit regions. The only major deviation in the relationship is that of the Southeast, which had a moderate energy deficit but a substantial increase in its relative income. This region is, of course, part of the rapidly growing Sunbelt, and in 1959 had the lowest income of any region (Miernyk 1984). This relationship between income growth and energy balance in the United States has also been tested by Manuel (1982). He found that there was a significant difference in the mean growth rates of per capita income between energy producing and energy consuming states after, but not before, the 1973 oil embargo. Furthermore, the difference was found to be more significant among Sunbelt states than among those to the north. Hoch (1980) tested the association between state per capita income growth and energy extraction over the 1972-75 period and found higher growth among energy extracting states. Hoch (1980, 259) concluded that "much of the estimated increase for fuel-producing states involves the direct effect of higher wages in fuel production."

The rise in energy prices during the 1970s also had a significant effect on state revenues (Landsberg 1982). As a consequence of higher prices, energy rich states have been able to generate more income, both directly through severance taxes and royalty payments on state energy properties and indirectly through corporate and personal income taxes and sales and property taxes. To some extent these revenues have been further supplemented by federal payments to states which are based on tax-effort formulae (Corrigan and Stanfield 1980). Although price decontrols have not raised the price of energy as much as expected and the demand for energy has declined lately, the increased revenues in energy producing regions should prove extremely beneficial in attracting businesses (through the use of larger incentives). In light of the current policy of transferring the responsibility for a number of programs from the federal government to the state governments this revenue generating capability may assume an even greater importance. Severance taxes constitute a large proportion of several state's revenues. In Louisiana, Wyoming, New Mexico, Alaska, Texas and Oklahoma they amount to about 20 percent of total revenues. Of the major energy producers, Kentucky has the lowest proportion of revenues coming from these taxes--7 percent (Cochran and Prestige 1981). The tax rates which generate these revenues vary quite considerably among the major producers, with highs of 12.5 percent for oil in Louisiana, 10 percent for gas in Alaska and 30.5 percent for coal in Montana (Cochran and Prestige 1981). Energy consuming states in the Midwest and the Northeast, needless to say, have gone to some lengths to limit such taxes, claiming that, since energy is a national resource, such revenues should be distributed nationally. Producing states, on the other hand, have argued that they are subjected to the costs of resource extraction and that such revenues reflect a desire to preserve something for their future (Corrigan and Stanfield 1980).

Energy producing states have also benefited from the influx of income, employment and capital through the revenues they generate via corporate and personal income taxes and from sales and property taxes. It has been estimated, for example, that Texas and Louisiana registered surpluses of about $10 billion annually during the mid-1970s through energy trading (Corrigan and Stanfield 1980). One consequence of these increasing revenues may be a further enhancement of these states' financial situations through increased federal aid. A number of federal programs to state and local governments are tied to a tax effort factor on the part of those governments. This has been the case with respect to the general revenue sharing and anti-recession fiscal assistance programs for example. In sum, rising energy prices should have enhanced the revenues of energy producing states at the expense of energy consuming ones. This certainly appears to have been the case in the mid- and late-1970s and in the beginning of the 1980s. In 1983-84 the revenues of most energy states were severely curtailed due to falling energy prices, but in the long run it seems all but impossible that these states won't continue to benefit.

Table 3 indicates which states possessed a surplus (production exceeding consumption) of energy resources by fuel for the years 1967, 1972, 1977 and 1980. There are three ways to measure a state's energy balance position (Bourque 1981). The production/consumption approach is most often used and is employed in this paper. An energy surplus state is simply one which produces more energy than it consumes. The balance for each fuel is computed here along with the energy input into power plants. Another method often used in determining a state's energy position is the export/import approach. This recognizes that energy forms vary in value and that energy prices provide a good measure of this value. Finally, the embodied energy trade can be computed using input-output analysis. This approach recognizes the energy embodied in the imports and exports of goods and services in addition to direct energy trade. Bourque (1981) found that although Washington State exported 17 million kilowatt hours of electricity in 1972, it was a net electricity importer when embodied trade was considered. For instance, the state purchased large quantities of goods from other states that required large amounts of electricity in their manufacture.

Table 3. State Relative Revenues Per Capita* and Energy Balance by Fuel, 1967-80

State	Relative Revenue Per Capita			
	1967	1972	1977	1980
AL	0.80 C O	0.76 C O	0.77 C O	0.90 C
AK	2.52 TCGO	1.99 TCGO	2.82 TCGO	7.28 TCGO
AZ	1.12 CG	0.96 C	0.95 C	0.91 C
AR	0.75 CGO	0.72 C	0.70 C	0.89
CA	1.36 GO	1.29 GO	1.25 GO	1.20
CO	1.13 CGO	1.04 TCGO	1.05 C	0.93 C
CT	0.98	1.05	0.96	0.96
DE	1.30	1.16	1.11	1.45
FL	0.89 O	0.83 O	0.81 O	0.72
GA	0.80	0.85	0.83	0.81
ID	0.95	0.86	0.86	0.94
IL	0.87 C O	1.04 C	0.96 C	0.93 C
IN	0.87 C O	0.82 C	0.78 C	0.76 C
IA	1.04 C	0.91	0.93	0.98
KS	0.95 TCGO	0.89 T GO	0.90 T GO	0.88 T G
KY	0.85 TCGO	0.77 TCG	0.80 TCG	0.99 TC
LA	1.03 T GO	0.92 T GO	0.91 T GO	1.10 T GO

Table 3(continued)

State	1967		1972		1977		1980	
ME	0.84		0.90		0.88		1.01	
MD	0.97	C	1.00	C	1.09	C	1.10	
MA	0.99		1.11		1.13		1.13	
MI	1.06	G	1.07	G	1.09	G	1.07	
MN	1.14		1.11		1.13		1.24	
MS	0.75	GO	0.81	GO	0.78	GO	0.95	
MO	0.84	C	0.81	C	0.76	C	0.72	
MT	1.08	CGO	1.07	TC	1.07	TC	1.16	TCGO
NE	0.95	GO	0.90		0.96		0.87	
NV	1.47		1.23		1.13		1.00	
NH	0.89		0.80		0.78		0.70	
NJ	0.87		0.98		1.05		0.93	
NM	1.16	TCGO	1.05	TC O	0.99	TC O	1.47	TCGO
NY	1.30		1.40		1.44		1.21	
NC	0.72		0.74		0.77		0.88	
ND	1.19	TCGO	0.98	TCG	1.07	TCG	1.34	TCG
OH	0.82	CGO	0.79	CGO	0.79	CG	0.73	C
OK	0.98	TCGO	0.85	TCGO	0.84	T GO	1.00	T GO
OR	1.16		0.99		1.11		1.13	
PA	0.84	CGO	0.92	C	0.89	C	0.92	C
RI	1.03		0.94		1.02		1.21	
SC	0.64		0.70		0.77		0.90	
SD	1.06		0.98		0.90		0.98	
TN	0.82	O	0.75	C	0.75	C	0.75	C
TX	0.81	T GO	0.78	T GO	0.80	TCGO	0.81	TCGO
UT	1.10	CGO	0.94	C	0.90		1.03	TC
VT	1.19		1.13		1.08		1.19	
VA	0.79	C	0.81	C	0.84	C	0.91	C
WA	1.19		1.09		1.11		1.13	
WV	0.84	TCGO	0.85	TC	0.82	TC	1.08	TC
WI	1.12		1.04		1.03		1.14	
WY	1.56	TCGO	1.28	TCGO	1.40	TCGO	1.64	TCGO

*State Revenue per capita divided by national average
+T=total energy surplus; C=coal surplus; G=gas surplus; O=oil surplus.

Source: U.S. Statistical Abstracts and Energy Information Administration

An overall picture of the distribution of changes in state revenues is also shown in Table 3, which presents data regarding the revenue per capita in each state relative to the national average for 1967, 1972, 1977 and 1980. It is quite clear that, with only two exceptions, states which consistently had surpluses with respect to all fossil fuels combined showed substantial improvements in their revenue positions, and that this improvement largely took place between 1977 and 1980. In virtually all surplus states revenue positions prior to 1977 were at best stable, and between 1967 and 1972 invariably worsened. For the great majority of the remaining states their revenue positions significantly declined in the later years. Most of the exceptions to this trend are to be found in southern states, whose populations grew much faster than average during the late 1970s. There is, of course, another important factor which cannot be accounted for here with any precision, that of changes in tax rates.

2. TESTING THE EFFECT OF RISING ENERGY PRICES ON STATE REVENUES

The effect that rising energy prices have on state revenues is governed by a state's energy balance. There will be an inflow of dollars to states that produce more energy than they consume and an outflow of dollars from those that consume more than they produce. The fact that a state has either a surplus or deficit with respect to energy is thus used in this study to account for a state's changing revenue status. Furthermore, since revenues may also be attributable to personal and corporate income taxes independently of the energy effect, these factors too need to be incorporated in a model of revenue growth. The following initial model is proposed:

$$\dot{R}_i = a + b\dot{I}_i + c\dot{V}_i + e_i \tag{1}$$

Where $\dot{R}_i$ represents the rate of change in relative state revenues per capita, $(R_{t+1} - R_t)/R_t$, in state i over the period t to t+1; $\dot{I}_i$ represents the rate of change in relative income per capita, $(I_{t+1} - I_t)I_t$, in state i over the period t to t+1; V_i represents the rate of change in relative value added in manufacturing per employee $(V_{t+1} - V_t)/V_t$, in state i over the period t to t+1; a, b and c are parameters to be estimated; and e_i is an error term. Since it is also assumed that the change in state revenues should be a function of a state's energy balance, the parameter a is expressed as a function of B_i, a dummy variable representing the fact that state i has an energy surplus (B=1) or deficit (B=0) during the time period t to t+1; hence

$$a = a_o + a_1 B_i.$$

Finally, we also hypothesize that the degree to which changes in both income and value added in manufacturing influence state revenues will differ between energy surplus and deficit states. It might be assumed, for example, that rising incomes, both personal and corporate, in an energy surplus state will result in a faster rate of revenue generation than in energy deficit states through tax-effort linked federal aid. Both parameters b and c are therefore expressed as a function of B_i; hence

$$b = b_o + b_1 B_i$$

$$\text{and} \quad c = c_o + c_1 B_i .$$

An expanded version of the model can then be developed by substituting the new expressions for a, b and c into equation (1) (see Casetti 1972). The final model is as follows:

$$\dot{R}_i = a_o + a_1 B_i + b_o \dot{I}_i + b_1 B_i \dot{I}_i + c_o \dot{V}_i + c_1 B_i \dot{V}_i + e. \qquad (2)$$

State data to estimate the parameters of this model were obtained for the years 1967, 1972, 1977 and 1980. In order to gauge the effect of various fossil fuels, the dummy variable, B, was defined in one of four ways: total energy balance, coal balance, gas balance and oil balance. In each case, a state was given the value of one if production exceeded consumption, and a value zero if consumption exceeded production. Furthermore, since the dependent variable, R_i, is expressed as the rate of change over a given period, a state was characterized as surplus only if it had a positive balance at both the beginning and end of the period.

3. RESULTS

The parameters of equation (2) were estimated by ordinary least squares for the periods 1967-72, 1972-77, 1977-80, 1967-77, 1972-77 and 1967-80. Alaska was originally included in the estimation, but the values of its variables proved to be so extreme that it clearly biased the results. The results presented here are based on OLS estimates using 48 states. Findings are shown with respect to total energy balance (Table 4), coal balance (Table 5), gas balance (Table 6) and oil balance (Table 7). Only parameter estimates significant at the 0.05 level are given in these tables.

When the total energy balance is used (Table 4) this factor became a significant positive influence only late in the period. Its coefficient is positive for the periods 1972-80 and 1977-80, with a larger value in the latter. Total energy balance is not a significant factor in either 1967-77 or in 1972-77. During the earlier period, 1967-72, it has a negative coefficient, and is indeed the only significant variable in that equation. Prior to the oil embargo and subsequent energy price rise of 1973-74 the evidence here clearly shows that states with a surplus of energy resources were at a significant disadvantage with respect to generating revenues. Toward the end of the 1970s, however, that situation was reversed. Furthermore, the difference in the absolute values of the energy balance coefficients for the periods 1967-72 and 1977-80 (0.072 and 0.214) indicates that the differential between energy surplus and deficit states was far greater after energy prices had risen. Given this reversal it is not surprising that the energy balance coefficient is not significant when estimated for the entire period 1967-80. Change in per capita income

Table 4. OLS Estimates for Equation (2) With Respect to Total Energy Balance*

Year	a_o	a_1B	$b_o\dot{I}$	$b_1B\dot{I}$	$c_o\dot{V}$	$c_1B\dot{V}$	R^2
1967-72	-0.024	-0.072					0.10
1972-77							
1977-80	0.012	0.214	-1.680		0.829	-0.847	0.44
1967-77							
1972-80	0.043	0.144					0.15
1967-80	0.007		0.596				0.10

*Only estimates significant at 0.05 level shown.
Source: Author's calculations.

Table 5. OLS Estimates for Equation (2) With Respect to Coal Balance*

Year	a_o	a_1B	$b_o\dot{I}$	$b_1B\dot{I}$	$c_o\dot{V}$	$c_1B\dot{V}$	R^2
1967-72							
1972-77							
1977-80							
1967-77							
1972-80	0.051		1.048				0.14
1967-80	0.007		0.596				0.10

*Only estimates significant at 0.05 level shown.
Source: Author's calculations.

Table 6. OLS Estimates for Equation (2) With Respect to Gas Balance*

Year	a_o	a_1B	$b_o\dot{I}$	$b_1B\dot{I}$	$c_o\dot{V}$	$c_1B\dot{V}$	R^2
1967-72							
1972-77							
1977-80	0.026		-2.223	5.167	0.759	-1.257	0.29
1967-77							
1972-80	0.051		1.048				0.14
1967-80	0.007		0.596				0.10

*Only estimates significant at 0.05 level shown.
Source: Author's calculations.

Table 7. OLS Estimates for Equation (2) With Respect to Oil Balance*

Year	a_o	a_1B	$b_o\dot{I}$	$b_1B\dot{I}$	$c_o\dot{V}$	$c_1B\dot{V}$	R^2
1967-72							
1972-77							
1977-80	0.045	0.230	-1.651				0.25
1967-77							
1972-80	0.051		1.048				0.14
1967-80	0.007		0.596				0.10

*Only estimates significant at 0.05 level shown.
Source: Author's calculations.

was the only significant variable for the entire period, with the expected positive coefficient. Toward the end of the 1970s, however, its coefficient actually becomes negative. Change in per employee value added is significant only in the more recent period, where its coefficient is positive. As one might expect, it indicates that improvement in a state's position with respect to its per employee value added in manufacturing led to improvement in that state's revenue position. The interaction term, BV, is negative, however, which implies that although there was a positive relationship between changes in revenues and changes in value added in manufacturing for energy deficit states, for energy surplus ones there was a small negative relationship (0.829 - 0.847 = -0.018).

The results using the coal balance variable (Table 5) show quite clearly that this has not been a significant influence on changes in revenue, either for the entire period or for any subperiod. Coal surplus states have varied enormously in their economic fortunes since 1967, from rapidly growing ones in the West to declining ones in Appalachia and the Midwest. Gas balance, too, does not appear as a significant independent influence in these equations (Table 6). In the latest period (1977-80), however, it does significantly interact with changes in income and value added. With regard to income, the results indicate that from 1977 to 1980 changes in per capita income were inversely associated with changes in per capita revenues in gas deficit states, but were positively related in surplus ones (-2.223 + 5.167 = 2.944). In other words, revenues responded to rising incomes in gas producing states, but not in consuming states. The reverse is the case with regard to value added. Revenues responded positively to changes in value added in consuming states, but these were inversely related in producing states (0.759 - 1.257 = -0.498). The last fuel, oil, is also only significant in the most recent period (Table 7), but unlike gas it appears to be an independent influence. As expected, revenues tended to rise significantly more in oil producing states than in consuming ones in 1977-80.

4. CONCLUSIONS

Rising energy prices during the past ten years have clearly had a discriminatory effect on the ability of states to generate revenues. The results of this study indicate that this effect was most pronounced at the end of the 1970s. Prior to the oil embargo of 1973, the relationship was the reverse. Energy surplus states were not able to improve their revenue position relative to the national average as much as energy deficit states. A state's position with respect to coal resources has not, in general, been a significant factor in revenue improvement, although it surely was in some cases. Gas and oil, on the other hand, were in general influential in the late 1970s. The fact that a state had an oil surplus was directly linked to its ability to improve its revenue position relative to the rest of the nation at that time. The link with regard to gas was less direct. A state's per capita revenue position was enhanced if its per capita income position improved only in gas surplus states. On the other hand, the revenue position of a state responded to changes in per employee value added only in deficit areas. In surplus areas, revenue generation appears to have been little affected by changes in value added in the manufacturing sector.

The relationship between changing energy prices, a regional economy's energy resource base, and that region's ability to improve its revenue position relative to the rest of the nation became much stronger in the United States during the late 1970s. Energy surplus states, however, had a setback in the early part of the 1980s due to the economic recession's impact on the demand for energy resources. This could be seen in the severely curtailed state budgets of growing economies such as Oklahoma as well as in declining economies such as West Virginia. In Texas, where 25% of state tax revenues are derived from oil and gas severance taxes, a $1/barrel decline in the price of oil corresponds to a $40 million drop in revenues (Zukosky 1985, 44). But in the long term the outlook for energy surplus regions is surely better than for deficit ones. This is not an insignificant issue when one considers the resources that states increasingly require to attract further economic activity and to pay for the various programs that are being steadily transferred from the federal government to the states. However, it should be emphasized that this study does not suggest that energy surplus states have improved their position compared to energy deficit states for other measures of economic vitality, including income growth, employment growth, and industrial relocation. State revenues, however, may be a precursor to long-term changes in state economic well-being but this awaits further analysis.

5. ACKNOWLEDGEMENT

We are grateful to the helpful suggestions of Dr. William H. Miernyk on an earlier draft of this manuscript and to Dwayne Porter who provided computer assistance. Naturally, the authors alone are responsible for any errors in the paper.

6. REFERENCES

Bourque, P.J. 1981. Embodied energy trade balances among regions. International Regional Science Review 6: 121-36.

Casetti, E. 1972. Generating models by the expansion method: applications to geographical research. Geographical Analysis 4: 81-91.

Cochran, T., and Prestige, J.R. 1981. The United American Emirates. Washington, D.C.: Northeast-Midwest Institute.

Corrigan, R., and Stanfield, R.L. 1980. Rising energy prices: what's good for some states is bad for others. National Journal 12: 468-74.

Garnick, D. H. 1983. Shifting patterns in the growth of metropolitan and nometropolitan areas. Survey of Current Business 63: 39-44.

Giarratani, F., and Socher, C.F. 1978. The pattern of industrial location and rising energy prices. In Regional impacts of rising energy prices, eds. W.H. Miernyk, F. Giarratani, and F. Socher, pp. 103-16. Cambridge, MA: Ballinger.

Goettle, R.J., Medine, E.S., and Kyle, S.C. 1977. Price-induced patterns of energy intensive industries of the Northeast. Brookhaven National Laboratory, Regional Energy Studies Program, Informal Report.

Hanham, R.Q. 1983. Energy price changes and the spatial realignment of the U.S. metropolitan system. Modeling and Simulation 14: 921-25.

Hansen, N. 1979. The new international division of labor and manufacturing decentralization in the United States. Review of Regional Studies 9: 1-11.

Hoch, I. 1980. Role of energy in the regional distribution of economic activity. In Alternatives to confrontation, ed. Victor L. Arnold, pp. 227-326. Lexington, MA: Lexington Books.

Landsberg, H. H. 1982. Energy 'haves' and 'have-nots.' In Regional conflict and national policy, ed. Kent A. Price, pp. 35-58. Washington: Resources for the Future, Inc.

Manuel, D. 1982. The effects of higher energy prices on state income growth. Growth and Change 13: 26-37.

Miernyk, W.H. 1977. Rising energy prices and regional economic development. Growth and Change 8: 1-7.

______. 1982. The differential effects of rising energy prices on regional income and employment. In High energy costs: assessing the burden, ed. H.H. Landsberg, pp. 297-330. Baltimore: Johns Hopkins University Press.

______. 1984. Energy and regional development. In Energy costs, urban development and housing, eds. A. Downs and K.L. Bradbury, pp. 226-86. Washington, D.C.: The Brookings Institution.

Moriarty, B. M. 1983. Hierarchies of cities and the spatial filtering of industrial development. Papers of the Regional Science

Association 53: 59-82.

Norton, R.D. and Rees, J. 1979. The product cycle and the spatial decentralization of American manufacturing. *Regional Studies* 13: 141-51.

Rees, J. 1979. Technological change and regional shifts in American manufacturing. *The Professional Geographer* 31: 45-54.

Richardson, H.W. 1984. Regional policy in a slow-growth economy. In *Regional development problems and policies in Eastern and Western Europe*., ed. G. Demko, pp. 258-81. London: Croom Helm.

Sawers, L and Tabb, W. K. 1984. *Sunbelt/snowbelt: Urban development and regional restructuring*. N.Y.: Oxford University Press.

Thompson, W.R. 1969. The economic base of urban problems. In *Contemporary economic issues*, ed. N.W. Chamberlin, pp. 1-47. Homewood Hills, Ill.: Richard D. Irwin Inc.

Zukosky, J. 1985. State budgets: Let the good times roll. *Business Week* January 28: 44.

CHAPTER 22

CONSTRAINTS ON REGIONAL COAL PRODUCTION IN THE U.S.: A TIME SERIES ANALYSIS

Thomas D. Georgianna
School of Public & Environmental Affairs
Indiana University
Bloomington, Indiana 47405

Barry D. Solomon
Office of Electric Power Regulation
Federal Energy Regulatory Commission
Washington, D.C. 20426

Kingsley E. Haynes
Center for Urban & Regional Analysis
School of Public & Environmental Affairs
and Department of Geography
Indiana University
Bloomington, Indiana 47405

ABSTRACT. This chapter investigates the impact of the 1970 Clean Air Act, the 1977 Amendments to the Act and the UMW strikes of 1977-1978 and 1981 as regional constraints on coal production. ARIMA time series analysis and ordinary least squares regression analysis are used. The results indicate that the Act had a negative effect on Midwestern coal and a positive effect on Western coal. Results with respect to the other events are less conclusive.

1. INTRODUCTION

Despite the projections of the mid-1970's that coal output would greatly expand in the United States, growth in the coal industry has not met expectations. This continued lack of strength in the United States coal industry, despite ambitious development goals and the proximal location of coal fields to large regional markets (Campbell et al. 1979), has been

F.J. Calzonetti and B.D. Solomon (eds.), Geographical Dimensions of Energy, 427–442.
© 1985 by D. Reidel Publishing Company.

perplexing (See Figure 1). Electric utilities, the major market for coal, now consume over 80 percent of the coal sold domestically. This steam coal market has been depressed since the early 1980's due to declines in the demand for electricity (see Energy Information Administration 1984).

Furthermore, coal as a source of electric power is limited by the difficulties of burning it in an environmentally acceptable manner. The lack of growth in coal production may be due in part to such environmental factors and to labor relations that have affected both supply and demand sides of coal markets. In particular, air pollution regulations impose a serious impediment to the use of coal in many areas. Regulations make it difficult to construct new coal-burning facilities. Labor disputes may also be a factor in the production of coal. Strikes by the United Mine Workers of America (UMW) may affect perceptions of reliability and thus demand. Mineworker strikes of course also have a short term effect of interrupting supply.

Each of these factors varies regionally. Environmental constraints for burning coal vary by region because of differing coal quality by region. Environmental constraints also vary by region because the regulatory environment varies as it seeks to reduce economic disruption in Eastern and Interior coal fields. Additionally, most UMW coal is produced east of the Mississippi River, although there has been some recent change in this distribution of unionized labor.

This chapter provides an empirical analysis of the regional impact of air quality requirements and labor-managements disputes on U.S. coal production for the years 1960-1981. These constraints are operationalized as the Federal Clean Air Act of 1970 and the 1977 Amendments to the Clean Air Act, and the 1977-1978 and 1981 strikes of the UMW. We first discuss these constraints in more detail, and generate research hypotheses. We then discuss the research methods used and data sources, and then present the results. Some conclusions and future research directions complete the chapter.

2. DEVELOPMENT CONSTRAINTS

2.1 Air Pollution

Air pollution is perhaps the major constraint to coal use. The Clean Air Act regulates air quality in several ways, although the most important controls that affect the coal industry apply to coal-fired electric power plants and industrial boilers. Important provisions of the law include new source performance standards (NSPS) for emissions from specific polluting facilities, and national ambient air quality standards (NAAQS) which set areawide emission standards for various pollution types. The actual standards have been established by the U.S. Environmental Protection Agency.

The Clean Air Act of 1970 has been considered by many observers as a landmark environmental law, since it was the first law to require national air quality standards. Previous Clean Air Acts in 1963, 1965,

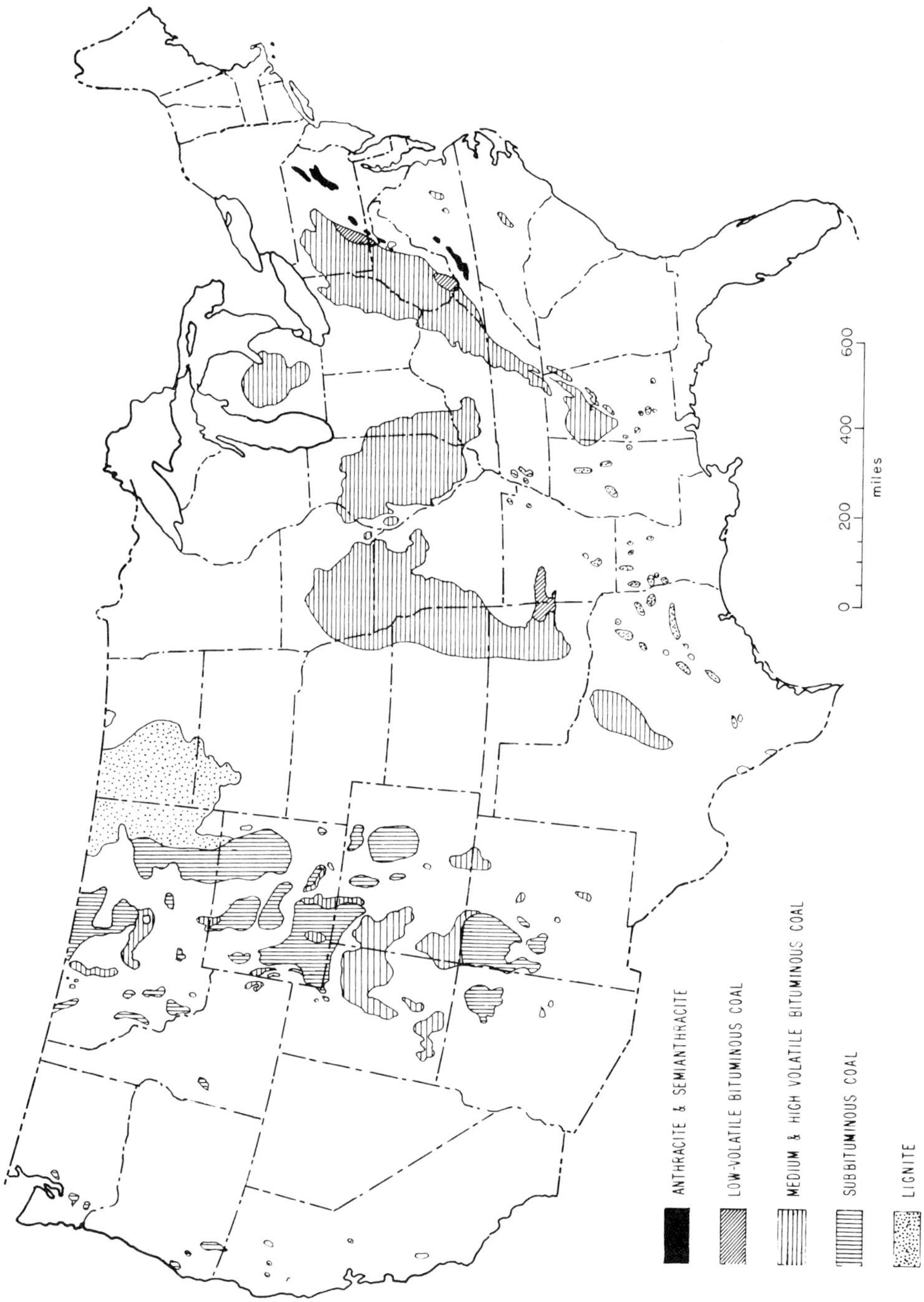

Figure 1. Coal Resources of the United States.

and 1967 only allowed for local air quality standards, which tended to be weak. Thus, not until the 1970 Act did air quality control have much national affect on industries such as coal mining and electric utilities. Moreover, the coal industry has been most concerned with the Clean Air Act's controls on sulfur dioxide (SO_2) emissions from power plants. It has been argued that the 1970 Clean Air Act standards (which were implemented in mid-1973) have depressed the market for high-sulfur coal, which is primarily found in the Midwestern and Appalachian regions. This would tend to expand the market for low-sulfur coal, which is mostly found in the West (Tyner and Kalter 1978), although the Southern Appalachian region has substantial low-sulfur coal reserves. However, the latter coal seams are primarily suited for the metallurgical market rather than for electric power production (which is of prime concern here). While Western coal production is often limited by its relatively high transportation costs to Midwestern and Eastern markets, as well as the coal's low heat content, one would expect that the 1970 Clean Air Act would result in a decrease in production of high sulfur, Midwestern coal, and an increase in production of low-sulfur, Western coal.

The air pollution requirements are fairly complex. The two most important controls are determined by the NSPS and NAAQS. In addition, prevention of significant deterioration (PSD) has been established for three types of areas. The most stringent of these categories applies to national parks and wilderness areas. PSD precludes further deterioration of ambient air quality where pollution levels are already below existing standards. PSD has been a hotly disputed issue, acted on favorably by the U.S. Supreme Court in 1973 and written into the Amendments in 1977. Western producers argued that PSD would be a stringent constraint on economic growth in the West (Tobin 1979).

The 1977 Clean Air Act Amendments strengthened the NSPS for SO_2 emissions. While the 1970 Act merely required that SO_2 emissions be limited to 1.2 pounds or below per million British thermal units (MMBtu), the 1977 Amendments required at least a 70% SO_2 removal rate <u>regardless</u> of BTU content. The new standards virtually required the use of flue gas scrubbers for most all coals burned in electric power plants constructed after September, 1978. If the SO_2 emissions are greater than 0.6 lbs. per MMBtu after scrubbing, the level of scrubbing efficiency must be 90 percent. If the emissions are lower than 0.6 lbs., the utility may choose a level of scrubbing between 70 and 90% (Zimmerman 1981).

According to Ackerman and Hassler (1981), the NSPS changes in the 1977 Amendments were a sellout of clean air to the high-sulfur coal industry. They further asserted that only through mandatory sulfur scrubbing would the high-sulfur coal industry regain its "lost" market from the low-sulfur coal industry in the West. Due to the regulatory delay, the 1977 Amendments were not implemented until mid-1979.

2.2 Labor-Management Relations

The coal mining industry has been characterized by great friction between labor and management in the U.S. and abroad. In the U.S., the coal mining labor force totaled about 222,000 in 1981, only a third of the level four decades earlier (U.S. Census Bureau 1982). Thus, coal miners have lost some of their power of earlier years. However, one of the worst strikes in the history of the UMW occurred in 1977-1978. But while about 75% of the U.S. coal miners are members of the UMW, the vast majority of the union works at underground mines in the Midwest and East, while relatively few work at the mostly surface (strip) mines in the West (Office of Technology Assessment 1979, 121-46). This would presumably lead to a regional effect of 1977-1978 and 1981 UMW strikes.

We investigate here the impact of labor disputes as perturbations in regional coal supply. We do not examine the arguments regarding the effects of UMW activity with respect to long term shifts in coal demand.

Zimmerman (1981) states that coal mining labor productivity declined sharply between 1969 and 1977 while real wages were increasing. The argument is made that because of the great labor intensity involved with underground mining a substitution takes place to less labor intensive strip mining. This suggests a counter hypothesis that UMW activity has little effect on coal production, although due to the prevalence of long term coal supply contracts such a shift would occur rather slowly.

The argument that is tested here is that the UMW strikes differ in the effect they have on coal production by region as an interruption in the production process. It may be argued that utilities may perceive production from unionized coal mines as less dependable. The counter arguments are that strikes are easily anticipated and the strategies for responding to an anticipated strike include stockpiling coal or, depending on the type of contract involved, switching for a short period to non-union Western coal.

3. METHODOLOGY AND DATA

3.1 Research Hypotheses

A set of hypotheses are examined here for selected coal basins in the United States. These hypotheses concern the relationship between coal production and a set of events that transpired in the 1970's and 1980's. The set of events are the Clean Air Act of 1970, the Amendments to the Clean Air Act of 1977, and UMW strikes in 1977-1978 and 1981.

As discussed above, the Act and subsequent amendments are thought to have influenced the regional competition between low-sulfur Western coal and high-sulfur Midwestern coal. The research hypotheses are operationalized in regional terms as follows: The Clean Air Act is negatively correlated to Midwestern coal production. The Clean Air Act is positively correlated to Western coal production. The 1977 Amendments are positively correlated to Midwestern coal production. The 1977 Amendments are negatively correlated to Western coal production. The UMW strikes are negatively correlated to Midwestern coal production. The UMW strikes are positively correlated with Western coal production.

The null hypotheses state that there is no correlation between the 1970 Act, the 1977 Amendments, the UMW strikes and coal production in any of the coal regions investigated.

3.2 ARIMA Models

The null hypotheses are first investigated by Autoregressive, Integrated, Moving Average (ARIMA) time series analysis.

ARIMA intervention analysis can be used to test the impact of a discrete intervention on a process such as coal production. In the ARIMA model the stochastic portion of the time series may be referred to as the noise component, N. The discrete intervention variable is referred to as I so that the impact assessment model is

$$X_t = w_i I_t + N_t \tag{1}$$

w_i is the quantity by which the time series increases or decreases at the onset of I. The subscript i denotes the order of the intervention component or the powers of the backshift operator employed, while t denotes the time period. Equation 1 represents an abrupt, permanent intervention. Gradual interventions require the addition of a third parameter, δ, for a lagged dependent variable:

$$X_t = \delta X_{t-1} + w_i I_t + N_t \tag{2}$$

The parameter δ takes on values between -1 and 1.

Finally, if the ARIMA component of the model is statistically adequate, the noise component can be subtracted to yield a deterministic intervention model:

$$X_t = \delta X_{t-1} + w_i I_t \tag{3}$$

Box-Jenkins time series analysis has been applied to regional problems in a variety of ways in the last ten years. Such models have been used extensively in modeling hydrologic characteristics of streams (Aguado 1982; Haynes et al. 1983; Beaumont 1979; Robinson and Rhode 1976). ARIMA time series investigation has also been applied to regional economic problems. Clark (1979) applied time series transfer function models to regional employment research. Similarly, ARIMA models have been applied by Fisher & Folmer (1982) to the measurement of the effect of Regional employment policy. Cook and Falchi (1981) demonstrate the family of Box-Jenkins techniques with an application for modeling spatial interaction in the formation of households. In energy geography research, Greene and Chen have applied the ARIMA time series models to the analysis of demand for gasoline by state (Greene and Chen 1983). One of the major purposes of using ARIMA models is to overcome problems of serial correlation in the error term, common to time series econometric models.

3.3 Econometric Models

Perhaps the most widely used model for estimation of economic and other social relationships in the linear statistical model

$$y = x\beta + e \tag{4}$$

where y is a vector of observations and x is non-stochastic matrix of explanatory variables. β is a vector of parameters to be estimated and e is a vector of unobservable random disturbances with a mean of zero. The parameters of this model have been estimated with least squares regression methods for an extremely large number of applications. The least squares criterion produces an estimate of the vector that minimizes the squared errors of the random disturbance e. Ordinary least squares (OLS) regression is the most widely used statistical method for empirically determining demand relationships (Baumol 1977).

The simplest form of the regression model used in energy research is probably the single equation. OLS regression model used in the study of gasoline consumption (Blomquist 1984). Here gasoline consumption is modeled as a function of the price of gasoline, real personal income, speed traveled, age of automobile and traffic density. The OLS results yield information on the influence of each of these variables on gasoline consumption.

Regional variation is introduced into these models in two ways. A separate regression analysis can be conducted for each spatial unit in the analysis. An example of the separate regression procedure is given in the estimation of returns to scale by Carlino (1982).

The other approach is that of dummy variables which takes on either the value of 1 or 0 to indicate the existence or absence of a particular condition. For example, if the data set include two or more regions, the regions could be represented with dummy variables. The variable D in equation 5

$$Y_t = B_0 + B_1 D + B_2 X_t + e_t \tag{5}$$

would take on a value of 1 for a model with two regions to indicate that the observation was from Region A and 0 if from Region B. The parameter B_1 indicates the increased value of the intercept in the presence of Region A. In equation 6:

$$Y_t = B_0 + B_1 D X_t + e_t \tag{6}$$

the dummy variable is used to indicate the change in slope of the regression line.

The regression model employed by the authors in this research is

$$X_{ci} = B_0 + B_1Y + B_2P_C + B_3P_O + B_4P_G + B_5D_1 + B_6D_2 + B_7D_3 + E \qquad (7)$$

This model assumes that coal production in region i (X_{ci}) is a function of demand (Y), fuel price (P_C, P_O, P_G,), environmental policy (D_1 ; D_2) and labor relations (D_3). The variables P_C, P_O and P_G are the price of coal, oil, and gas respectively. The intent in including these three prive variables is to test the effect of own price and price of possible substitute fuels on coal production. The variables D_1, D_2 and D_3 are dummy variables included to test the effect of the interventions of the 1970 Act, the 1977 Amendments and the UMW strikes. The variable Y is real personal income and it is used as a proxy for demand (Perryman 1984; Jambekar and Campbell 1984). It is expressed in terms of constant 1958 dollars. Demand as used here is a measure of economic activity that affects the production of coal-related energy in the national marketplace. We are suggesting that coal from different regions are goods that are substitutes in the national market.

3.4 Data

Monthly coal production statistics from four coal basins are used (U.S. Department of Interior 1960-1976). The coal basins are: (1) the Illinois coal basin in Indiana, Illinois and Western Kentucky, (2) the Powder River Coal Basin in Montana, (3) a sub-bituminous coal basin in Oklahoma and Arkansas and (4) Colorado and New Mexico. The first two coal basins are important because they represent high-sulfur coal of the Midwest and low-sulfur coal of the West.

The production data for 1977 through 1981 were collected from the U.S. Department of Energy. Prior to 1974 the coal production data were published by the coal district. In 1974 and after the reports were made by state. The basins used for data here are basins which were reported prior to 1974 that were discrete within state boundaries, thus providing data from 1960 to 1981 without changes in reporting regions. Because of changes in reporting areas, coal basins often do not have consistent data over the relevant time period.

The four included coal basins plus one other small basin in the State of Washington are the only ones which provide consistently measured data over the period of time in question. Thus, they are the only coal basins with an unbroken time series of data for the period 1960-1981. The basins used here are basins which may be aggregated in their pre-1974 form to an exactly corresponding basin in the 1974 and after form.

The variables for the interventions are zero before the intervention is in place and one after the event is initiated. For the UMW strikes the variables also take on a value of zero after the strike is over.

Additional data used in the regression model are price of coal, price of natural gas, price of fuel oil, and real personal income (U.S. Department of Energy 1981; U.S. Census Bureau 1960-1982). The price data are reported yearly and are national averages. For most of the regions in the analysis, most of the coal produced is transported to other regions for use. The national prices more properly reflect the potential market for the product.

4. RESULTS

4.1 ARIMA Analysis

Time series analysis is carried out for each of the four coal production regions. The sequence of identification, estimation and diagnosis is applied to the coal production of each region. The results are presented in Table 1.

Table 1. Time Series Analysis

	Coal Production Region			
	Indiana/W. KY/ Illinois 1	Oklahoma/ Arkansas 2	Colorado New Mexico 3	Montana 4
ARIMA Model	$(1,0,1)(0,1,0)^{12}$	$(1,0,0)(0,1,0)^{12}$	$(1,0,1)(0,0,0)^{12}$	$(1,1,1)(1,1,0)^{12}$
Interruption Variable		t values		
Act w	0.06	9.94*	9.49*	0.23
δ	.03	5.37*	4.26*	.01
Amends w	-3.94*	-3.58*	-3.66*	- .98
δ	-3.24*	-1.03	-2.12*	- .04
Strike w	-0.26	-8.02*	-2.77*	7.53*

* indicates significance at α = .05 KEY: Amends = Amendments

The ARIMA models shown in Table 1 indicate that the analyses of the coal production time series yield a different model for each coal production basin. The intuitive interpretation of this result is that the production stream from each of these basins is a result of a set of different factors. The basins vary widely in the sulfur content of the coal and in the BTUs per ton in yield. They also vary widely in the type of market that they serve.

The intervention analyses yield different results for the coal basins as well. In Region 1 there is only one significant interruption variable. The significant variable is the 1977 Amendments. Region 1 is the Midwestern region. The expectation in the Midwestern basin originally was for a negatively significant coefficient for the 1970 Act, to support the hypothesis that the Act had a negative impact on the producers of high-sulfur coal. The results show the Act to be insignificant in its correlation with production.

In addition, the expectation for the w coefficient for the Amendments was a positive, significant coefficient. The coefficient is significant but it is negative, indicating that the Amendments had a negative effect on the basin. Finally the largest negative effect for the strike was expected in the Midwest. The result, however is an insignificant coefficient for the UMW strike. The delta coefficient is significant for the Amendments indicating that the effect of the amendment in this region was indeed gradual rather than abrupt. Region 2, Arkansas and Oklahoma, provided significant w coefficients for all three interruption variables. The t value of w for the Act in Region 2 is large and positive indicating that the 1970 Act actually caused an increase in coal production from the initiation of the air quality standards. The delta parameter for the Act in Region 2 is strongly positive indicating that the effect of the Act was gradual rather than abrupt in this region.

The t value of w for the Amendments is negative and significant indicating a negative impact from mandatory scrubbers on coal production. These results are more in line with expectations from western states with low-sulfur coal. Arkansas and Oklahoma are in the Interior Western coal basin with a substantial amount of low volatile bituminous coal, which is typically high in sulfur content. Region 2 also displays a large negative impact from the UMW strike.

Region 3, Colorado and New Mexico, in Table 1 yields a set of results that are expected of Western states with low-sulfur coal. The exception is the significant negative coefficient of w for the Act, which is in conformance with the expectation that the Act would provide an incentive for Western coal production. The negative and significant coefficient of w for the 1977 Amendments is in conformance with the expectation that the Amendments would have the opposite effect.

Region 4, Montana, did not perform entirely according to expectations either. The only significant variable in Region 4 is w for the UMW strike. In this case the sign is the one expected, since the Western coal mines are surface mines and generally characterized as non-union. The expectation was that the strike would cause Western coal production to increase while Midwestern coal production declined. The expectation is borne out in Region 4. The Amendment and the Act are not significant, counter to expectations. The signs on these two variables are the expected signs.

It is also of interest to note that w for the Act has a positive sign across all four basins, regardless of significance level. The Amendments have a negative sign for w across all four basins and it is significant in three of four cases. The strike produces mixed results in its signs when compared across basins. The sign for this variable is negative in three instances and positive in one.

4.2 Econometric Analysis

OLS regression has been applied to econometric models of coal production in each of the four study regions. These results are summarized in Table 2, and are consistently superior to those from the ARIMA analysis. However, many of the signs on the coefficients initially appear counterintuitive, but can probably be attributed to regional structural differences in the U.S. coal industry. Overall, the adjusted $\bar{R}^2$ are satisfactorily high, and autocorrelation was not detected in the error term in any region.

With coal production as the dependent variable Region 1 (the Midwest) shows a respectable adjusted $\bar{R}^2$, with both disposable personal income and coal price strongly significant and exhibiting the expected signs. Neither oil nor gas price is significant in this region. The 1970 Clean Air Act (D_1) shows the expected negative sign and is significant. The 1977 Clean Air Act Amendments (D_2), surprisingly, is highly significant with a negative sign. This may indicate that the mandatory requirement for scrubbers at coal-fired electric power plants hurt coal production in the Midwest with the necessary capital investment costs not causing much or any regional shift in production. This may reflect the fact that coal-fired power plants with construction permits before 1977 could continue to choose the low-sulfur coal option, and thus the shift wouldn't occur immediately. Finally, the UMW strike (D_3) exhibits the expected negative sign, but is insignificant in the unionized Midwest.

Region 2 of Oklahoma and Arkansas also has a high $\bar{R}^2$, but coal price is insignificant, as is personal income. However, this region is a very large producer of natural gas, and gas price is strongly significant here. The positive coefficient on gas price indicates the expected cross price elasticity relationship. Once again, all three dummy variables have negative signs, but only the UMW strike is significant (and strongly so).

Region 3 of Colorado and New Mexico has a high $\bar{R}^2$, but with only three significant explanatory variables. All three price variables have the expected sign but they are not significant. In the case of coal price, as price declines production rises as one would expect if demand was driving production. This indicates, perhaps, that there is a cushion between the national market and production in this region due to the captive market for electric utilities. The Clean Air Act increases coal production in this region as expected, and significantly, while the 1977 Amendments also result in increased production, counter to expectations but not significantly so. The UMW strike significantly increases coal production in this relatively non-unionized region as expected and is significant. Personal income is a highly significant, positive explanatory variable as expected as well.

Region 4, Montana, also has an exceptionally high adjusted $\bar{R}^2$, with only two significant variables, the price of coal and the 1970 Act. In the case of coal price the positive sign may reflect an outright preference for Western coal in times of rising energy prices. It should be noted that much of the Western coal is exported to other regions in contrast with the fate of Midwestern coal. D_1 is positive and is significant as in Region 3, while D_2 is insignificant with a negative sign. The most surprising result in Region 4 is that D_3 has an unexpected negative sign, although this variable is insignificant in this region. The reason for this result is unclear.

Table 2. Regression Results of X_c †

Explanatory Variables	Coal Production Region: Indiana/W. KY/ Illinois	Oklahoma/ Arkansas	Colorado/ New Mexico	Montana
constant	19,694.0 (1.92)	-896.8 (1.748)	-14,010.3 **(7.55)	-10,446.6 *(2.523)
Personal Income	-209.1 **(9.5)		47.82 **(7.56)	11.866 (1.312)
Price of coal	-1,173.0 *(2.13)	151.79 (1.41)	-467.89 (1.637)	1,139.7 *(2.791)
Price of oil			407.299 (1.578)	256.17 (0.69)
Price of gas		*3007.25 (2.169)	1,996.1 (0.734)	-3,847.4 (0.99)
1970 Act (D_1)	-12,139.98 *(2.138)	-358.72 (0.466)	4296.729 *(2.591)	6421.34 *(2.709)
1977 Amend- (D_2) ment	-24,254.2 **(4.378)	351.42 (0.431)	124.894 (0.065)	-2114.67 (0.765)
UMW Strikes (D_3)	-10,930.7 (1.651)	-2,789.1 *(2.416)	7364.58 **(3.129)	-858.155 (0.255)
$\bar{R}^2$	.896	.858	.975	.964
DW	1.7136	2.4337	2.5365	1.5428
N	22	22	22	22

† t-values in parentheses (2-tailed test)
* indicates significance at the 95% confidence level (α = .05)
** indicates significance at the 99% confidence level (α = .01)

To summarize the regression results with regard to the research hypotheses state earlier, the results are interesting but quite mixed. First, the OLS model shows that for the 1970 Clean Air Act the expected shift of coal production from the Midwest to the West occurs. The 1977 Clean Air Act Amendments show an unexpected negative sign in the Midwest. The sign is negative as expected in Montana but the variable is not significant. The 1977 Amendment is insignificant in the other two regions as well. Finally, the UMW strike results in a large decline in coal production in the unionized Midwest, and Oklahoma and Arkansas, a gain in non-unionized Colorado and New Mexico, but a decline (albeit insignificant) in non-unionized Montana.

5. CONCLUSION

5.1 Results

The one result that stands clear in terms of analysis of policy is that of the impact of the Clean Air Act of 1970. The OLS analysis indicates that, as expected, the Act had a negative impact on Midwestern coal production and a positive impact on coal production from the large Western coal district in Montana. This is validated by the results in the Colorado/New Mexico Western coal district.

The results are less clear with regard to the 1977 Amendments and the UMW strikes. The only basin in which the Amendments are significant in the OLS analysis is the Midwestern coal basin, and the results show that the Amendments had a negative impact instead of the expected positive impact. The results are the same in the ARIMA analysis for the Midwestern coal basin. The two smaller basins also have significant negative results in the ARIMA analysis. The fact that the required scrubbers were imposed on new construction and not on existing power plants probably has a great impact here. The lag time between the 1977 Act and its impact would be long enough that the time period 1977-1981 would not give much evidence of the effect of scrubbers on regional coal production.

5.2 Limitations

Both models used are subject to limitations and these limitations have to be weighed in the process of interpreting the results. The ARIMA model is limited because of its relatively simple view which is that the determinant of production in time t is coal production in time t-1. This view produced modest results in terms of goodness-of-fit for the ARIMA models tested.

In considering the relative limitations of the two types of models, the regression models produced consistently better goodness-of-fit than did the Box-Jenkins interruption time series models. The danger to be avoided in the regression model is that of a spurious regression due to misspecification of the error structure. It is not unusual to find econometric work with high $\bar{R}^2$ and very low Durbin-Watson values. (See

Granger and Newbold 1977, for a discussion of this problem).

In the regression analysis here, the $\bar{R}^2$ measurements were high and the Durbin-Watson statistics were not unduly low. In comparison, the univariate ARIMA models fit to the coal production data produced moderate results in reducing the error terms to a white noise process.

A useful extension of this research would be a multi-equation model of the process that would investigate the independent variables in a more disaggregated fashion. Disaggregation would also be beneficial in terms of the coal fields. To some degree there is a homogeneity problem built into the analysis in that within each of these coal basins the coal produced is not likely to be as homogenous a product as one would like. Despite the limitations, it would be fruitful for geographers to contribute their regional perspective toward a better understanding of the impacts of coal development constraints.

6. ACKNOWLEDGEMENTS

The authors acknowledge the support of the National Science Foundation Grant #48-967-12. The views expressed in this paper are those of the authors and not necessarily those of the above institutions.

7. REFERENCES

Aguado, E. 1982. A time series analysis of the Nile River low flows. Annals of the Association of American Geographers 72:109-19.

Akerman, B.A., and Hassler, W.T. 1981. Clean coal/dirty air. Binghamton, New York: Vail Ballou Press.

Baumol, W. J. 1977. Economic theory and operations analysis. 4th ed. Englewood Cliffs, New Jersey: Prentice Hall.

Beaumont, C. 1979. Stochastic models in hydrology. Progress in Physical Geography 27:363-91.

Blomquist, G. 1984. The 55 m.p.h. speed limit and gasoline consumption. Resources and Energy 6:21-40.

Campbell, T. C., Hwang, M. J., and Shahrokh, S. 1979. Market delineation in the coal industry. The Review of Regional Studies 9:6-17.

Carlino, G. A. 1982. Manufacturing agglomeration economies as returns to scale: a production function approach. Papers of the Regional Science Association 50:95-108.

Clark, G.L. 1979. Predicting the regional impact of a full employment policy in Canada: A Box-Jenkins approach. Economic Geography 55: 213-26.

Cook, T., and Falchi, P. 1981. Time series modeling in a regional economy: An exposition of Box-Jenkins techniques. Environment and Planning A 13: 635-44.

Energy Information Administration. 1984. Annual energy review, 1983. Washington, D.C.: U.S. Department of Energy.

Fisher, M.M., and Folmer, H. 1982. Measurement of the effects of regional policy by means of time series analysis. Papers of the Regional Science Association 49: 133-50.

Granger, S.W.J., and Newbold, P. 1977. Forecasting economic time series. New York: Academic Press.

Greene, D.L., and Chen, C.K. 1983. A time series analysis of state gasoline demands, 1975-1980. The Professional Geographer 35: 40-51.

Haynes, K.E., Georgianna, T.D., Miller, T.K., Whittington, D. 1983. Time series analysis of low flow stages of the Nile River: 1622-1469. Modeling and Simulation 14: 905-09.

Jambekar, A.B., and Campbell, G.A. 1984. The use of the system dynamics in studying the behavior of the mineral industries: The case of the U.S. zinc smelting industry. Modeling and Simulation 15: 1271-76.

Office of Technology Assessment. 1979. The direct use of coal: prospects and problems of production and combustion. Washington, D.C.: U.S. Congress.

Perryman, M.R. 1984. An integrated econometric modeling strategy for projecting energy demand in small utility service areas: Theory and application. Modeling and Simulation 15: 1253-60.

Robinson, S.A., and Rhode, F.G. 1976. Stochastic modeling of monthly stream flow in the Federal Republic of Germany. Nordic Hydrology 7: 43-56.

Tobin, R.J. 1979. The social gamble: Determining acceptable levels of air quality. Lexington, Massachusetts: D.C. Heath & Company.

Tyner, W.E., and Kalter, J.J. 1978. Western coal: Promise or problem. Lexington, Massachusetts: D.C. Heath & Company.

U.S. Census Bureau. 1960-1982. Statistical abstract of the U.S. Washington, D.C.: U.S. Department of Commerce.

U.S. Department of Interior, Bureau of Mines. 1960–1976. *Minerals yearbook*. Washington, D.C.: U.S. Government Printing Office.

U.S. Department of Energy. 1981. *Energy data report; cost & quality of fuels for electric utility plants*. Washington, D.C.: U.S. Government Printing Office.

Zimmerman, M.B. 1981. *The U.S. coal industry: the economics of policy choices*. Cambridge, Massachusetts: The MIT Press.

CHAPTER 23

REGIONAL ECONOMIC TRADE-OFFS IN SULFUR EMISSIONS CONTROL POLICY

Mark Berkman
National Economic Research Associates
123 Main Street
White Plains, New York 10601

Peter Blair
Office of Technology Assessment[1]
U.S. Congress
Washington D.C. 20570

ABSTRACT. The current debate over acid rain controls is chiefly a regional conflict. The distribution of employment impacts may be the most critical obstacle to establishing a control policy. Despite the importance of this issue, there have been no thorough employment impact studies. This chapter reports on the development and application of a regional modeling system designed to fill this gap. The system combines several existing models including a multiregional input-output model and a regional electricity model. Initial results indicate that net employment losses from acid rain control are typically overstated as a result.

1. INTRODUCTION

While the country's coal supply is found in many regions, the cost of extraction, coal quality and proximity to market vary substantially by region. As a result, Federal coal and environmental policies affect regions differently--a policy which benefits one region may harm another. The current debate over acid rain control is the prime example of this conflict. Acid rain--rain contaminated with sulfur dioxide and nitrogen oxides--is thought to cause serious damage to the environment. It results largely from burning hydrocarbons (coal, fuel oil and gasoline). Coal-fired electric utility plants are responsible for a large share of the emissions causing acid rain. Consequently, legislation proposed to control acid rain concentrates on coal-fired plants.

A wide array of legislative proposals have been formally introduced in Congress or publically promoted during the last two sessions. Eleven bills were considered during the 98th Congress. Six bills were introduced during the 97th Congress. The Congress has not succeeded in passing

F. J. Calzonetti and B. D. Solomon (eds.), Geographical Dimensions of Energy, 443–470.
© 1985 by D. Reidel Publishing Company.

legislation because of two fundamental disputes. The first dispute is over the magnitude of the acid rain problem. The second dispute is among those who agree on the magnitude of the problem but who disagree over how acid rain should be controlled.

The legislative proposals generally fall into three categories reflecting these differences--those requiring technology-based controls, those requiring market-based controls, and those requiring further research rather than control. The technology-based proposals require the installation of coal scrubbing equipment to meet utility emissions standards. The market-based proposals establish similar emission standards, but allow utilities to meet them by the least costly method available (certain bills limit the methods available to some extent). For many utilities, this method will be burning low sulfur coal rather than adding costly scrubbing equipment. Proposals in the third category parallel the Reagan Administration's position that further research regarding the environmental consequences of acid rain must be completed before a costly control policy is implemented.

Even without Presidential support, there appears to be sufficient Congressional support to enact some form of control legislation. The real obstacle to passing such legislation is the dispute among control proponents. While the market-based approach appears far more cost effective on an emission-reduction-per-dollar basis, there is strong opposition by eastern and midwestern representatives and coal interests because of the coal differences across regions noted above. The market based approach creates a premium for low sulfur coal which occurs predominantly in the west (although sizeable low sulfur coal reserves exist in central Appalachia). Opponents to the market-based approach claim that the resulting shift in coal markets from east to west will create serious unemployment among eastern coal miners. These direct job losses, along with indirect losses, are projected to cause widespread economic disruptions in eastern coal mining regions. For example, the United Mine Workers (UMW) forecast annual economic losses of $3.0 to $4.6 billion under a market approach. The Office of Technology Assessment (OTA) estimates annual losses in the high-sulfur coal areas of $0.6 to $2.7 billion.

Proponents of the market-based approach acknowledge that there will be some job losses in certain eastern coal regions, but they also note that increased demand for eastern low sulfur coal will expand job opportunities in other eastern coal regions. They also point out that increased demand for western coal will create jobs so that nationwide, job losses will not be large. In fact, one analysis found that when jobs in the transportation sector are considered, the market approach results in more jobs nationwide than the technology approach (ICF 1983). OTA, for example, estimates that despite losses in the midwest, there will be no substantial economic losses in the aggregate (OTA 1984). OTA also estimates that the technology approach would cost approximately $1.5 billion more than the market approach. Given the substantial cost differences between the two control policies, market-based proponents have even supported the addition of some form of job assistance program to mitigate job dislocation, believing that the additional costs will still result in a control program well below those incurred under a technology-based program (Portney 1983).

This debate over employment impacts is an extension of the regional conflict over sulfur emissions that began with the Clean Air Act Amendments of 1970 and resulted in the Clean Air Act Amendments of 1977. As Ackerman and Hassler (1980) note, job protection rather than environmental concerns have dominated the policy debate. Despite the importance of the employment issue, however, there have been no comprehensive employment impact studies. Studies conducted by OTA (1984), ICF (1983), the United Mine Workers (1983), and the Congressional Research Service (1983) rely on simple multiplier analysis to estimate the regional employment and income impacts of acid rain control policies. For example, ICF relies on miner productivity statistics to forecast mine employment as coal production levels vary by mining region (coal production changes are multiplied by productivity measured in tons per miner). Transportation employment resulting from increased coal shipments (primarily by rail) between regions is also estimated based on productivity statistics. However, no attempt is made to forecast where these jobs will occur. OTA and UMW follow similar procedures. A preliminary report by Jonash (1984) suggests a more comprehensive and systematic approach has been taken. Pollution control employment is considered in addition to coal and transportation. However, the modeling procedures used are not described in the report.

2. THE NEED FOR REGIONAL MODELING

Given the importance of the employment issue, there is an obvious need for a more thorough analysis utilizing a comprehensive regional model. As noted above, most (if not all) coal employment impact analyses conducted during the acid rain debate have relied upon the simple multiplier procedure to estimate economic impacts. This method, however, does not account for many of the possible employment impacts which may accompany a regional shift in coal production. The simple multiplier method provides what economists term a partial equilibrium analysis. This form of analysis does not capture supply and demand responses of other economic sectors or regions. A general equilibrium analysis is required to account for these responses.

The lack of interregional linkages appears to be the most critical omission of the simple multiplier method. Most regions are not self-sufficient in all goods and services and, therefore, rely on each other for the provision of certain goods and services. Because of these interregional linkages, changes in one region's economy have direct consequences in other regions.

Such linkages may result in important employment gains in certain sectors of a region despite losses in other sectors. Ignoring these offsets will overstate employment losses. For example, while there may be a loss of coal mining jobs in Illinois as coal production shifts to Wyoming, there may also be an increase in manufacturing jobs in Illinois because of increased demand for mine equipment in Wyoming.

There is evidence that such offsets will occur. A study by Ballard and Wendling published in Growth and Change (1980) found that interregional linkages are important in assessing coal production impact on employment. Ballard and Wendling traced the impacts of increased coal mining in Montana across all regions of the country using a multiregional model which enabled them to conduct a general, rather than partial, equilibrium analysis. They found that next to Montana, the East North Central region (Illinois, Indiana, Michigan, Ohio and Wisconsin) benefitted the most from increased Montana coal production in terms of employment, income and regional output. The positive impacts in the East North Central region were attributed primarily to increased demand for manufactured goods.

There is also data showing the significance of the link between western coal production and eastern and midwestern employment. For example, Table 1 shows the regional flow of construction equipment. (Construction equipment includes most mining equipment except for motor vehicles.) The Mountain region, which includes Montana, Wyoming and Colorado, relies heavily on the East North Central region (Illinois, Indiana and Ohio) for this type of equipment. Almost 40 percent of construction equipment used in the Mountain region is produced in the East North Central region. Consequently, increased mining activity in the Mountain region will create demand for midwestern equipment production and, therefore, jobs. Table 2 shows mining equipment manufacturing by plant and region. Again, the East North Central and Northeast regions account for most employment.

The interregional effects, however, are complex and care must be taken in measuring them. Employment offsets cut both ways and are impossible to trace thoroughly without the use of a model. Factors such as existing mining capacity and the frequency of equipment replacement must be accounted for. Also, the employment estimates under the projected regional coal production shifts must be compared to those made assuming that a technological fix is imposed. Demand for scrubbing equipment may have important positive employment effects in some regions. On the other hand, the addition of such equipment will result in higher electricity costs which may dampen regional activity, and therefore employment, in certain regions.

3. THE MULTIREGIONAL COAL POLICY IMPACT MODEL

The modeling framework developed for this work was designed with the following capabilities.

1. Estimation of the regional coal supply as a function of mining cost and cumulative production of coal in the region.
2. Estimation of regional electricity demand and non-electric demand for coal as a function of relative prices for alternative fuels and a derived weighted average price of energy; coal exports and demand for metallurgical coal are specified exogenously.
3. Estimation of coal demand for electric power generation as a

Table 1. Regional Trade Flows: Construction and Related Machinery

Origin	Destination					
	Northeast	East North Central	Central	South	Mountain	Pacific
	Percent					
Northeast	26.4	9.9	4.4	10.5	5.1	4.9
East North Central	47.3	60.3	28.4	52.2	39.6	33.0
Central	12.2	17.1	55.0	16.2	39.3	21.8
South	9.2	6.7	3.9	19.3	2.7	3.7
Mountain	2.6	1.9	1.4	0.6	6.0	1.6
Pacific	2.3	4.1	6.9	1.2	7.3	35.0
	100.0	100.0	100.0	100.0	100.0	100.0

Note: Northeast includes New England and Middle Atlantic Census regions. Central includes West North Central and West South Central Census regions. South includes South Atlantic and East South Central Census regions.

Source: U.S. Department of Transportation, Commodity Transportation Survey, Census of Transportation, 1980.

Table 2. Mining Equipment Plants By Region 1981 - 1982

Region	Number of Plants	Percent of Total	Number of Employees	Percent of Total
1. Northeast	15	18.1	8,000	27.2
2. East North Central	19	22.9	6,300	21.4
3. Central	13	15.7	4,600	15.6
4. South	21	25.3	5,800	19.7
5. Mountain	11	13.3	3,600	12.2
6. Pacific	4	4.8	1,100	3.7
Total	83	100.0	29,400	100.0

Source: Marketing Economics Institute, "Marketing Economics Key Plants," 1981.

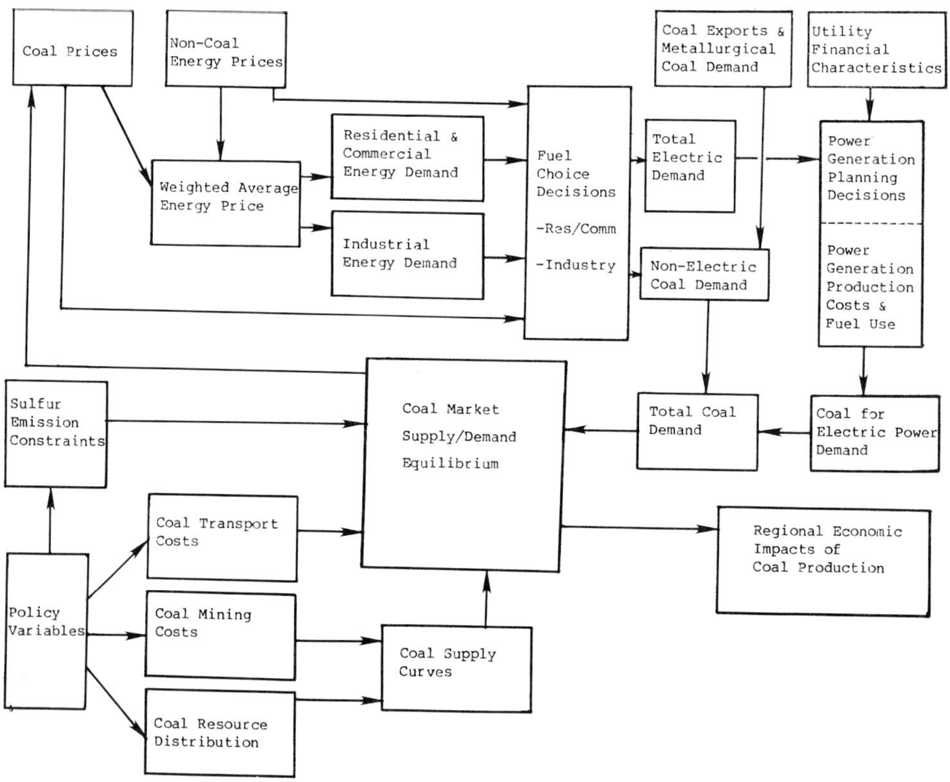

Figure 1. Multiregional coal impact assessment model.

function of regional electric power system load and financial characteristics as well as the relative costs of competing power generation technologies.

4. Calculation of regional production levels and interregional shipments of coal given supply characteristics specified in (1), coal demands specified in (2) and (3), and constraints on sulfur emissions of various coal categories.
5. Determination of regional primary and secondary impacts of coal production levels computed in (4) in terms of regional product, employment and income levels.

This model was developed in part by modifying and combining a number of existing models. In particular, (1) and (2) above follow the methodology of Zimmerman (1981) who developed regional coal supply curves as a function of a number of observable geological characteristics; we discuss this methodology in more detail later. Component (2) is an econometric demand model derived from a version of the Baughman et al. (1979) model. Component (3), also based on the Baughman et al. approach, predicts new coal-fired generation planning and production costing tools. Component (4) is a linear programming supply/demand equilibrium model much like that employed by Zimmerman (1981), ICF (1977), and other coal market models. Component (5) employs an aggregated version of the U.S. Multiregional Input-Output Model (MRIO), developed by Polenske (1980). The overall modeling framework is shown schematically in Figure 1.

3.1. Coal Supply

A national-level coal model based on Zimmerman (1981) is used to simulate the market for coal. This model is an integrated supply and demand model designed for coal policy analysis. Six coal producing regions and nine coal demand regions are distinguished by state.[2] The supply model is based on twenty cumulative cost functions per state (for eighteen coal producing states). The Zimmerman cost functions are specified by type of mining (strip and deep mining) and coal sulfur content (ten sulfur levels) accounting for the twenty equations. For computational purposes of this study, the states are aggregated to six regions, the cost curves for strip and deep mining are summed, and the sulfur categories are reduced from ten to three.

The deep mine cost function is derived from a model of the production process because cost data by mine is not available. The basic cost function is given as:

$$C = (Th, Q) + \varepsilon \qquad (1)$$

where C is cost, Th is seam thickness, Q is mine output and ε is a disturbance term reflecting other mining conditions. To model the production process, the mine is described in terms of production units or mining sections. The output of these units is described as:

$$\frac{Q}{u} = f(Q, G) + \varepsilon \qquad (2)$$

Where Q is the output of the mine, u is the number of producing units in the mine, G is the collection of measured geological characteristics and ε is the disturbance term. Capital and labor expenses are then assumed functions of the number of producing units or

$$E_j = g(u), j = 1,2,\ldots, N \tag{3}$$

where E represents a particular type of expenditure j, capital, labor, materials, etc. Substituting equation (2) into (3) gives a derived cost function where expenditures E are a function of output Q and observable and unobservable geological characteristics.

$$E_j = \left(\frac{Q}{f(Q, G) + \varepsilon}\right) \tag{4}$$

Following a statistical specification of mine productivity, by section, reflecting these geological characteristics and expenditures, by type, per mining section, a total cost function is derived.[3] Total cost is then used to estimate the minimum average cost and the output rate that meets this cost (assuming that in the long run marginal cost equals minimum average cost). Marginal cost can now be defined in terms of geographical characteristics[4] (seam thickness or overburden); the distribution of coal by the cost is obtained. This is estimated for each state. Finally, the distribution of sulfur content is introduced, resulting in cost function (supply curves) for each state by mining type and sulfur content.

The supply curves estimated by Zimmerman (1981) take the following form:

$$S = aRn^{\alpha} + \beta \text{ for surface mining and} \tag{5}$$

$$S = \frac{K}{(Th)^{\alpha}\,\varepsilon} \text{ for deep mining}$$

where a, β, S, α, and K are estimated parameters; R is the overburden ratio (feet of overburden to feet of coal); Th is the seam thickness; and ε and n represent underlying geologic conditions; S is the average cost per ton produced. Based on regional distributions of the observed geological conditions of coal reserves (over-burden and seam thickness), supply curves are restated in terms of production cost and cumulative production. In our version of the coal model we specify 18 supply curves, i.e., 3 sulfur categories for each of 6 supply regions.

For computational purposes, the supply curves are linearized into a sequence of supply steps as depicted in Figure 2. The curves in this form can be introduced to a linear programming problem integrating coal supply and demand which we discuss later.

To estimate the delivered price of coal, a transportation model is linked to the coal supply model. Transportation costs are estimated by two econometric equations, one for eastern and midwestern coal and one for

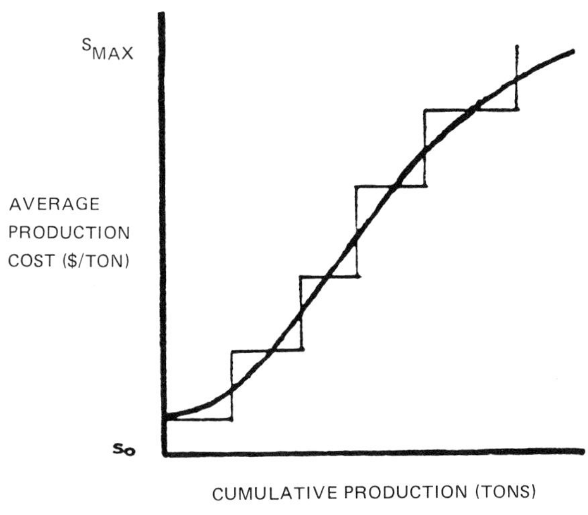

Figure 2. Coal supply curves.

western coal (supply and demand regions are identified based on existing potential routes). The difference in equation specifications reflects historical differences in rates in the two regions. The model produces a matrix of coal shipping rates (dollars per ton) by supply and demand regions. This information is introduced to the overall supply-demand model integration framework described below.

3.2. Coal Demand

The demand submodel has three components, an electric utility demand model, an industrial demand model, and export and metallurgical demand. The electric utility model is based on the Regionalized Electricity Model (REM) developed by Baughman, Joskow, and Kamat (1979). REM is an engineering-economic model that incorporates fuel choice by residential, commerical, and industrial sectors. A logit specification is used to model these decisions. Demand for electricity (one of three fuel types) by sector is then integrated with an electric utility supply model. The electricity supply model predicts utility response to estimated demand in terms of which type of generating plant is used (coal, natural gas, oil, nuclear, and hydro). This generation decision is based on projected load duration and financial and operating characteristics of available generators. Additional plant capacity decisions are also modeled. Investment in new plant is based on projected load duration and the economic parameters of alternative plant technologies (there are five technologies available: peaking units, coal fired units, oil and gas fired units, light water nuclear reactors, and liquid metal fast breeder reactors). Utility coal demand from REM is based on utility decisions regarding existing coal fired capacity, coal retrofit and on the construction of new coal fired capacity. In turn, utility decisions are based on electricity demands of two sectors, residential-commercial and industrial.

Residential-commercial demand is specified as a function of the price of energy relative to the price of other goods and services, P, income per capita, Y, and other demographic or geographic variables, Z. These variables include a measure of population density, average temperature for the warmest three months of the year, and average temperature for the coldest three months of the year. The basic energy demand equation for region i can be written as:

$$D_i = f(Y_i, Z_i, P_i, E) \tag{6}$$

where E is a disturbance term.

Electric energy demand is derived from this estimation by a logit specification of energy type choice. Energy choice is described as a function of relative prices, household income and certain geographic variables (the same temperature related variables are used). The fuel choice model specification is written as

$$\ln\left(\frac{S_1}{S_3}\right) = \alpha_0 + \alpha_2 \ln\left(\frac{P_1}{P_2}\right) = \alpha_3 Y_{it} + \alpha_4 Z_i + \alpha_5 \ln\left(\frac{S_1}{S_3}\right)_{i,t-1} + \varepsilon_{it} \tag{7}$$

$$\ln\left(\frac{S_2}{S_3}\right) = \beta_0 + \beta_2 \ln\left(\frac{P_2}{P_3}\right) = \beta_3 Y_{it} + \beta_4 Z_{it} + \beta_5 \ln\left(\frac{S_2}{S_3}\right)_{i,t-1} + \varepsilon_{it} \tag{8}$$

$$S_1 + S_2 + S_3 = 1 \tag{9}$$

Where S are the market shares, P are the fuel prices, Y is income per capita, Z are geographical variables, and ε_{it} is a disturbance term. The α and β terms are adjustment parameters to the lagged dependent variables indicating the speed of fuel adjustments (movement from actual to desired fuel consumption). The fuel choice equations are linked directly to the demand equation because the price variable in the latter is weighted by the fuel split choices.

The industrial energy demand model of the REM model has three components: (1) an aggregate energy demand equation, (2) a set of location decision equations, and (3) a set of fuel choice locations by state. The aggregate demand equation assumes that total demand is a function of value added, and average energy price.

The location equations determine the fraction of total demand consumed in each state (excluding Hawaii and Alaska and including the District of Columbia). State industrial energy consumption is based on a logit formulation which assumes that industrial decisions to locate and purchase energy in a given state are a function of state-specific attributes, a weighted average fuel price and state population. The latter is a proxy for existing and potential labor supply. Other possible state attributes such as cost of capital services were either not available or assumed not to vary significantly across states.

The fuel choice equations are specified in logit form by state. Fuel prices for the four fuels available (oil, natural gas, coal, and electricity) are the explanatory variables. Fuel choice decisions are used to estimate direct industry coal demand and indirect coal demand via demand for electricity. Electricity price is introduced to the fuel choice equations through a financial and regulatory model that produces two prices, one for industrial sector and another for the residential-commercial sector. The model calculates these prices based on estimated rate bases and regulated rates of return for utilities in nine census regions. The financial regulatory model partially integrates the supply and demand model because it requires outputs from both to determine utility operating costs.

The REM model is used to specify two of the three components of coal demand, i.e., by industry and electric utilities. The third component, metallurgical and export coal, is specified exogenously. Demand is allocated based on historical patterns and projected by trend analysis. Figure 1 summarizes the relationship of the energy demand and fuel choice equations to the estimation of total coal demand.

3.3. Coal Market Equilibrium

Integration of coal supply and demand is accomplished by posing a linear programming problem. The objective function is to minimize the cost of mining and transporting coal subject to several sets of constraints. The demand constraint forces the model to meet demands by region. The sulfur constraints reflect the sulfur standards binding in a particular region and for particular power generating plants. There are four relevant standards in constructing these constraints: (1) the new source performance standards (NSPS) for plants beginning operation after 1975, (2) the scrubbing requirements for plants under construction after 1978, (BACT), (3) state emission standards, and (4) nondegradation requirements.

A third set of balance constraints guarantee the consistancy of the model integration, i.e., quantities produced are set equal to quantities shipped. The coal supply curves are linearized, and non-negativity is established. The linear programming formulation is the following:

$$\text{Minimize TC} = \sum_{i=1}^{6} \sum_{j=1}^{6} t_{ij} \sum_{k=1}^{3} Z_{ijk} + \sum_{i=1}^{6} \sum_{j=1}^{3} \sum_{k=1}^{6} c_{iks} Y_{iks} \tag{10}$$

Subject to

$$\sum_{i=1}^{6} \beta_i \sum_{j=1}^{6} Z_{ijk} \geq D_j ; \; j = 1,2, \ldots, 6 \tag{11}$$

$$\frac{\alpha \sum_{i=1}^{6} \sum_{k=1}^{3} \rho_k Z_{ijk}}{\sum_{i=1}^{6} \beta_i \sum_{k=1}^{3} Z_{ijk}} \leq S_j ; \; j = 1,2, \ldots, 6 \tag{12}$$

$$\sum_{s=1}^{6} Y_{iks} - \sum_{j=1}^{6} Z_{ijk} = 0 ; \; i = 1,2, \ldots, 6 \quad k = 1,2,3 \tag{13}$$

$$Y_{iks} \leq U_{iks};\ i = 1,2,\ \ldots,\ 6 \quad k = 1,2,3 \quad s = 1,2,\ \ldots,\ 6 \tag{14}$$

$$Z_{ijk},\ Y_{iks} \geq 0;\ i = 1,2,\ \ldots,\ 6 \quad j = 1,2,\ \ldots,\ 6 \quad k = 1,2,3 \quad s = 1,2,\ \ldots,\ 6 \tag{15}$$

where

TC is total mining and transportation cost across all supply and demand regions and sulfur categories.

Z_{ijk} is the coal in tons of sulfur category type k shipped from supply region i to demand j.

Y_{iks} is the level of coal production in region i of sulfur category type k in supply step s.

t_{ij} is the transport cost per ton of shipping coal from supply region i to demand region j.

c_{iks} is the coal mining cost per ton of producing coal in supply region i of sulfur category k in supply step s.

B_i is the average heat content (Btu) per ton of coal produced in supply region i.

ρ_k is the sulfur content lbs. per 10^6Btu for sulfur category k.

α is the scaling factor to convert percentage sulfur specifications to lbs. per 10^6Btu.

U_{iks} is the amount of coal production of sulfur type k produced in region i included in supply step s.

Finally, it must be noted that such a model is essentially a long-term equilibrium model. Incremental changes in coal supply and demand are estimated based on long run costs. To account for rapid changes, however, the demand constraints reflect the estimated depletion of mines in operation at the time the model begins a period by period, i.e., year by year, simulation. Therefore, optimization in each period is based on incremental demand which accounts for net increases and mine closures.

3.4. Economic Impact Assessment

The third and final stage of the analysis methodology is the assessment of the economic impacts of the proposed revisions, using a multiregional input-output analysis model (MRIO). The MRIO model is used to derive regional economic activity (employment and income) estimates from regional coal production values. Changes in regional production levels and patterns will affect the relative level of economic activity across regions.

The MRIO model was selected for our analysis based on a review of regional models with energy policy applications.[5] The review indicated that no ideal model exists and that there were tradeoffs between important model characteristics. The more dynamic models, for example, also tended to have fewer energy policy "handles." Models providing significant regional and industrial detail were found to be too complex to be useful in the current study.

The MRIO, despite its static nature and dated data base, seems the most attractive model for this research. Two of its characteristics make it the model of choice, namely its straightforward and comprehensive accounting structure and its high level of sectoral and spatial disaggregation. The detailed accounting structure makes it possible to compensate for its static behavioral assumptions; dynamic submodels have been introduced to modify technical and trade coefficients. Updates have also been completed to improve the data base. Information is available to accomplish this for a number of important sectors.[6] The high level of disaggregation increases the likelihood that submodels can be linked (it is easier to aggregate a disaggregate model) and allows for the development of more policy "handles."

The MRIO model can be succinctly stated by the following:

$$X = (I - C\bar{A})^{-1}\ CY \tag{16}$$

where

- X is an mn element vector (m regions, n sectors per region) of total outputs for each sector in each region, i.e., X_j^k is the total output of commodity j in region k (j=1,2,...,n; k=1,2,...,m).
- Y is an mn element vector of final demands for each sector in each region, i.e., Y_j^k is the final demand for sector j in region k.
- A is a block diagonal matrix (mn x mn) of technical input coefficient for each region where the direct input coefficients for each region A^k (n x n) appears as m diagonal blocks of A. An element of A, a_{ij}^k, represents the dollar worth of industry i's output required per dollars worth of industry j's output in region k. i.e.,

$$a_{ij}^k = z_{ij}^k \;/\; x_j^k \tag{17}$$

where Z is the matrix of transactions for region k; z_{ij}^k is an element of Z.

- C is a matrix (mn x mn) of trade coefficient matrices where the trade coefficient matrices appear along the principal diagonals of m^2 blocks of C partitioned as in the following:

$$C = \begin{bmatrix} C_{11} & C_{12} & C_{13} \cdots\cdots & C_{1n} \\ C_{21} & C_{22} \cdots\cdots & & \vdots \\ \vdots & \vdots & & \vdots \\ C_{n1} & C_{n2} \cdots\cdots & & C_{nn} \end{bmatrix}$$

where C is the vector of trade coefficients of all commodities shipped from region k to region l (an m x l vector) placed along the diagonal of an m x m matrix (the off diagonal elements are all zero), i.e., a particular coefficient is defined by:

$$c_i^{kl} = z_i^{kl} / x_i^l \qquad (18)$$

where x_i^l is the flow of good i from region l.

For the purposes of impact analysis, either a change in total output X (or final demand Y) is estimated based on changes in the other parameters, e.g., X or Y, $\bar{A}$ and C. Equation (16) is solved accordingly. For example, the impact on outputs of a change in final demand can be stated as follows:

$$\Delta X = (I - C\bar{A})^{-1} C \Delta Y \qquad (19)$$

The matrix $(I - C\bar{A})^{-1}$ is the matrix of direct and indirect requirements where I is an identity matrix.

In this analysis changes in energy, coal and electricity demands projected by the coal model are introduced through the Y and C terms. The regional impacts in terms of output, income, and employment resulting from changes in coal policy can then be estimated.

4. ANALYSIS

The coal policy impact model was employed to examine the consequences of a market-based emissions control policy. Two policies were simulated--a market-based control policy restricting sulfur emissions to 1.2 lb. per million Btus of coal burned and a relaxed emissions policy restricting emissions to 2.5 lb. per million Btu. The relaxed emissions policy serves as a rough approximation for the scrubbing requirement by eliminating the premium for low-sulfur coal. A six region, twelve industry version of the model was used. Coal trade flows are traced across six supply regions representing the major coal producing districts of the U.S. and six demand regions based on the nine U.S. Census regions. (The supply and demand regions are summarized in Table 3.)

Table 4 shows the regional coal trade flows under the sulfur emissions policies outlined above between 1980 and 1985. Under the strict emissions standard, coal production from the midwest (supply region 3) drops substantially compared to a relaxed emissions policy from 226 million tons to 169 million tons annually. Conversely, coal production from Southern Appalachia (supply region two) and Montana-Wyoming (supply region four) is notably higher under the strict emissions standard reflecting the low sulfur coal reserves available in those regions.

These results are consistent with other coal models used in coal policy analysis (ICF 1984; Amit 1982). Predicted coal productions shifts of this magnitude are in fact the basis for the concern of eastern and midwestern representatives and coal interests. For example, a

Table 3. Supply and Demand Regions in the Multiregional Coal Policy Impact Model

	Supply Regions[1]
1.	Northern Appalachia: Pennsylvania, Northern West Virginia and Ohio
2.	South and Central Appalachia: Southern West Virginia, Eastern Kentucky and Tennessee
3.	Midwest: Western Kentucky, Illinois and Indiana
4.	Montana/Wyoming
5.	Colorado/Utah
6.	New Mexico/Arizona
	Demand Regions[2]
1.	Northeast/Middle Atlantic: Connecticut, Maine, Massachusetts, New Hampshire, Rhode Island, Vermont, New York, New Jersey and Pennsylvania
2.	East North Central: Illinois, Indiana, Michigan, Ohio and Wisconsin
3.	West North Central/West South Central: Iowa, Kansas, Minnesota, Missouri, Nebraska, North Dakota, South Dakota, Arkansas, Wisconsin, Oklahoma and Texas
4.	East South Central/South Atlantic: Alabama, Kentucky, Mississippi, Tennessee, Delaware, District of Columbia, Florida, Georgia, Maryland, North Carolina, South Carolina, Virginia and West Virginia
5.	Rocky Mountain: Arizona, Colorado, Montana and Wyoming
6.	Pacific: California, Oregon, Washington, Alaska and Hawaii

[1] These regions are aggregations of coal mining districts defined by the Bureau of Mines, U.S. Department of Interior.

[2] These regions are based on regions defined by the Bureau of the Census, U.S. Department of Commerce.

Table 4. Coal Trade Flow By Policy: 1980-1985
Average Annual Production

Supply Regions	Demand Regions 1	2	3	4	5	6	Total Supply
	(millions tons per year)						
Strict Sulfur Constraint							
1	68.9	55.3	0.3	25.0	*	*	149.5
2	9.0	42.5	0.6	138.1	*	*	190.2
3	*	76.4	49.4	42.6	0.2	*	168.6
4	*	24.9	91.6	*	45.6	5.7	167.8
5	0.0	2.7	3.7	1.1	15.6	20.0	43.1
6	0.0	0.0	21.8	0.0	23.9	4.0	49.7
Total Demand	77.9	201.8	167.4	207.8	85.3	29.7	769.0
Moderate Sulfur Constraint							
1	68.9	45.0	0.3	25.0	*	*	139.2
2	9.0	42.5	0.6	108.0	*	*	160.1
3	*	87.6	63.9	74.7	0.2	*	226.4
4	*	24.9	77.4	*	44.5	5.7	152.5
5	0.0	2.7	3.7	1.1	15.6	16.9	40.0
6	0.0	0.0	18.5	0.0	24.8	7.2	50.5
Total Demand	77.9	202.7	164.4	209.8	85.1	29.8	769.0

*Indicate trade flows of less than 100,000 tons per year (figures may not add due to rounding).

production decline in the midwest of 58 million tons or more annually will result in fewer mining jobs--approximately 11,500 jobs assuming a productivity factor of 5,040 tons/man-year. Applying employment and income multipliers to such losses suggest that overall economic impacts will be as large as those predicted by the United Mine Workers.

However, as shown in Table 5, a general equilibrium analysis accomplished through the use of the multiregional input-output model reveals that the regional losses associated with the decline in mining are at least partially offset by growth in other sectors of the economy. Increased coal production in other regions results in greater demands for midwestern products. Gross product increases in the construction and in stone, clay and glass reduce the losses in gross product from reduced coal production. These increases are the result of increased demand from other regions experiencing coal production gains.

These changes in gross output indicate that net job losses will not be as great as predicted when simple multiplier methods are used. As shown in Table 6, coal mining job losses are partially mitigated by job increases in sectors where gross output grows. These results suggest that the estimates of economic losses in the midwest because of coal production declines are overstated because offsetting gains are ignored.

5. LIMITATIONS AND IMPLICATIONS FOR FURTHER RESEARCH

Clearly, the results presented above do not provide sufficient information to gauge the overall impact of a strict sulfur emissions policy. While they do show the importance of interregional effects in measuring impact, they do not measure the impacts of reduced coal production at the local level. For example, certain counties in the midwest may suffer relatively large employment and income losses when high-sulfur coal mines are shut, despite small net losses at the state or regional level. Determining the magnitude of this problem can be accomplished in part through a reaggregation of the coal impact model. However, empirical analysis of the counties faced with potential harm is also needed.

As a first step, the importance of coal employment at the state level is reviewed. As shown in Table 7, coal mining accounts for less than 0.5 percent of total employment in Illinois, Indiana and Ohio (the midwestern states facing the greatest direct coal employment losses). At the county level, coal employment is more important, but as shown in Tables 8, 9, and 10, accounts for 10 percent or more of total county employment in less than half of the coal mining counties of these states (22 of 58 counties). Consequently, dramatic impacts from coal production declines are possible in a relatively small number of counties. A similar analysis conducted by the Congressional Research Service (1983) emphasized the severity of impacts in certain counties.

Another limitation of the above analysis is that the effects of electricity price on factor substitution and industrial location are not fully captured. There is evidence that electricity prices are important in firm input and location decisions (Anderson 1979; Carlton 1982).

Table 5. Sulfur Emission Standards Impact on Gross Output in the East North Central States[1] -- 1985

Sector[2]	Strict Sulfur Standards	Relaxed Sulfur Standards	Impact of Strict Standard
	(millions of 1980 dollars)		
			(1)-(2)
	(1)	(2)	(3)
Agriculture	$ 25,002.99	$ 25,002.13	0.8
Coal Mining	2,673.98	3,235.62	-561.6
Other Mining	4,816.04	4,815.21	0.8
Construction	41,515.03	41,516.45	-1.4
Food, Fabrics, Wood	80,304.55	80,301.84	2.7
Machinery and Equipment	174,175.50	174,041.20	134.3
Petroleum Products	36,336.81	36,335.51	1.3
Stone, Clay, Glass	9,576.18	9,573.96	2.2
Primary Metals	47,244.04	47,213.52	30.5
Services	212,913.00	212,921.10	-8.1
Transport	22,003.99	22,000.78	3.2
Electric Utilities	14,629.52	14,640.79	-11.2
Total			-406.4

[1] East North Central Census Region includes Illinois, Indiana, Michigan, Ohio and Wisconsin. Gross output is defined as all intermediate and final sales.

[2] Industrial sectors are aggregations of the standard industrial classifications. Various levels are used, based on the 19 sector model presented in K. Polenske, The U.S. Multiregional Input-Output Accounts and Model, Table A-3, p. 230. The 10 sector aggregation is as follows:

Sector	MRIO Sector
Agriculture	1,2
Coal Mining	3
Other Mining	5
Construction	6
Food, Fabrics, Wood	7,9
Machinery and Equipment	8,15
Petroleum Products	4,10,11
Stone, Clay, Glass	12
Primary Metals	13,14
Services	16,18
Transport	17
Electric Utilities	19

Table 6. Sulfur Emission Standards Impact on Employment in the East North Central States -- 1985

Sector	Gross Output Change[1] (millions of 1980 dollars)	Employment Factor[2]	Employment Impact[3]
Agriculture	0.8	5.4	4.6
Coal Mining	-561.6	9.8	-5,492.3
Other Mining	0.8	14.0	11.6
Construction	-1.4	18.6	-26.4
Food, Fabrics, Wood	2.7	14.1	38.2
Machinery and Equipment	134.3	12.4	1,668.1
Petroleum Products	1.3	2.6	3.3
Stone, Clay, Glass	2.2	16.0	35.4
Primary Metals	30.5	1.8	53.4
Services	-8.1	81.0	-659.3
Transport	3.2	22.0	70.6
Electric Utilities	-11.2	5.2	-59.4
Total	-406.4		-4,352.2

[1] Estimated change in gross output from strict sulfur emission standards (see Table 5).

[2] The employment factor is defined as employees divided by gross output (millions of dollars). These factors are weighted average values taken from U.S. Department of Labor, Bureau of Labor Statistics, Time Series Data for Input-Output Industries, Bulletin 2018, Washington, D.C.: Government Printing Office, March 1979. Values are presented as of 1976. Employment and output were adjusted to 1980 based on trends between 1958 and 1976.

[3] Estimated as the product of the change in gross output and the employment factor.

Table 7. Coal Mining and Total Employment by State, 1981

State	Coal Miners	Total Employment	Percent Coal Miners
	(000's)	(000's)	(percent)
	(1)	(2)	(3)
Alabama	10,239	1,488	0.69
Kentucky	48,050	1,522	3.16
Maryland	1,202	2,007	0.06
Ohio	13,752	4,595	0.30
Pennsylvania	33,340	5,018	0.66
Tennessee	3,364	1,919	0.18
Virginia	16,595	2,442	0.68
W. Virginia	55,921	702	7.97
Arkansas	89	936	0.01
Illinois	17,444	5,103	0.34
Indiana	5,227	2,353	0.22
Iowa	147	1,320	0.01
Kansas	321	1,138	0.03
Missouri	1,052	2,139	0.05
Oklahoma	1,665	1,391	0.12
Texas	2,131	6,701	0.03
Alaska	33	1,488	0.00
Arizona	918	1,187	0.08
Colorado	4,062	1,445	0.28
Montana	1,227	355	0.35
N. Mexico	1,679	533	0.32
N. Dakota	846	294	0.29
Utah	4,166	603	0.69
Washington	615	1,798	0.03
Wyoming	5,217	240	2.17
Total	229,302	48,717	0.47

Note: Total employment includes all civilian employees.

Source: Col. (1): United States Department of Energy Energy Information Administration, Coal Production, 1981, p. 58.

Col. (2): United States Department of Commerce Bureau of the Census, Statistical Abstract of the United States, 1982-83, p. 378.

Table 8. Employment in Coal Producing Counties Ohio - 1981

County	Coal Mining Employment Range	Coal Mining Employment Midpoint	Manufacturing Employment	Total[1] Employment	Coal as a % of Total (2)/(4)	Manufacturing as a % of Total (3)/(4)
	(1)	(2)	(3)	(4)	(5)	(6)
Belmont	2500-4999	3,750	3,828	18,700	20.1	20.5
Carroll	20-99	60	1,453	3,404	1.7	42.7
Columbiana	286	286	9,329	24,508	1.2	38.1
Coshocton	337	337	5,073	10,300	3.3	49.3
Gallia	100-249	175	1,117	7,007	2.5	15.9
Guernsey	20-99	60	4,624	11,271	0.5	41.0
Harrison	1000-2499	1,750	1,119	4,405	39.7	25.4
Hocking	20-99	60	2,182	4,437	1.3	49.2
Holmes	50	50	2,462	5,713	0.9	43.1
Jackson	100-249	175	1,724	5,237	3.3	32.9
Jefferson	500-999	750	6,569	22,130	3.4	29.7
Lawrence	20-99	60	1,878	7,403	0.8	25.4
Mahoning	98	98	21,895	84,718	0.1	25.8
Meigs	1000-2499	1,750	312	3,481	50.3	9.0
Monroe	1000-2499	1,750	3,750	7,500	23.3	50.0
Muskingum	453	453	8,485	25,330	1.8	33.5
Noble	100-249	175	633	1,693	10.3	37.4
Perry	500-999	750	1,593	4,713	15.9	33.8
Stark	173	173	54,472	136,794	0.1	39.8
Tuscarawas	601	601	8,311	23,166	2.6	35.9
Vinton	500-999	750	780	2,110	35.5	37.0
Washington[2]	656	656	5,999	18,243	3.6	32.9
Wayne[2]	660	660	13,688	29,100	2.3	47.0
Total		13,257	140,809	411,910	3.2	34.2

[1] Excludes government and railroad employees and self-employed persons.

[2] U.S. Department of Commerce Employment Statistics not broken down beyond "Mining" occupation --however, due to the predominance of coal mining in these counties, all mining employees are assumed to be involved in the mining of coal.

Source: U.S. Department of Commerce, Bureau of the Census, County Business Patterns, 1981, Ohio.

Table 9. Employment in Coal Producing Counties Illinois - 1981

County	Coal Mining Employment Range	Coal Mining Employment Midpoint	Manufacturing Employment	Total[1] Employment	Coal as a % of Total (2)/(4)	Manufacturing as a % of Total (3)/(4)
	(1)	(2)	(3)	(4)	(5)	(6)
Christian	0-19	10	1,503	7,227	0.1	20.8
Clinton	500-999	750	765	5,467	13.7	14.0
Douglas	500-999	750	1,650	5,735	13.1	28.8
Franklin	2500-4999	3,750	529	8,479	44.2	6.2
Fulton	500-999	750	2,137	8,535	8.8	25.0
Gallatin	500-999	750	166	1,539	48.7	10.8
Hamilton	250-499	375	100	1,267	29.6	7.9
Jackson	250-499	375	1,551	13,774	2.7	11.3
Jefferson	1000-2499	1,750	2,449	11,786	14.8	20.8
Macoupin	1000-2499	1,750	1,196	9,025	19.4	13.3
Montgomery	500-999	750	1,815	7,503	10.0	24.2
Peoria	100-249	175	30,049	92,646	0.2	32.4
Perry	1943	1,943	1,137	5,764	33.7	19.7
Randolph	859	859	3,504	9,563	9.0	36.6
St. Clair	1000-2499	1,750	8,404	54,028		
Saline	500-999	750	352	7,099	10.6	5.0
Vermilion	20-99	60	11,976	29,189	0.2	41.0
Wabash	500-999	750	1,460	4,808	15.6	30.4
Washington[2]	47	47	397	2,276	2.1	17.4
Williamson	500-999	750	3,351	12,573	6.0	26.7
Total		18,836	74,491	298,283	6.3	25.0

1 Excludes government and railroad employees and self-employed persons.

2 U.S. Department of Commerce Employment Statistics not broken down beyond "Mining" occupation --however, due to the predominance of coal mining in this county, all mining employees are assumed to be involved in the mining of coal.

Source: U.S. Department of Commerce, Bureau of the Census, County Business Patterns, 1981, Illinois.

Table 10. Employment in Coal Producing Counties Indiana - 1981

County	Coal Mining Employment Range	Coal Mining Employment Midpoint	Manufacturing Employment	Total[1] Employment	Coal as a % of Total (2)/(4)	Manufacturing as a % of Total (3)/(4)
	(1)	(2)	(3)	(4)	(5)	(6)
Clay	20-99	60	858	3,560	1.7	24.1
Daviess	20-99	60	1,304	5,086	1.2	25.6
Dubois	126	126	7,507	15,019	0.8	50.0
Fountain[2]	0-19	10	2,478	4,630	0.2	53.5
Gibson[2]	321	321	3,084	8,553	3.8	36.1
Greene	100-249	175	1,108	4,348	4.0	25.5
Knox	20-99	60	1,891	10,426	0.6	18.1
Martin	0	0	375	1,437	0.0	26.1
Owen[2]	20-99	60	339	1,143	5.2	29.7
Pike	500-999	750	89	1,693	44.3	5.3
Spencer	154	154	852	3,619	4.3	23.5
Sullivan	500-999	750	572	3,144	23.8	18.2
Vermillion	250-499	375	1,660	3,525	10.6	47.1
Vigo	57	57	12,330	39,208	0.1	31.4
Warrick	1000-2499	1,750	3,750	10,340	16.9	36.3
Total		4,703	38,196	115,731	4.1	33.0

1 Excludes government and railroad employees and self-employed persons.

2 U.S. Department of Commerce Employment Statistics not broken down beyond "Mining" occupation --however, due to the predominance of coal mining in these counties, all mining employees are assumed to be involved in the mining of coal.

Source: U.S. Department of Commerce, Bureau of the Census, County Business Patterns, 1981, Indiana.

Although the Regional Electricity Model (REM) accounts for location decisions as a function of fuel price, the results are aggregate. Consequently, they are not fully measured by the MRIO Model. According to work by Anderson (1979), an elasticity of approximately -.14 exists for industrial employment with respect to electricity prices. Consequently, higher electricity prices because of scrubbing requirements or the purchase of premium low sulfur coal may result in substantial job losses not fully captured by the model.

Further analysis comparing the market-based and technology-based acid rain control strategies is necessary once these price effects are incorporated into the model framework at a more disaggregate level. This is especially important in light of recent work by Jonash (1984) which presents results contrary to those presented here.

6. FOOTNOTES

1 This paper refers to material in published OTA studies. The views expressed in the paper, however, are solely those of the authors and do not necessarily reflect those of the Office of Technology Assessment or the Technology Assessment Board.

2 Two states, West Virginia and Kentucky, are partially represented in two regions.

3 A similar procedure is followed for strip mining, but the producing unit definition and geologic characteristics are different.

4 Several distributions were used based on available information. A lognormal distribution was assumed for deep mining and western strip mining. United States Geological Survey data and techniques were used for eastern strip mining.

5 See, for example, Olsen et al. (1977), Fromm, et al. (1981), and Polenske (1980).

6 The MRIO model was recently updated by Jack Faucett Associates. Future analysis will employ the updated data base.

7 The model can be rewritten as well to compute the results of a mixture of Y's and X's; see Miller and Blair (1985).

7. REFERENCES

Ackerman, B. and Hassler, T. 1981. Clean coal, dirty air. New Haven: Yale University Press.

Amit, R. et al. 1982. A methodology for analyzing the impact of regulations on the coal industry. In Perspectives on resource policy modeling: energy and minerals, edited by Raphael Amit and Mordecai Ariel. Cambridge, Mass: Ballinger Publishing Co.

Anderson, Kent. March 1979. Industrial Location and Electric Utility Price Competition. New York: National Economic Research Associates.

Baughman, M.L., Joskow, P.L., and Kamat, D. 1979. Electric power in the united states: models and policy analysis. Cambridge: MIT Press.

Berkman, M., Blair, P. and Kathan, D.. November 1982. Regional Impacts of Federal Coal Policy. Working Paper No. 74, Department of Regional Science, University of Pennsylvania.

Carlton, Dennis. August 1983. The location and employment choices of new firms: An econometric model with discrete and continuous endogenous variables. The Review of Economics and Statistics Vol. LVX No. 3: 440-449.

Congressional Budget Office, U.S. Congress. April 1983. The Clean Air Act, the Electric Utilities and the Coal Market. Washington, D.C.: Government Printing Office.

Congressional Research Service. 1983. Mitigating acid rain: Implications for high-sulfur coal regions. In Congressional Research Service, A Survey of Data and Current Analyses. A report prepared for the use of the Subcommittee on Health and the Environment of the Committee on Energy and Commerce, U.S. House of Representatives. Washington D.C.: GPO, pp. 910-923.

Energy Information Administration, U.S. Department of Energy. June 1983. Impacts of the Proposed Clean Air Act Amendments of 1982 on the Coal and Electric Utilities Industries. Washington, D.C.: U.S. Department of Energy.

Fromm, D., et al. 1981. The Wharton EFA multiregional econometric model. In Modeling the multiregional economic system, eds. F.G. Adams and N. Glickman. Lexington, Mass: Lexington Books.

ICF. February 1984. Analysis of Cost-Effective, Phased-In Reductions of Sulfur Dioxide Emissions. Prepared for the Alliance for Clean Energy, Washington, D.C.: ICF.

ICF. February 1983. Analysis of a Senate Emission Reduction Bill (S-3041). Prepared for the U.S. Environmental Protection Agency, Washington, D.C.: ICF.

Jonash, R. 1984. Economic impacts of alternative acid rain control strategies. Paper presented to Acid Rain: Economic Assessment Conference. Arthur D. Little Inc. and Energy Ventures Analysis Inc.

Massachusetts Institute of Technology Energy Laboratory. October 1981. The ICF, Inc. Coal Model and Electric Utility Model: An Analysis and Evaluation Vols. 1-7 (prepared for the Electric Power Research Institute). MIT EL 81-015.

Massachusetts Institute of Technology. May 1979. Independent Assessment of Energy Policy Models (prepared for the Electric Power Research Institute). EA-1071.

Miller, Ronald and Blair, Peter. 1985. Input output analysis: Foundations and extensions. Englewood Cliffs, New Jersey: Prentice Hall.

Morrison, M. 1983. A Linear Programming Model to Examine Sulfur Dioxide Control Strategies for U.S. Coal-Fired Plants. Center for Energy and Environmental Studies Carnegie Mellon University, Pittsburgh, Pennsylvania.

Office of Technology Assessment, U.S. Congress. 1984. Acid Rain and Transported Air Pollutants: Implications for Public Policy. Washington, D.C.: Government Printing Office.

Olsen, R., et al. 1977. Multiregion: A Simulation Forecasting Model of BEA Economic Area Population and Employment. (ORNL/RUS-25) Oak Ridge National Laboratory.

Price, Kent, editor. 1982. Regional conflict and national policy. Washington, D.C.: Resources for the Future.

Polenske, K. 1980. The U.S. multiregional input-output accounts and model. Lexington, Mass: Lexington Books.

Portney, Paul. October 1983. Making Sensible Environmental Policy: The Case of Acid Rain. Prepared for the Alliance for Clean Energy.

Temple Barker and Sloan. September 1983. Evaluation of H.R. 3400 The "Sikorski/Waxman" Bill for Acid Rain Abatement. Prepared for the Clean Air Act Issues Group. Lexington, Mass: Temple Barker and Sloan.

United Mine Workers. 1984. Employment impacts of acid rain. In Congressional Research Service, A Survey of Data and Current Analyses. A report prepared for the use of the Subcommittee on Health and the Environment of the Committee on Energy and Commerce, U.S. House of Representatives Washington D.C.: GPO, pp. 910-923.

U.S. Department of Energy. February 1979. An Assessment of National Consequences of Increased Coal Utilization. Executive Summary, TID-29425-V2.

Van Kriken, J. and Krohm, G. 1980. The Argonne coal model. In *Simulation of Energy Systems: Part I*, edited by Kenneth Watt, pp. 69-76. La Jolla, CA: Simulations Councils Inc..

Wendling, R. and Ballard, K. October 1980. Projecting the regional economic impacts of energy development. *Growth and Change* Vol. 10, No. 1.

Zimmerman, B. 1981. *The U.S. coal industry: the economics of policy choice*. MIT Press: Cambridge.

CHAPTER 24

HYDROELECTRIC ENERGY: AN AGENT OF CHANGE IN AMAZONIA (NORTHERN BRAZIL)

Rolf Sternberg
Department of Environmental,
Urban and Geographic Studies
Montclair State College
Upper Montclair, New Jersey 07043

ABSTRACT. Perceptions of the Amazonian resource base are changing, accompanying unprecedented energy development. Government policies are directed toward the integration of Amazonia into the national economy. Infrastructure formation is a fundamental step toward this objective, and hydroelectric energy forms an essential part of such a program. The costs of hydroelectric development in Amazonia, as well as Brazilian perceptions of the region, complicate resource management. Without hydroelectric development, Amazonia would continue as a resource-yielding region without any return to the national economy. Hydroelectric power may draw different kinds of industries into the region and may gradually redraw the economic map of Brazil.

1. INTRODUCTION

Perceptions of the Amazonian resource base are changing as different resource utilization technologies are introduced into the region. In the past, the Amazonian resource base has been associated with specific products, such as valuable woods, Brazil nuts and rubber. Livestock raising is a recent addition to the region, but its permanence remains to be proven. Hydroelectric energy fosters a reassessment of the Amazonian resource base that introduces complexity previously not encountered. It will impose the systematizing of human activities, thereby marking a radical departure in the use of the Amazonian resource base and the organization of Amazonian society, space and economy.

This chapter examines Amazonian hydroelectric energy projects and considers how these will contribute to regional and economic shifts in economic activities. To what extent will hydroelectric energy serve as an integrative component? Will the wedge created by the Belem-Brasilia highway become a broad base rooted in hydroelectric energy? Do these changes presage the effective integration of Amazonia into the national economy? Resource extraction now requires costly infrastructure

F. J. Calzonetti and B. D. Solomon (eds.), Geographical Dimensions of Energy, 471–494.
© 1985 by D. Reidel Publishing Company.

formation and a spatially planned economic system. The environmental impact of these facilities and the associated development must also be considered.

Brazilian Amazonia is experiencing far-reaching changes. One of the driving considerations for change is the "interiorization process," which aims to make Amazonia an effective part of the nation, something that has not been the case in the past. As part of the "interiorization process," infrastructure formation occupies a prominent position on the national planning agenda. A dependable energy supply system will play a key role in such a scheme. Hydroelectric energy will be a leading energy source for the region which will power many of the changes to be introduced into Amazonia.

2. THE AMAZONIAN ENVIRONMENT AND ITS HYDROELECTRIC POTENTIAL: SOME GENERALIZATIONS

Sioli (1980, 257-68) claims that Amazonia contains the world's most complex ecosystem. Its complexity and fragility have been assessed by a growing body of diverse specialists, including Denevan (1981, 25-44) and Goodland (1980, 9-26). Hydroelectric power projects by their very character impose notable changes upon the environment. Subsistence farmers, commercial agriculturalists and livestock ranchers as such modify the local environment, but not as radically as a hydroelectric project.

Hydroelectric projects require very detailed surveys and long range planning. Even if the Transamazon highway became an effective transportation link for selected segments in the region, its original planning was rudimentary and its execution rustic. Hence travel on it is difficult during the rainy season. Erosion scars pit the road and tubeless barriers to drainage channels are readily breached by the rushing mass of water. Each new project will have to be assured long term functional viability with minimal environmental disruption to conserve the total infrastructure and physical environment.

Hydroelectric energy will imprint itself notably in the Amazonian landscape. It requires space for the dams, the reservoirs, and the transmission systems. Projects of this kind also require their own environmental conservation. Eletrobras[1] and Eletronorte[2] have a growing number of environmental specialists on their staffs to address environmental problems. Eletronorte also has engaged INPA[3] to evaluate the environmental impacts of the Tucurui dam.

2.1 Hydroelectric Energy and Geography: Some Generalizations

Little research has been conducted by American geographers on hydroelectric energy. Murphy (1954, 256) noted that "almost nothing has been done about the geographical aspects of forest utilization, and very little with regard to the geography of hydroelectric power." Whitbeck and Finch (1941, 218-22) review water power in general, including an instructive map of "Estimated Potential Water Power Resources" (p. 219). That review was directed toward potential power and available installed power in relation to existing or future consumption. Hydroelectric power was presented in the context of a larger design, not as the geography of hydroelectric energy. Whitbeck and Williams (1940, 243, 395-96) replicate this approach, giving a half page to Argentine water power potential and about one page to that of Brazil. The estimated potentials then projected were based on existing transmission technology in place rather than the number of possible dam sites for each river basin and the nameplate potential output of each dam (name-plate potential is the maximum KW output of dam). Alexander and Gibson (1979, 180-4) treat the subject with greater lucidity than the others, but they do not advance the subject matter significantly. Hydroelectric energy has not received the attention of coal, gas, petroleum, or natural gas.

Fels (1954, 82-95) addresses the topic of reservoirs and their formation in some detail. Fels reviews the problem of reservoir size, which mirrors in part land availability in different regions of the world. This may also explain why reservoir areas in Central and Western Europe tend to be well below 100 km^2 in area, while those in the USSR and USA are readily over 100 km^2, and often well over 1000 km^2. Obst (1965, 528-84) discusses hydroelectric energy from the point of view of water availability, mass and distance of fall, to give insight into the measurable energy availability. Obst then turns to hydroelectric energy and its development in fossil-fuel poor nations such as Italy, France, Norway, and so on. Hydroelectric energy was recognized as a substitute for fossil fuels.

Several authors have discussed hydroelectric power in larger studies of energy geography (Manners 1964; Guyol 1974; Cook 1976; Smil and Knowland 1980). Analysis of hydroelectric energy and its spatial character is given more specific attention in Soviet energy surveys. Shabad and Mote (1977, 27-33) review its role in the context of Siberian changes, especially the BAM (Baykal-Amur-Mainline) system, designed to provide access to hitherto unused resource deposits. Electricity was to ease access to these resources, and allow resource transformation to occur locally using inexpensive electricity. The aluminum plant at Shelkhov, about 16 km from Irkutsk, is an example of this. In a more broadly based study, Dienes and Shabad (1979, 131-150) present an analysis of the role of hydroelectric energy in the Soviet Union. The authors regionalize the power production and its character, differentiating base-load and peak-load, explaining why hydroelectric power serves more in the latter role to meet peak demand and cases of emergencies.

Cuff and Young (1980, 311-38) in a survey of hydroelectric energy in the United States emphasize that two-thirds of hydroelectric potential of the U.S. will in most likelihood never be developed because of numerous socio-environmental considerations. They note that U.S. hydroelectric potential can be increased by about 60%, and that the energy output at existing dams can be increased by about 1.5 times that installed. Pryde (1983, 76-99) provides a general survey of hydroelectricity, with an emphasis upon the U.S. By placing hydroelectricity in the context of "nonconventional energy resources," the potential increases in this sector are critically evaluated, as are pumped storage options and small dams i.e., units of 15 MW or less. Environmental constraints in the U.S. were reviewed. Multiple uses of water were considered, as were numerous potential problems with dams. The examples given were based on storage dams that generally are lacking in spillways.[4] The analysis closes with the presentation of a "comparative regional potential" for the U.S., in a way similar to the work of Whitbeck and Finch.

2.2. Hydroelectric Energy and Geography: The Brazilian Setting

Brazil has embarked upon the quest for energy autarchy. Hydroelectric energy is assuming an increasingly important role and by 1983 contributed 30.7% to the national energy mix. Hydroelectric energy in Brazil commands a presence it never held in the U.S. Its impact is far more pronounced, and weighs even more heavily in Amazonia where most changes are predicated upon its availability. Energy availability changes the perception and value of space, hence when hydroelectric energy becomes the key source of energy, the land use system may sustain substantial changes. The Tucurui dam on the Tocantins is an example, as it is designed to power aluminum production, iron and steel industries, chemical plants, and numerous lesser industries. It will influence and shape resource and land utilization in Amazonia for years to come.

2.3. Amazonia's Hydroelectric Potential: A Preliminary View

When reference is made to the Brazilian Amazon basin's hydroelectric potential, the potential hydropower of the master stream is excluded from any consideration here. The river's grade is too low. From Tabatinga to Macapa the river has a 55 meter grade over a distance exceeding 2500 km. At this time the stream flows of such major tributaries as the Tapajos, Madeira, Purus, Jurua, Tefe, are being inventoried (see Figure 1), but large segments of the basin are inadequately surveyed, especially for the rivers west of the Xingu. Most secondary tributaries have not been listed in the surveys of their hydroelectric potential for future use. Available data projects 48840 MW of firm power for the basin to date. Firm power as used by Eletrobras Eletronorte refers to one half of the nameplate of installed power, or 50% of the potential. Douglas S. Luz (1983), president of Eletronorte, took this as a conservative measure, and indications are that this could be increased by 15 to 20%. Even in this incomplete survey, the Amazon basin (Figure 1) contains nearly 50% of the nation's potential hydroelectric power.

The streamflow analysis is structured into sides and major river-

Figure 1. Northern Brazil Locations.

basins. The left-hand margin, the north side of the master stream, has a potential of 7,770 MW of firm power. The right-hand margin has an estimated potential of 28,400 MW of firm installed power. Most of these values remain to be confirmed, and it is virtually certain that the data available now will change substantially once the field work is completed. In the left-hand data cited, nine streams are listed but only the Uatuma and the Trombetas are considered inventoried. Similarly, on the right-hand side only the Xingu and the Curua Una (Figure 1) are deemed inventoried, while the Madeira's firm power is estimated at 7,240 MW, yet only 60 MW are listed as proven for this river. Eletronorte engineers refer to one power site on the upper Madeira with a potential for 18,000 MW (Eletrobras 1980). The hydroelectric potentials are thus extremely large.

2.4. Surveys and Projections

Power potential surveys for selected rivers in the eastern Amazon basin have been made. Prominent among these rivers are the Xingu, the Trombetas, the Uatuma, and the Jamari. The other important river basin inventoried for its power potential is the Tocantins-Araguaia system. The Tocantins system is not part of the Amazon river, though it is included in the region of Amazonia (see Figure 1; Table 1). Changes in these values upon project completion have to be anticipated; the true range may lie around 10% above or below present data. The data shown here for the Trombetas apply only to this river, not to its tributaries.

Table 1. Hydroelectric Potential of Selected Rivers - Northern Brazil

River	Number of Dam Sites	Firm Power (MW)	Maximum Potential (MW)
Jamari	2	68	275
Trombetas	9	1,068	2,553
Uatuma	2	203	406
Xingu	7	9,800	19,600
Tocantins-Araguaia	22	8,223	16,446
			39,280

Other river basins have been surveyed and these data are more tentative. Among these are the Tapajos, Madeira, Negro, Branco, and several smaller streams (see Table 2). These 48,194 MW comprise 23% of the hydroelectric potential of the nation. If the values of Table 1 are included, the percent increases to 41%, but this percent will change once more complete inventories become available that are essential prior to actual project installation (see Figure 1).

3. HYDROELECTRIC PROJECTS--COMPLETED, IN PROGRESS, AND PLANNED

Hydroelectric projects require considerable lead time before entering energy production. The siting, planning, construction and connection takes ten or more years, the variations in length of time reflecting the magnitude of the projects. In Amazonia the three phases of completed, in progress, and planned hydroelectric projects will be common for several decades or more. At this time the number of projects in progress is limited by financial constraints, but indications point to increased numbers of small dams to service isolated communities. An addition to these projects already exists--a small "run-of-the-river-turbine" that generates 1000 Kw to 10,000 Kw (1-10 MW). These projects only serve local areas.

Table 2. Tentative Hydroelectric Potential Projections of Selected Rivers - Northern Brazil[5]

River	Estimated Maximum Potential (MW)
Tapajos	19,200
Madeira	16,340
Negro	4,280
Branco	3,778
Cotingo	548
Erepecuru	1,648
Jari	1,096
Ituxi/Endimari	200
Jatapu	580
Araguari	524
Total	48,194

3.1. Completed Projects

Tucurui was inaugurated in November, 1984 and came on-stream in early 1985. It is a major regional power source, compared to Curua Una and Coaracy Nunes. Curua Una serves the Santarem region, and perhaps will serve a nearby cement plant. As late as 1981, its largest customer was the Hotel Tropical.

Coaracy Nunes has many customers and serves Macapa, the ICOMI manganese pellet plant at Porto Santana, and other communities in the southern part of the state. Curua Una had two 10 MW turbines operational in 1980, and preparation to place the two remaining units were underway. The Coaracy Nunes plant on the Araguari (Amapa)[6] has two turbines for 39.2 MW in place, and provisions for one more exists, bringing its maximum to 69 MW.

3.2. Projects in Progress

Two projects are underway, Tucurui and Balbina. Work began on Balbina in 1982 and it will come on-stream by 1986 or 1987. It is designed to substitute for thermally-generated power. The plant is comprised of five 50 MW units, and is tied to a 230 KV transmission line. The plant is 172 km from Manaus, but will not be large enough to completely eliminate the thermal units in use.

Tucurui Dam is a two-phase project. When completed by 1990 or so, it will have 24 turbines at 332.5 MW each, each with a 369 MW maximum. The first phase includes 12 units, and the last turbine of the first phase is to be placed in service in 1987 (see Figure 2). The second phase can be completed after the dams of Santa Isable (Araguaia River) and Carolina (Tocantins River) are in place and control upstream flow. Tucurui's size and location allow it to serve a large mineral-rich region. Tucurui attracts investment to the region and provides the energy for infrastructured improvements.

3.3. Future Projects

Many large projects will be placed into the Eletronorte territory, in the following order:

1. Samuel--Dam on the Jamari River, to service Porto Velho and Rio Branco, cities in the center-west. Projected maximum output is set at 216 MW (see Figure 2).

2. Santa Isabel--Dam on the Araguaia River will serve as an energy source to the Maraba-Carajas region. It is projected to generate 1258 MW with a nameplate potential of 1761 MW. It is one of two projects to control the Tocantins' streamflow variations, and will assure a better regulated streamflow into the Tucurui reservoir, as well as provide a dependable base-load output for its many power consumers. Santa Isabel will not enter service until 1991 or 1992.

3. Sao Felix on the upper Tocantins will serve as power a source and major storage reservoir that controls much of the Tocantins' streamflow. With this dam in place, the Tucurui reservoir will benefit from upstream regulation of its water sources and provide a more assured baseload capacity to meet its varied regional energy obligations. Tucurui reservoir holds 45.8 km^3 of water, of which 25.4 km^3 serves the turbine peak.[7] Sao Felix will be a 1330 MW dam with 16.8 km^3 of water for its power peak. Its completion date is anticipated for 1992-93.

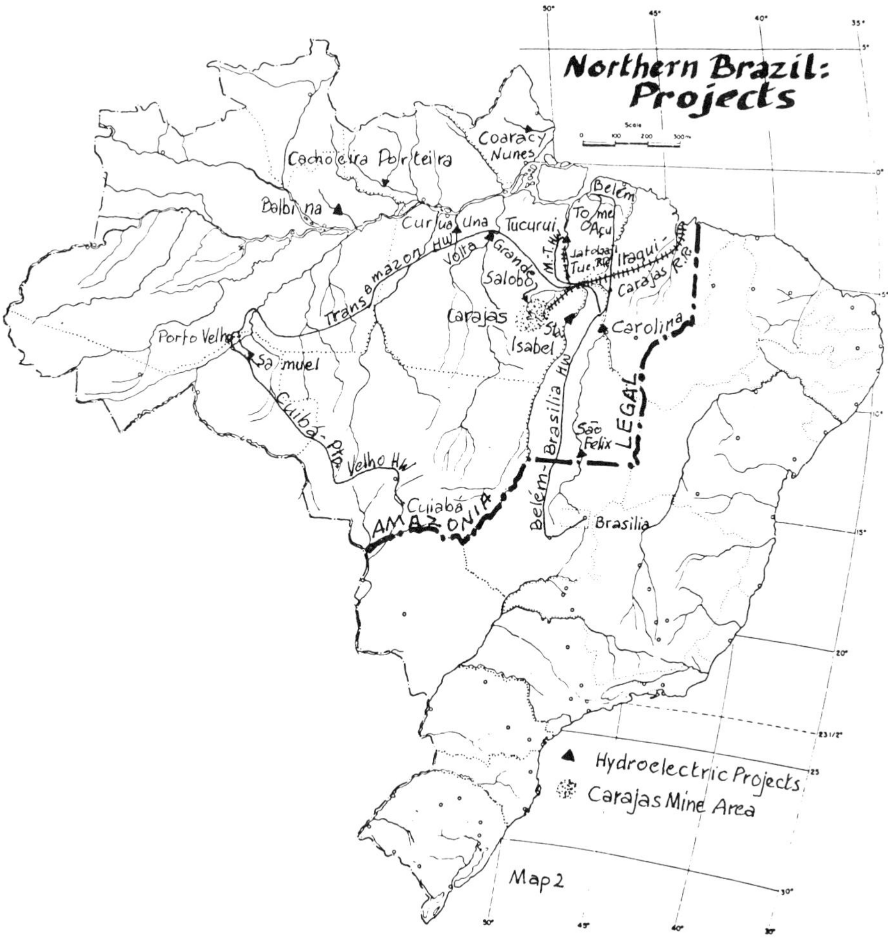

Figure 2. Northern Brazil Projects.

4. Volta Grande--Dam on the Xingu River will be a complex of two dams, one above Altamira, and the other just above Vitoria, comprising a combined nameplate potential in excess of 14,000 MW. This project may be initiated by 1990 and may take 15 years to complete. Most of the energy is to be exported to the Southeast. This complex places a very substantial energy source in the heart of the Amazonian region, allowing access to resources previously unavailable. It is another way to push the Tucurui experience west.

5. Cachoeira Porteira--Dam on the Trombetas river system has been under consideration for over 30 years. It is projected to enter service by 1994. It is planned as a project for two dams, greatly reducing reservoir areas and achieving nearly comparable electricity generation. Notably lower investment requirements are anticipated as well. This project is of growing significance because it will service the nearby bauxite mining community of Porto Trombetas, as well as many other left margin communities in the Amazon valley, ultimately including Manaus. The availability of energy at the mine may encourage alumina production instead of exporting bauxite.

4. HYDROELECTRIC PROJECTS AS AGENTS OF ENVIRONMENTAL CHANGE

Hydroelectric projects introduce numerous changes in the environment. Most conspicuous among these is the reservoir, the dam and the transmission system. Environmental changes must be anticipated in stream regime's flora, fauna, and economic activities. Hydroelectric generation is virtually pollution free, but the industries that depend upon it may cause severe environmental stresses. It is this latter aspect that merits close consideration.

4.1. Observed Changes - Physical Attributes

Hydroelectric projects are a recent addition to the Amazonian lands ape. In the second half of the 1970's two small hydroelectric power plants came on stream, Curua Una, 70 km east of Santarem, and Coaracy Nunes on the Araguari River in central Amapa state. Future projects will be much larger than these. Table 3 compares land requirements of projects and illustrates the variations in installed Kw/ha. The range in these values and reservoir size in relation to nameplate potential indicates the scale of potential problems. For example, Salta Osorio on the Iguacu River has minimal land requirements, compared to Balbina on the Uatuma. It may well be that Curua Una and Balbina serve local needs in contrast to Tucurui, which will service regional and national demands. Land quality and other considerations also influence the level of impact. Curua Una and Balbina provide little if any base for permanent human occupance. The dominant soils in either case are oxisols.

Table 3. Selected Dams - Some Characteristics[8]

River	Dam Name	Year Begun	Year Completed	KW (1000)	Reservoir Area - ha	Kw/ha
Curua Una	Curua Una	1968	1977	40	8,600	4.6
Araguari	Coaracy Nunes	1956	1976	69	2,300	30.0
Uatama	Balbina	1982	-	250	254,000	1.0
Tocantins	Tucurui	1976	1984	8,000	243,000	32.9
Volta (Ghana)	Akosombo	1962	1964	833	850,000	1.0
Iguacu	Salto Osorio	1970	1975	1,053	6,100	172.6

Curua Una experienced swift deterioriation of aquatic and floral life in the reservoir. The reservoir was not cleared prior to lake formation, provoking serious environmental problems. These include very high fish mortality, macrophyte infestation (water hyacinths), dying trees, and active formation of sulfuric gas that causes underwater explosions in the intake basin. A prolonged drought would reduce some of the problems naturally into the macrophyte concentration, decimate the snail population and greatly reduce mosquito breeding grounds and populations. Although power production would suffer, this may be the fastest way to improve the reservoir environment.

Health-related factors are extremely important. Malaria and schistosomiasis can root in the reservoir and infect populations. In this part of Para malaria is endemic, and its control complicated. The snail carries schistosomiasis, and black flies are a health hazard because of ulcerations and blindness that their larvae produce in humans. Possible solutions to these problems exist, but these may prove incompatible with the prevailing culture. Information services and education would be the most beneficial, the least costly, and a highly efficacious approach, provided the populace responds. The other approach would be a two kilometer wide land reserve surrounding the reservoir, which would serve as a "cordon sanitaire" and an erosion control measure.

The Araguari river differs from the Curua Una in that it traverses a podzolic soil zone on the Amapa savanna. Prior to flooding the Coaracy Nunes reservoir, 50% of the area was deforested, a sharp contrast to the Curua Una condition. The dam's crest is at 120 meters, but all land up to 114 meters was cleared of forest, and land in proximity to the dam was cleared up to and including the 120 meter level. The general environmental conditions surrounding this reservoir are good. The river sports active fishing and water hyacinths are absent. Malaria is the main disease in the area, while snails and black flies and their associated diseases are lesser problems because of the cleaner environment.

Traditional resource extraction depends upon the floral resource base yielding such products as wood, nuts, rubber and essences. Resource extraction did not require much investment save the energy needed to obtain the product. Such activities may have depleted resources such as sarsaparilla, clove, rosewood oil, and altered the existing floral patterns by destroying the productive plants. As Moran (1981, 61-63) says, "extractive economic exploitation has been a common experience throughout the tropics."

In contrast, considerable investments are needed to exploit minerals (iron ore, manganese, bauxite) and develop hydroelectric energy. The early resource extraction practices left little if any capital in the region nor did they create a long-term spatial order. New capital-dependent activities impose such an order prior to the beginning of the production phase. A case in point is the hydroelectric project in Tucurui, which imposes a regional transmission system. This will intensify land use demands.

An INPA scientist identified water quality as one of the key problems with hydroelectric power projects. Associated with this is ichthyology and dams, because of the separation of feedings and breeding zones for different species of fishes. A third concern is climatic modification which is associated with larger reservoirs, such as the one at Tucurui (2430 km^2) or the one at Balbina (2540 km^2). A fourth issue is the speed of an unobstructed river compared to a blocked one. At Tucurui, water will be slowed to about 1/7 its original (pre-dam) velocity.

The lack of dependable data on stream flows and precipitation complicates analysis in Amazonia where little authoritative information was assembled until very recently. This is especially the case in the hydrological field. Stream gauges broke, and were monitored irregularly by non-professionals. For example, at Tucurui prior to 1976 the stream flow data indicated an average annual rate of 9300 m^3/s. In 1984, the average annual rate was 11,400 m^3/s, a 23% increase. A problem is the peril of false estimations which may bring on an unanticipated flood that could damage the project, as nearly happened at Tucurui in March of 1980. As actual construction of numerous dams in the Amazonian region has been delayed for economic reasons, important and dependable hydrologic and climatic data can be assembled. Consulting firms are deemed unreliable by this INPA scientist, who thinks that they dole out too much misinformation. As an example, reference was made to a consultant's report in which the Tucurui area contained 46 m^3 of Brazil nut wood/ha--a very questionable measure according to this source.[9]

4.2. Anticipated Changes

Reservoir formation will produce numerous environmental changes while the power plant itself is an instrument of regional development. Changes in stream patterns, lake margins, lake surface and evaporation, water quality, local climate, and pisiculture are some physical impacts that draw attention. The associated population increase encourages greater attention to land uses, health needs, and economic activities in the project and reservoir areas. Societal needs command more systematic and prudent resource utilization than had been practiced in Amazonia until very recently.

Stream patterns above Tucurui are expected to change because of slowed water mobility. How this will affect sediment deposition in the master stream remains to be identified. It may help to know that the Tocontins derives much of its sedimentary load from the crystalline plan-alto region of central Brazil. The Araguaia, main tributary to the Tocantins, carries clear water similar to the Rio Negro.[10] Most of the sediment then is Tocantins-borne and will alter its regime in years to come. It is likely that sand bars will mark the entrance to Lake Tucurui, and thus cause delta formation in tributaries at the junction with the master stream above the reservoir.

The lake margins are a general concern. Currently Brazilian law sets only a 100 meter belt surrounding reservoirs aside for lake protection (Cals de Oliveira Filho 1977), but the idea to widen this belt to form an ecological reserve is under discussion. Actual width of this belt has not been fixed, though two kilometers on the larger units has been mentioned. Such belts could also serve as "cordon sanitaire" for malaria and schistosomiasis controls.

Protection of reservoirs has been the concern of Eletrobras since 1975, as has been environmental protection in a wider context. The current Minister of Mines and Energy, Cesar Cals de Oliveira Filho (1976), a former member of the Eletrobras directorate, specialized in environmental protection in the watersheds to control sedimentation and pollution as key factors. He included pisiculture and health as well. Taulois, a member of the Eletrobra's environmental assessment department, has studied the problem of sedimentation and dam protection for a decade. Besides these studies from Eletrobras, consulting and construction firms (Budweg 1979) and their staffs have been taking an active interest in fostering environmental concerns and share this by publishing detailed studies. Sensitivity to environmental considerations is making notable gains--as summarized by Menezes and Pereira (1979). Reservoirs present notable changes in the environment and their creation and management is part of the project planning process.

Pollution and navigation are two problems of reservoirs.

1. Pollution--if large communities form above the dam on the waterbody, the effluents could present serious problems for waste quality, sedimentation of pollutants, and ultimate corrosion of turbines; fertilizers from agriculture that can reach the reservoir present another possible source of turbine corrosion and there are possible metallic industries that fail to control their effluents; all these cause concern for the energy planners (Menezes and Pereira 1979).

2. Navigation--though the rivers in Amazonia are not navigable thoughout because of rapids or falls and sharp seasonal variations in water levels, the greater segments are navigable for the relatively flat-bottomed river boats. Lake formation will increase winds on the lake which is absent on winding and tree-lined rivers. Boats using lakes must be constructed to withstand wind.

Health is another concern. Two areas of immediate attention are disease vectors and waste. Malaria and schistosomiasis are the diseases that give greatest cause for concern. Unless the lake margins are widened significantly, snail and mosquito populations will have their hosts to perpetuate and spread "plasmodium vivax" and "sanbom". Large reservoirs complicate insect and mollusc controls significantly. To give an idea about the challenge to disease control, consider if 264 snails are counted in 1 meter^2, or 1 snail/38 cm^2, control of such numbers becomes difficult, if not impossible.[11]

Only by separating snails and people can schistosomiasis be controlled. Mosquitos also need to be separated from potential hosts to reduce and control malaria in the reservoir area. Mosquito control is complicated by the insect moving from host to plant, not dwelling. This in turn breaks the cycle of control. Incidence of malaria could grow at Tucurui because of the extent of the waterbody, especially if the population density near the lake increases. One way to curb high population density near a waterbody would be declaring reservoir belts as a national security zone. Another disease that is difficult to control is chagas, which is transmitted by a large black fly.

4.3. A Comparative Assessment of Hydroelectric Projects: Function and Scope

Two kinds of dam have been built in Brazil: those that serve as water storage units in the arid Northeast, and those designed for hydroelectric power generation. Few dams have been planned or built for multiple functions. Regional water needs led to construction of single function dams. Energy shortages prompted the implantation of dams that enter and entered power generation promptly. Few other functions on hydroelectric dams have been included until recently. For example, Jupia (1969) included a navigation lock (still not in use). A more ambitious scheme was planned for Sobradinho (on the Sao Fransisco) to: a) regulate the stream flow for downstream power plants; b) provide water for irrigation agriculture in the Sao Francisco Valley; and c) generate energy. In Amazonia, to date, only Tucurui dam includes a second function, namely, future navigation via its major locks. The first of these is half completed. This leaves ichthy-o-fauna as the area in which efforts are being made to compensate for the possible change in riverine fish life, an important local food source. Experiment stations and fish stocking are used to attenuate any dislocations caused by dams.

In Amazonia, the operating and planned hydroelectric dams are only designed to generate electricity. The government has decided not to add costly features that will serve few if any users. At this time there is nothing comparable in Amazonia to the Tiete River in Sao Paulo state, on which the dams include locks for river navigation (locally known as the "alcohol route"). Until Tucurui dam entered service (1985), there was no major regional energy system functioning in Amazonia, and except for Curua Una and Coaracy Nunes, biomass and fossil fuels served local energy needs. It is technically possible to build multipurpose dams in Amazonia, but at present it is not politically or economically feasible. Much of the energy generated at Tucurui is destined for the northeast (Eletrobras 1982, 40). The 14,000 MW Volta Grande project on the Xingu at Altamira is slated to send its energy to the industrial Southeast, and current plants refer only to energy generation, nothing about locks for navigation. A regional development effort is more appropriate for the Sao Fransisco river basin, or the Iguacu or Uruguay river basins than those under construction and planned in Amazonia.

Conservation-consciousness is supported in the scientific and technical communities, but must yield to perceived national needs. Indications are that large projects will not be as readily accepted in the future as in the past and face growing challenges. Objections to large projects because of size alone would betray more obstructionism than understanding of regional and national needs. In the perspective of time and national energy needs, Electrobras has provided additional energy with minimal adverse environmental changes. Single-purpose dams, for hydroelectric power in this case, are benign. Careful differentiation among Brazilian river basins is necessary to focus upon unique regional conditions and needs.

Amazonia of 1985 resembles the U.S. prairie lands and their emerging infrastructure formation in the 1850's, in that occupying the land was more pressing than asking how best to use it. The current process of land occupance proceeds without the needed scientific-technical preparations to assure long-range stability on the land. Much more is known about the deleterious effects of careless land use today than in the 1850's. The thrust for immediate results persists in Amazonia. Large hydroelectric projects require comprehensive regional analysis of the physical base prior to construction and are much more carefully conceived than other land use patterns in the region. Agro-projects and their impact upon the land should not be compared with hyroelectric projects in Amazonia. For hydroelectric projects to be productive, conservation of the environment is fundamental. It is the scale of the investments and the sectors serviced that accounts for this pronounced difference in environmental response and perception.

5. HYDROELECTRIC PROJECTS AS AGENTS OF SOCIETAL CHANGES

Hydroelectric power projects can be identified either as subsequent additions to development or as antecedent instruments of change. Curua

Una may fit the former characteristic, but Tucurui is clearly an antecedent agent of change. The hydroelectric projects in Southeastern Brazil were built to overcome serious electricity shortages experienced for over twenty years (1948-1970). Numerous industrial firms built their own hydroelectric or thermal power plants to insure a regular energy supply and avoid damage to machinery (Tendler 1968, 7-42). These projects in Southeastern Brazil can be characterized as subsequent power plants. Many small hydroelectric units will enter service in small, isolated communities. These are locally significant. Regionally, large projects will be needed to contribute to the energy infrastructure in Amazonia. Amazonian hydroelectric energy projects will create access to resources that were formerly unavailable. Hydroelectric power projects usually have modest space requirements, but can provide energy for large population concentrations in limited spaces.

5.1. New Towns--Hydro-project Related

Amazonia is undergoing urbanization. In the early 1980's Manaus grew at an annual growth rate in excess of 9% (SEPLAN, Para 1979). In Para state, 22 new communities were established between 1970 and 1977, with a total population exceeding 100,000 persons, or an average population of 4,680 each. Deforestation is already underway before forestry management plans can be completed. Small holders--50 and 100 ha lots--may spare considerable areas of Amazonia from hasty deforestation. Even though the population growth in part of the region exceeds 2.5% a year for the moment (in Rondonia and Mato Grosso in-migration reached 8 to 10% per year in the last five years), much of the growth pressure is being diffused by the established urban centers and the formation of new urban centers. The above-cited SEPLAN study refers to these changes as "de-ruralization" (desruralizacao). It is to these trends that the urban centers that are organized to serve dam construction have to be added. Other towns are already being planned and built as mining and industrial centers. These latter centers are evolving in concert with the formation of the Tucurui energy growth rate.

Hydroelectric project-related new towns have been built at Curua Una and Tucurui. Tucurui is substantial (Tucurui 1981, 2); in 1979 the town housed over 26,000 persons and by May 1981 over 55,000. Tucurui has its own urban administration, has the key urban functions such as a large hospital, an inter-city bus terminal, post office, airport with a runway for B-737's and a primary and secondary educational system. There are numerous shopping centers, bus service, utility infrastructure, and recreational facilities to meet the needs of the population. The city is divided into distinct neighborhoods, and the manager's section contains houses of $180m^2$ each overlooking the river and project. The rest of the town will be relinquished to the next wave of dwellers who are expected to come here to work in metallurgical plants. Planners are arranging for a smooth transition in Tucurai from hydroelectric projects to industrial activities.

As a consequence of the construction of Tucurui, other urban centers will be established and existing ones will change. Several new communities are under construction in the center of the Carajas mineral deposit region and a 70,000 person urban center is in progress outside of Vila do Conde, the new aluminum city of Barcarena on the Tocantins, near Belem. Pargominas can be expected to sustain active growth to service the bauxite mines nearby; Marba is projected to grow to 177,000 people by 1990. Maraba is also being rebuilt on higher ground to avoid the perennial March floods. It already has a bridge across the Itacaiunas as part of the Transamazon highway. A bridge across the Tocantins is forthcoming in the near future for the railroad that links Carajas to Itaqui in Maranhao, the deep seaport. Maraba is being converted into a major transport node and plans project several heavy industries to be sited there.[12] All of this raises questions about agriculture and grazing in the region, which have been introduced without adequate scientific and technological foundations. Farmers and ranchers may move to the cities, creating local food shortages.

5.2. The Evolving Mining Industry

The discovery of the Carajas iron ore deposit was instrumental in spreading change. At this time, the deposit is estimated to contain 18 billion tons of 66%+ grade iron ore. Other ores include at least a billion tons of copper ores, in excess of 60 million tons of manganese, 40 million tons of bauxite, to mention only the more widely used metals. Gold, as by-product, is estimated to exceed 400 tons. This is a highly mineralized area which may become the core of a new state. It was this iron ore discovery in 1967 that deflected the construction of the Cachoeira Porteira dam on the Trombetas which was originally to be built by the early 1980's. Political-economic considerations were thus instrumental in initiating these changes. Furthermore, it may be anticipated that once the fundamental infrastructure is in place, most Brazilian iron ore exports will be from Carajas and no longer from Minas Gerais. This will conserve those deposits for the domestic market and keep transportation costs to a minimum as long as possible. The Tucurui Dam can be considered a creater of Carajas, and this part of Amazonia will be shaped in large part by the energy generated at Tucurui for a very long time to come.

5.3. Transportation - A Changing Panorama

Transportation in Amazonia depends heavily on the river system. As such, the river is deeply rooted in practice as well as in the psyche of the people of Amazonia. The first major change was the Belem-Brasilia highway. Next came the Transamazon Highway, followed farther south by the Cuiaba-Port Velho road. Each of these roads required large investment and great sacrifice. These roads were sliced through vast unpopulated rainforests. They subsequently became corridors of settlements. The implantation of Tucurui was substantially furthered by the Belem-Brasilia and the Transamazon Highway and its branch road to Tucurui. The motivations to build each were very different, but the sum of their impact is impressive as it has sped the "interiorization" process in Amazonia.

Overland transportation has become very popular and is much quicker in many cases than riverboat travel. There remains active riverine passenger service in the immediate area of Belem in the para delta. As the metropolitan region expands its industrial activities, the water routes will serve until bridges and causeways change the current travel patterns. Land transportation has found ready general acceptance.

How people will respond to the new railroad from Itaqui to Maraba and Carajas will depend in part on the settlement pattern changes that are planned. Planners in Sao Luis, Maranhao and staff of Compania Vale do Ric Doce as well as in their publication (1981, 132-34) (the Brazilian government mining firm) claim that at least seven farm communities are planned along the railroad in Maranhao, each with a station. This line is 860 km long, with a projected spur to Paragominas of an estimated 150 km (see Figure 2). In the vicinity of Paragominas exist proven bauxite reserves of 2.4 billion tons, enough to support a branch-line. Other spurs to other mineral deposits are possible and the most important will be the link to the copper mine at Salobo. The reason for linking Carajas to Itaqui instead of Belem or some other place in Para state is that Itaqui offers berthing facilities without dredging for vessels of up to 300,000 tons.

The main line has engendered considerable disfavor among planners and politicians in Para. Planners discuss this from the point of view of Belem as a port city and the Tocantins as the transport artery. The key point is that the Tocantins from Jatobal to Tucurui (see Figure 1) was never a navigable river because of rapids and falls. Even if the river were constantly carrying $35,000m^3/s$, navigation would be hazardous, but its actual annual average is $11,200$ to $11,400m^3/s$. On a field trip in 1984, when the river carried 3,700m3/s, a 6.5 meter aluminum boat had to be navigated with great skill in daylight above the dam to avoid the basalt outcrops. Below the dam the river takes on a far more placid character and barges for bulk freight are an indicated transport medium. Pinto (1980) has been the chief publicist in the local newspaper (O'Liberal) writing against large projects in the region, including the railroad. What is generally forgotten in this discussion is the Jatobal-Tucurui railroad (see Figure 2) which was built precisely to resolve the transport problem of the Brazil nuts collected in the Maraba region and processed in Belem for export. As the river proved too hazardous as a water route, a parallel land route had to be developed. The railroad entered service in 1941 and ceased operations by the mid-1970s when trucking replaced it.

Out of this opposition to the railroad and its route emerged construction of locks at Tucurui. One has already been built. This can be viewed as a conciliatory gesture to the State of Para on the part of the federal government at this time. The locks may justify the investment in the future. At this time they cannot be considered useful and are a diversion of costly capital more effectively invested in transmission lines for the hydroelectric power services. There is no freight in quantity to compensate for this investment between Maraba and Belem for the next decade. For the locks at Tucurui to be effective, the dam at Santa Isabel (see Figure 2) and its locks will have to be in place. If the agricultural production takes on the patterns projected for Goias and Mato Grosso states in the Araguaia valley, then there may be the cargo volume that justifies the investment in these locks. For the Carajas iron and copper ores the river route lacks logic. The projected river route is more for agricultural than mineral products. The railroad thus became an instrument for a political end, namely the locks.

The railroad then has been built foremost as a carrier. As these farm colonies evolve, substantial tonnages of agricultural products are planned to be sent via this system. This would be especially the case if the babacu palm Icernel were to be transformed into metallurgical coke for iron ore smelting plants in Maraba and Tucurui. Preliminary plans project annual tonnages on the line to be 72 million tons, 35 million tons iron ore, 7 million tons agricultural products, and the balance in other products (Compania Vale do Rio Doce 1981-B).

6. HYDROELECTRIC ENERGY AS AN INSTRUMENT OF CHANGE: CLOSING OBSERVATIONS

Hydroelectric energy served as the moving force in establishing industries in the southeast, south and northeast of Brazil. The additional hydroelectric projects that are being constructed or planned in that part of Brazil will continue to service an established clientele, and in part serve as replacement energy for thermal power plants. In the Amazonian setting hydroelectric energy assumes a position of a leading contributor and promoter of change. Tucurui can be considered an instrument of "resource enrichment" in a region that had proven to be "economically uninviting" for major projects. Hydroelectric energy is made into an agent of change in Amazonia because it provides the instrument to reorder the attitudes and objectives of the people in the region as well as those from without.

Without Tucurui or a comparable project, Amazonia would continue as a resource-yielding region without any return to the national economy. Tucurui is a major investment that will serve to attract other large investments into the region, as for example the alumina and aluminum producing facilities at Barcarena, Para (see Figure 2). It was the discovery of the massive iron ore deposits at Carajas in 1967 that led to the decision to construct Tucurui as soon as possible and postpone Cachoeira Porteira. The spatial consequences of constructing Tucurui (for the moment the fourth largest hydroelectric facility in the world) are far reaching, for it will furnish the energy for the infrastructure at Carajas, and in time power the railroad system that delivers the iron ore at dockside in Itaqui (see Figure 2). Brazilian iron ore exports will be relocated from Minas Gerais to Para-Maranhao and thereby conserve the iron ore deposits of southeast Brazil for the regional domestic markets. The Brazilian aluminum industry will be pulled toward the resource-rich Amazon where inexpensive power is available. Aside from repeating itself in the future, this process will gradually redraw the map of economic activities in Amazonia and Brazil.

The harnessing of the Amazonia hydroelectric potential induces the formation of industrial centers in empty quarters. This stands in stark contrast to the role of hydroelectric energy in the Brazilian southeast. In either case the hydroelectric power projects serve foremost for the generation of energy, by contrast comparable projects in the United States were designed to serve multipurpose functions in which hydroelectric energy was chiefly a secondary consideration (Eckstein 1958, 237-58, see especially 237). In the Amazonian case, hydroelectric power will free the region of predominant external energy dependence and serve as major factor in tying the region into the Brazilian economic system. Amazonia will become part of the "effective national economy" and as such part of the "effective national territory" (James 1956, 887-88), something it has not yet been. To evaluate the consequences that will flow from these changes requires time, but one also needs to understand the processes that shape the future. And that is exactly what is happening at Tucurui. What is being observed here is how an idea took hold, namely former president Kubitschek's concept of infrastructure formation. Jaguaribe (1968 152-4) summarizes Kubitschek's administrative goals (1956-1961):

> "The Target Program, incorporating existing plans in a broader and more systematic framework, set up thirty-one targets grouped under six broad heads: energy, transport, food, basic industries, education and the construction of the country's new capital, Brasilia."

In terms of estimated expenditures, the energy sector (chiefly hydroelectric power and oil production) was slated to receive 43.4% of Brazil's capital resources, followed by transportation 29.6% for transportation. Kubitschek stressed energy and roads, and he could hardly ask for a more effective way to be remembered then by the realization of the program that he set in motion then for all of Brazil.

7. ACKNOWLEDGEMENTS

This study has benefited from much understanding, help and encouragement, from a long list of professionals in Eletrobras, Eletronorte, INPA, OAS, CAEMI, IDESP, SEPLAN, IBDF, and many others. To them all I owe a great deal. I can only hope that what is reported here is clearly recognizable to them. A note of special thanks for released time granted by Montclair State College.

8. FOOTNOTES

1. Eletrobras - the national power company in Brazil, a holding company that coordinates electric energy planning and financing. Hydroelectric energy is its primary responsibility; some thermal electric power units planned and financed by Eletrobras.

2. Eletronorte - a regional power firm in Brazil, responsible for major hydro-electric projects in the Amazon region; subsidiary of Eletrobras.

3. INPA - National Research Institute for Amazonia, located in Manaus (Instituto Nacional de Pesquisas para Amazonia).

4. Reference to potential earthquakes of water-storage reservoirs rather than hydroelectric facilities indicates that for the moment no direct link to hydroelectric projects has been documented.

5. Luz (1983), UHE TUCURUI, app. 2.

6. The Araguari has potential for up to 262 MW as inventoried to date, or another two dams of comparable scale could be placed.

7. Each turbine at Tucurui generating at nominal capacity will discharge 576 $m/^3s$ of water, or if 24 units were in operation, 13,824 m^3/s. The Tocantins at Tucurui averages roughly 11,400 m^3/s, so about five turbines would have to be idled. This may provide some insight into the concepts of base and peak load, as well as the significance of stream flow control in relation to energy generation.

8. Luz (1983), UHE TUCURUI, app. 2.

9. Interview, INPA, Manaus, June 22, 1981.

10. Interview, INPA, Manaus, June 22, 1981.

11. Interview, INP, Manaus, June 22, 1981.

12. Maraba has engaged the attention of several planning agencies. Notable changes can be anticiapted as iron-ore movements, railraod maintenance, and metallurgical industries develop interest in regional markets.

9. REFERENCES

Alexander, J.W. and Gibson, L.S. 1979. Economic geography. Second Edition. Englewood Cliffs, NJ: Prentice-Hall.

Budweg, F.M.G., ed. 1979. Barragems, reservatorios e o meio ambiente: A practica Brasileira. Sao Paulo: Themag.

Companhia Vale do Rio Doce. 1981a. Eastern Amazon: Preliminary development plan. 2 vols. Rio de Janeiro: C.V.R.D.

______. 1981b. Projecto ferro cavajas. Rio de Janeiro: C.V.R.D.

Cals de Oliveira and Filho, C. 1976. Politica de protecao das grandes represas Brasileiras. Rio de Janeiro: Eletrobras.

______. 1977. Consideracoes sobre a utilizacao multipla de grandes represas Brasileiras. Rio de Janeiro: Eletrobras.

Cook, E. 1976. Man, energy, socety. San Francisco: W.H. Freeman and Company.

Cuff, D.J. and Young, W.C. 1980. The United States energy atlas. New York: The Free Press.

Denevan, W. 1981. Swiddens and cattle versus forest: The imminent demise of the Amazon rain-forest reexamined. In Where have all the flowers gone? Deforestation in the third world. ed. V.H. Sutlive, Williamsburg VA: William and Mary.

Dienes, L. and Shabad, T. 1979. The Soviet energy system. New York: John Wiley and Son.

Eckstein, O. 1958. Water resource developments: The economics of project evaluation. Cambridge: Harvard University Press.

Eletrobras. 1979a. Plano de atendimento aos requisitos de energia eletrica ate 1995. Rio de Janeiro: Eletrobras.

______. 1979b. Potencial hidroeletrico da Regiao Amazonica. Vol. VII. Rio de Janeiro: 1979.

______. 1980. O potencial hidroeletrico do Brasil. Rio de Janeiro: Eletrobras.

______. 1982. Plano de suprimento aos requisitos de energia eletrica ate o ano 2,000. Rio de Janeiro, Eletrobras.

Fels, E. 1954. Der Wirtsch aftende Mensch als Gestalter der Erde.

Stuttgart: Franckh'sche V.

Goodland, R.J.A. 1980. Environmental ranking of Amazonian development projects in Brazil. Environmental Conservation VII: 1.

Guyol, N.B. 1971. Energy in the perspective of geography. Englewood Cliffs NJ: Prentice-Hall.

Jaguaribe, H. 1968. Economic and political development: A theoretical approach and a Brazilian case study. Cambridge: Harvard University Press.

James, P.E. 1956. Latin America: State patterns and boundary problems. In The changing world: Studies in political geography. eds. W.G. East and A.E. Moodie. New York: World Books.

Luz, D.S. 1983. UHE TUCURUI. Sua Importancia para a regiao Amazonica e para o Brasil. Tucurui: Eletronorte.

Manners, G. 1964. The geography of energy. London: Hutchinson.

Menezes, L.C., and Pereira, A. de A. 1979. O sector eletrico e o meio ambiente. Rio de Janeiro: Eletrobras.

Ministerio do Interior, PRODIAT. 1982. Diagnostico da bacia do Araguaia - Tocantins. Brasilia: PRODIAT.

Moran, E.F. 1981. Developing the Amazon. Bloomington: Indiana University Press.

Murphy, R.E. 1954. The fields of economic geography. In American geography: Inventory and prospect. eds. P.E. James and C.F. Jones, Syracuse: Syracuse University Press.

Obst, E. 1965. Allgemeine wirtschafts- und verkehrsgeographie. Berlin: de Gruyter.

Paiva, M.V. 1977. The environmental impact of man-made lakes in the Amazonian region of Brazil. Rio de Janeiro: Eletrobras.

Pinto, L.F. 1980. Amazonia: No rastro do saque. Sao Paulo: Hucitec.

Pryde, P.R. 1983. Nonconventional energy resources. New York: John Wiley and Son.

Scazzocchio, F.B. 1980. Land, people and planning in contemporary Amazonia. Cambridge: Center for Latin American Studies, University of Cambridge.

SEPLAN, Para. 1979. Politica de desenvolvimento urbano. Belem: SEPLAN.

______. 1981. Perfil urbano 1981: MARABA. 3 vols. Belem: SEPLAN.

Shabad, T. and Mote, V.L. 1977. Gateway to Siberian resources. New York: Scripta.

Sioli, H. 1980. Foreseeable consequences of actual development schemes and alternative ideas. In Land, people and planning in contemporary Amazonia, ed. F.B. Scazzocchio, Cambridge: Center for Latin American Studies, University of Cambridge.

Smil, V. and Knowland, W.E. 1980. Energy in the developing world. New York: Oxford University Press.

Taulois, L. 1976. Protecao de reservatorios: Erosao e assoreamento. Rio de Janeior: Eletrobras.

Tendler, J. 1968. Electric power in Brazil. Cambridge: Harvard University Press.

Tucurui. 1981. Resumo Estadistico. Tucurui, A.M.

Whitbeck, R.H. and Finch, V.C. 1941. Economic geography. New York: McGraw Hill.

Whitbeck, R.H. and Williams, F.E. 1940. Economic geography of South America. New York: McGraw Hill.

PART VI

Prospects in Energy Geography

CHAPTER 25

GEOGRAPHY AND ENERGY: THE QUEST FOR ROLES AND MISSIONS

Thomas J. Wilbanks
Energy Division
P.O. Box X
Oak Ridge National Laboratory
Oak Ridge, Tennessee 37831

ABSTRACT. Energy has emerged as an important field of specialization in geography, offering geographers the opportunity to have an important impact on public policy as the world struggles through the transition to non-fossil sources of energy services. To take advantage of this potential, energy geographers need to focus on the real problems, to establish their credentials as experts on energy as well as geography, and to communicate effectively with those who take actions to resolve issues and solve problems.

1. INTRODUCTION

For several reasons, this book is a benchmark in research by geographers on energy questions. Appearing a decade after the oil embargo of late 1973, it documents the response of an academic discipline to the sense of crisis that followed that event. At least as importantly, it appears at a time that represents a kind of watershed for energy as the focus of a subfield of geography, both because the main stimulus for research on energy questions has faded, at least for the time being, and because geographers now know enough about energy to have an impact on public policy, if they so choose.

The change during the past decade is quite clear. In 1974, geography as a discipline had strong external reasons to pay attention to energy issues, but it had very limited experience to offer. In 1984, the situation is reversed.

What makes the present situation especially interesting is what it promises to tell us about geographers. Faced with external needs but internal inadequacies in 1974, there was never any question how geography would respond. It would learn and do, taking advantage of a cross-cutting perspective which is the captive of no particular policy issue. But faced now with external apathy and internal strengths, there are very real questions about how geography will react.

In some ways, from a point of view of the current relevance of the

F.J. Calzonetti and B.D. Solomon (eds.), Geographical Dimensions of Energy, 497–510.

discipline, this is encouraging. The perspectives and skills of geographers are indeed adaptable, and geographers can be very agile in applying them. On the other hand, one senses the lure of a sort of faddishness, as issue-oriented scholars pursue the crisis of the moment, lacking the long-term view that more often leads to major contributions to knowledge and effective policy.

This final chapter is an attempt to sketch the outlines of such a long-term view. In general, I will argue, if geography can sustain its attention to energy as a policy issue, maturing in its sense of how to participate in a broad policy arena, it can put itself in a position to play a central role in one of the great intellectual and institutional challenges of the next several generations: the transition to non-fossil sources of energy services.

2. A QUESTION OF FOCUS

The central question has to do with the way we decide where to direct our efforts as scholars: what is most worth doing and for whom it is done. Looking back to the past decade, it is probably not too much of an exaggeration to suggest that geographers writing about energy have mainly been concerned with reaching an audience of other geographers. In their attention to energy issues, to paraphrase Herman Postma, they have usually been jumping on the bandwagon rather than leading the band. More often than not, the original contributions have been in the development and demonstration of methodologies rather than the illumination of issues. Certainly, it can be argued, this is the sort of orientation which pays off in the academic world: a link with a "hot" policy issue as a tactic to attract research support while the focus remains on what disciplines value (which is not necessarily what will help to resolve the issue).

These comments are not meant as criticisms, because such an orientation has a certain rationale, epistemological as well as practical; and, besides, this book contains a number of notable exceptions to it. But these tendencies do suggest some questions about the future, as we consider our roles as geographers with energy as a specialty. These questions, it seems to me, center on (1) relevance as a measure of the value of research and (2) effective communication beyond the boundaries of the research community as such.

2.1. Relevance as a Measure of Value

On occasion, I have asserted that the main problem with geography--or at least economic geography--is that it is too often irrelevant (e.g., Wilbanks 1982b). This does not mean that geography is wrong, and it does not necessarily mean that it is uninteresting (at least to those of us who love it). Likewise, it certainly does not mean that accumulating knowledge for its own sake is without social value. But we need to face the fact that, in a society with limited resources, these verities are not enough. We can see the evidence in enrollments for geography courses, the infrequent use of our research by people grappling with the major problems of our society and economy, and often the virtual invisibility of our subject within society at large.

In looking for solutions to this problem, we need to look both at ourselves and our society. Clearly, few people--at least in the United States--value geography for its own sake. What most people are concerned about are such things as job prospects, economic productivity, and the utility of knowledge for confronting challenges in the real world. To the extent that we speak effectively to these concerns, we are valued--whatever our label. To the extent we do not, we are ignored, regardless how happy we may be within our own little circles. Is there any question where the balance falls right now?

We need to worry about this "revealed preference" for two reasons. One is that such utilitarianism in society can be dangerous, even anti-intellectual. Along with our colleagues in the humanities, we need to persuade our fellow citizens that there is more to life than this. But the other is far less comforting. Society's viewpoint toward our subject tells us to take a close look at ourselves. What have we been learning and teaching? How have we decided what is worthwhile? Have we been so busy following our noses and filling our niches that we have missed what is going on in society around us? Have we neglected the important in our exposition of the familiar? Certainly, our literature indicates that we have contributed very little to such major issues for American society in the 1980's, such as the following:

1. Avoiding nuclear war.

2. Increasing the productivity of the U.S. economy in an era of intense international competition.

3. Providing for critical social needs without asking government to play roles (or spend money at levels) we consider undesirable.

4. Increasing the effectiveness of our institutions and structures for decisionmaking, both in solving current problems and doing the right things to prepare us for the future.

5. Improving "North-South" relationships in international development.

Ten years ago, Torsten Hagerstrand, speaking about contributions by geographers to social issues, suggested that the first prerequisite is to worry less about disciplines and more about society (Hagerstrand 1975). "Like the Biblical injunction, perhaps we must be prepared to lose ourselves in order to be saved" (Wilbanks 1985).

Research focused on social issues and problems is usually tagged with the vaguely pejorative label "applied geography," which implies an activity fundamentally different from (and somehow inferior to) the search for new knowledge (e.g., Berry 1981)--perhaps analogous to the distinction between engineering and science. To the extent that this impression is correct, however, it has little to do with epistemology. It reflects observed differences in the questions that are posed, not intrinsic differences in how a research question is answered.

It is quite possible, in fact, to conceive of what might be termed "mission-oriented basic research" (Roy 1979): fundamental research extending the frontiers of knowledge in geography, arising from questions about society rather than questions about geography itself, opening new doors for building theory and methodology. The potential for truly original work, different from what we usually do, is associated with three characteristics of this kind of outward focus. First, it often suggests the importance of new variables (or the unimportance of familiar ones). A recent effort to understand the spatial behavior of electric utilities, for example, has demonstrated the inadequacy of conventional location theory as an approximation of complex institutional decisionmaking. For utilities, the _expansion_ decision is viewed as more important than the _siting_ decision for new power plants; in siting a power plant, minimizing _delay_ is more important than minimizing _distance_; and, to reduce costs and risks, many new power plants are owned by consortia of utilities supplied by a national power grid, which tends to blur the process of identifying an optimal location (Hillsman and Alvic 1985). In this kind of analysis, geographers often benefit from intensive contact with non-geographers, who challenge our disciplinary biases and our jargon--requiring us to understand the whole phenomenon that we observe, not just those elements that we would normally consider geographic (Wilbanks 1979). Second, a "mission orientation" encourages us to seek concepts that inspire the imagination and uplift the spirit, as contrasted with bodies of facts (readily accessible to non-geographers through information retrieval systems) and collections of techniques (which are often limited to situations where we cannot realistically channel our inquiry to fit technical specifications). In debates about social issues, nothing carries more weight than good, interesting, insightful ideas. They distinguish the gem from the journeyman, the leader from technician (Wilbanks 1981a). Third, looking outward helps us to understand the difference between _truth_ and _meaning_. The primary concern of science is truth, and the scientific method is designed to uncover it. In contrast, the primary concern of society (and philosophy, incidentally) is with meaning: a non-trivial "bottom line". We know that meaning without truth is dangerous, the tool of the demagogue, but we also need to recognize that truth without meaning is empty. A sensitivity to social issues reminds us to ask "so what?" as we lay out a research plan, to the benefit of geography in the long run (see, for example, King 1976; Berry 1980; and Gould 1982).

Being relevant, therefore, does not mean that we must be prepared to sacrifice professional standards or to give up our concerns with theory and methodology. But it does require that we be willing to start from a different kind of awareness in deciding what research questions are most important, at the risk (at least initially) of shaking ourselves loose from secure and comforting disciplinary paradigms.

In our energy work as geographers in the future, are we willing to turn our attention to the research questions which are most important for resolving energy issues, even if our most elegant tools cannot be applied? Are we willing to strike out in new directions in our basic research in order to improve our ability to deal with these issues, even if we must set aside some of the expertise that we have labored so painfully to get?

2.2. Effective Communication with a Broader Audience

Once we have original knowledge relevant to critical energy issues, if that is what we decide we want to create, the next challenge is to communicate it effectively to the appropriate audience. But just what (or who) is the audience? For research which is not mission-oriented, we know that we are aiming the results at fellow members of the research community and at learners in the educational system. For mission-oriented research, however, the objective is to contribute not only to _learning_ but to _doing_. And this requires communicating as well with those whose business it is to _do_ things to resolve issues and solve problems.

One of the reasons that we (like most other academic disciplines) have generally failed to meet this challenge is that we are too often unaware of some of the prerequisites for reaching certain audiences. For example, communicating with policymakers depends heavily on earning recognition as a substantive expert (i.e., an _energy_ expert), as distinguished from a disciplinary expert, and producing materials that are timely and incisive, not just technically competent (Wilbanks 1981b; Wilbanks 1982a; Wilbanks and Lee 1985). And communicating effectively with an interdisciplinary audience usually requires more attention than usual to issues that _others_ agree are important and to using the regular English language in place of familiar jargon (Wilbanks 1979).

More fundamentally, it appears that most uses of social research in policymaking are indirect rather than direct. Ideas and information from the research world are absorbed by the lay public, or at least the technocrats within it, and then communicated in diverse ways by that public to policymakers as general wisdom (Weiss 1977). This suggests the importance of still another audience, one that we tend especially to neglect. Perhaps our impact as a discipline on energy policy over the next several decades will depend mostly on our success in reaching not a professional but a _general_ audience with insights that are fresh, sensible, clearly articulated, and verifiable through the experience of a wide range of non-scholars. Such fields as economics and psychology have met this challenge more effectively than we, year after year.

In our energy work in the future, are we willing to pay the price to become effectively "multilingual" in communicating our work? Are we willing to make a special effort to be understandable as well as good, salient as well as correct, even "popular" at the expense of some precision?

3. THE CHALLENGES AHEAD

Clearly, during a time when energy seems to have a relatively low priority as a policy issue and a funding target, many geographers will choose to respond to other challenges than these--even though one can cite evidence that mission-oriented energy research is rewarding in a great many ways (e.g., Wilbanks 1984). They will contribute occasionally from the periphery of the policy arena--offering new tools, methodological concepts, and case studies--and they will continue to be a disciplinary sounding board for those geographers who are _energy_ specialists: all roles which are needed and appreciated. But they will not solve the world's or the nation's energy problems.

Where geography can play a central role in solving these problems is by providing first-rate professionals who immerse themselves in energy as a field of study and learn to communicate eloquently with the energy policymaking system (broadly speaking). From an individual point of view, the main incentives are two: (a) good science can make energy policy better, to the larger benefit of society in the long run, and (b) the effort is interesting intellectually. The first incentive helps to tide one over when the second is not so true, and the second helps to sustain one when progress with the first is hard to see (Wilbanks and Lee 1985).

If we produce enough such professionals, and if they direct their efforts so that valid insights are ready when energy issues again become too conspicuous for society to ignore, then geography will truly make a difference in the lives of millions of people during the lifetimes of most of us. This I believe. I believe it because, during the next several decades, the world will have to get serious about the transition from natural liquid and gas fossil fuels to alternative sources of energy, and we have a long way to go before we are prepared for such a transition. I believe it because energy is too important to the quality of life to treat casually as a field of public policy. And I believe it because the skills and perspectives of geographers are especially useful in some connections with important energy issues.

3.1. The Great Energy Transition

Few terms are more hackneyed in the energy policy arena than "the energy transition"--from petroleum and natural gas to whatever will eventually take their place. Every knowledgeable person agrees in principle that such a transformation will come; the only disagreements are about when and what will happen along the way.

The very familiarity of this idea makes it difficult to appreciate the mystery and wonder associated with a fundamental change in the system which provides energy services. Historians write of the far-reaching effects of the shift from wood to coal associated with the Industrial Revolution in Europe. Even more dramatic have been the impacts in the 20th Century of the movement away from the delivery of energy in solid form to the delivery of energy as a hydrocarbon liquid or as electricity (e.g., Wilbanks 1983). The automobile, the airplane, the telephone, the elevator, television, the computer--these powerful instruments of social change are almost unimaginable without the great energy transition of the early 20th Century.

The coming transition is likely to be less profound in its social effects than the last one, because it will not revolutionize the forms in which energy is supplied. Instead, it will change the way these forms are produced. But because it seems to involve a substitution of more problematic options for relatively cheap, convenient, clean oil and natural gas, the way will not be easy.

Consider the basic options, as we know them now. Coal is dirty. Nuclear energy is risky. Solar energy is expensive. Even energy conservation is complicated, because it depends on action by so many different individual parties. At this point, none of the major alternatives to oil and natural gas offer us a smooth path to the future.

At the same time, population growth and aspirations for economic development will be calling for more and more energy services to be provided. In particular, closing the gap between North and South in the quality of life will require that a greater proportion of the world's energy services be channeled to the developing countries.

In my view, meeting these needs while at the same time replacing our predominant sources of primary energy is one of the great challenges of our time. Perhaps only the challenge of avoiding nuclear war is more important.

The central problem is not that we lack resources and technologies to meet global energy needs without much petroleum or natural gas. Resources and technologies, or at least technology concepts, are available in abundance. The problem is that major energy options take decades to develop, in the sense of creating institutions to provide energy services cheaply, reliably, and acceptably, while society seems increasingly shortsighted. Under pressures of government budget deficits and high interest rates, for instance, the national investment in the U.S.--public and private--in energy options that lie between current commercial markets and the far horizons of basic research has diminished sharply, even though it was never very substantial. With the flattening of oil prices in the early 1980's, the sense that institutional innovations may be needed to improve our capability to deal with energy problems has waned, even though it was never very resolute.

As a result we can anticipate a time, perhaps as early as the 1990's, certainly no later than the first several decades of the next century, when a true energy crisis will appear. Because supplies of petroleum and ready substitutes for it will have dwindled while demand has increased, the price of these premium fuels will climb--most likely precipitously, triggered by some sort of political action which creates a sudden awareness of energy insecurity (Kash et al. 1983). Unprepared to respond relatively quickly with a shift to other energy sources, most countries will face economic stress, and many will experience sociopolitical instability. The events of the post-1973 period were but a rehearsal for what lies ahead.

This, at least, is what lies ahead unless the research community can find a way to be more farsighted than the society which supports it, which is no small task. The challenge to us is to develop and test ideas and to generate information so that alternatives are available, surrounded by as little uncertainty as possible about their requirements and impacts, when the crisis comes.

Some of this effort lies beyond the purview of most professional geographers: seeking technological breakthroughs in solar cells, biotechnological improvements in biomass production, and fusion energy; clarifying the world's potential for natural gas production and use; developing and demonstrating transportation technologies that are not dependent on petroleum; and the like. But many of the key issues are more accessible, for instance:

(1) The environmental impacts of coal development and use: e.g., clarifying the relationship between coal use and such environmental concerns as acid rain and increased atmospheric concentrations of carbon dioxide;

(2) Location, scale, and technology potential: e.g., developing a better understanding of the ways that technology potentials are related to variations between places (for example, heat pump performance and climate, solar energy and local energy conditions)--particularly the role of economies and diseconomies of scale in energy production and use;

(3) The process of deploying new technologies: including technology transfer, conflict resolution, access to capital, waste disposal, and institution-building;

(4) Approaches to energy distribution: getting energy from its sources to its uses, involving questions of movement systems, delivery forms, political and economic control, and organizational performance;

(5) Potentials for energy efficiency improvement: the degree to which efficiency improvements can substitute for supply capacity increases in delivering energy services, and the prospects that advances in the efficiency of energy use can help to keep the price of energy services at an acceptable level as the cost of energy production rises;

(6) Strategies for quick response: policy alternatives for an uncertain energy future, ranging from emergency preparedness to a greater stress on energy options that can be put into place relatively quickly, such as moderate-scale standardized units for electricity supply (as contrasted with larger units having much longer lead times).

None of these issues is exclusively geographical, nor is the intellectual territory ours because of our label. But each is important, complex enough to be intellectually rich, and part of the difficult job of assuring a smooth transition to a global energy future which is sufficient, just, sustainable, and participative.[2]

3.2. Energy in Context

A key reason for directing our attention to energy, in my opinion, is that it is not just another commodity. It has a special relationship with the quality of human life. Because of the virtual "invisibility" of energy in most of what we do (Aronson et al. 1984), however, this relationship is not well understood. Whether the question pertains to, say, connections between energy conditions and economic growth in developing countries or the role of electricity as an energy delivery form in modern societies, our present understanding is remarkably incomplete. In order to put energy in its proper context in the next century, avoiding misdirection as well as neglect, it seems to me that any undertaking to solve energy problems should be coupled with an undertaking to understand those problems better: why and how we need energy services.

The starting point is that, for society, energy is a _means_, not an _end_. For example, we seek comfort in life, for which we use energy to control climate in artificial environments. We seek convenience and personal productivity, for which such energy-dependent devices as machines, appliances, and computers are useful. We seek mobility, and energy lets us expand our potentials for movement. It is the connection of energy with human aims that make it important.

One needs to be careful not to debase the argument by exaggerating it but, as an illustration, consider one's requirements in a situation that calls for starting from the basics: the proverbial desert island or, somewhat more realistically, the conditions after a truly major disaster. Survival requires food, water, and shelter. Unless the surrounding physical environment is unusually benevolent and/or the level of life is rock-bottom, energy is needed for space conditioning in the shelter and for food preparation. At a higher level of existence, energy is needed for lighting, most likely for systems to provide sustainable supplies of food and water, and possibly for the storage of perishable medicines. At still a higher level, energy is essential for communications (e.g., a radio) and for converting available materials into useful articles, ranging from cloth to implements. By itself, energy is of little importance, but it appears as an aspect of nearly every item that is a clear requirement. Unsurprisingly, most scenarios (and plans) for surviving a disaster on the scale of an asteroid strike or a nuclear war place a considerable emphasis on the availability of energy. We get just a hint of this whenever our home power supply fails or gasoline becomes scarce.

In a profound sense, energy is one of the fundamental "building-block commodities" of the quality of life--along with food, materials, and information. Like these other building blocks, its main function in society is as a facilitator: an enabler of activities and accomplishments that people value. But this supportive role, realized almost entirely in combination with other factors, has made it difficult to establish unambiguous relationships between (a) energy as a physically measurable substance or force, something which is produced or developed, and (b) energy services, those aspects of our lives which require energy as a price of opportunity (Wilbanks 1983, 438). Energy seldom stands alone as an independent variable affecting human actions.

Conceptualizing about the role of energy in human affairs is further complicated by the facts that:

> it can be supplied to us by the human body, draft animals, falling water, the sun, solid materials, liquid or gas materials, engines, wires and microwaves. It can, if seen as a commodity, be defined by the economic and political processes that regulate trade and be associated with our views about how such processes should operate. Or, if seen as a resource, it can be defined by our views about needs, potentials, and alternative ways to reach our goals, and thus be associated with our views about ecological stability and change. Or it can be seen as a right, an entitlement, a necessity for human beings if they are to lead a decent life in the society in which they find themselves, and thus be associated with our views about justice and equity. Or it can be seen as a strategic material, a part of the arsenal for geopolitical economic competition, and thus be associated with our views about national power and vulnerability. It is usually defined in terms of measurable physical units, such as barrels and Joules and calories and kilowatts, but we also talk in terms of 'mental energy', a very real source of an 'ability to do work' (Wilbanks 1983, 438, after Aronson et. al. 1984).

In spite of these difficulties, however, our likelihood of solving future energy problems depends considerably on our ability to identify and understand the workings of energy in our societies and economies. A pragmatic reason is that documenting the importance of energy can be instrumental in acquiring support for energy-related research. A more exalted reason is that, especially when the pool of resources for energy research is relatively small, our intellectual and financial resources are too precious to waste them on efforts that are off-target. For instance, to what purpose do we increase total energy supplies if they cannot be delivered in forms that people can readily use or at prices they can afford to pay? How much help are new technologies if we cannot agree whether to use them? Where will the money come from to develop the energy potentials of the developing countries? What do onrushing technological revolutions, and perhaps major social trends as well (e.g., Harman 1979), mean for future requirements for energy services? For particular social aims, where is energy a necessary condition? How does it interact with other factors? What are the possible alternatives?

Again, we cannot claim that such questions belong uniquely to geography, but neither do they belong solely to any other academic discipline. They await the attention of smart, curious, persistent people who want to focus on issues rather than on defining boundaries; and they offer chances to get close to the essence of how human systems work.

3.3. Contributions from Geography

Clearly, I believe we should concentrate on what _geographers_ can do, working as issue-oriented professionals with a sense of mission that transcends disciplinary boundaries, not on what _geography_ can do as a discipline. But once we redirect ourselves in this way, this book--together with other recent experience--suggests that we have some special qualities and distinctive insights to offer.

Part of the reason is that geography tends to attract people with certain attributes. One example is our spatial awareness, operating in a public policymaking system which is (in the U.S.) exquisitely geographic (Wilbanks 1985). Location, spatial structure, and movement are important issues in this system, for energy policy as well as other concerns; and maps are powerful tools. Another example, perhaps, is the value of our particular kind of curiosity as a catalyst for good policy analysis. More than many others, one observes, we are likely to look for the central essence of a situation, without regard to conventional subject matter boundaries or distinctions between the physical and human worlds--not because of the nature of geography, in my opinion, but because our regional tradition and our cross-cutting locational perspective bring many _people_ with this sort of orientation to choose geography as their professional home (if they have a chance to get to know about geography). Still another factor in many cases is our practicality, which relates to our traditional grounding in the empirical. Compared with other field of science, geographers are often more willing to fit the approach to the problem, rather than trying the converse, benefiting from the fact that most of us get exposed to a variety of methodologies for research. Although we share the common compulsion to prove ourselves as a scientist, many of us seek out situations where we can work more like engineers, which is rare among the social sciences. Finally, geographers tend to be unwilling to assume away diversity and complexity, whether topical (e.g., variations in elasticities affecting economic decisions) or areal (i.e., the rest of the world). Such blinders often make analysis more tractable, but they have little to do with the realities of policymaking. Accordingly, an analyst who remembers their significance is more likely to produce useful and realistic advice for policymakers.

But we are not just useful because of the kinds of people we are. Individually and collectively, we bring concepts and tools from geography to bear on four subjects which pertain directly to energy choices for the future. Obviously related to each other, these subjects are:

(1) Locational decisions: including facility siting (e.g., nuclear waste repositories), conflict resolution, spatial economic efficiency, control/vulnerability, and relationships between the scale of an activity and the ability to arrive at consensual decisions in a participatory society (Wilbanks 1981b);

(2) Movement systems: movement needs and constraints related to energy systems (...pipelines, transmission grids, rail networks,...), energy-using systems (...highway transportation...), energy systems support requirements (...capital, labor...), energy system impacts (...pollutants...), and energy options themselves (...diffusion of innovations...);

(3) Spatial structures: energy efficiency as an implication of the spatial structure of activity (e.g., movement requirements, efficiencies in building envelopes, cogeneration potentials), concentration vs. dispersion in energy supply systems, evolving spatial distributions of activities as they relate to the socioeconomic and environmental impacts of energy policies;

(4) Relationships between people and their environment: not only resource potentials and environmental impacts but also environmental appraisals--in the sense of the much-abused concept of "carrying capacity"--under a variety of assumptions about resource/technology/ society conditions for the development and use of energy options, especially for rural areas in the developing world.

In these connections and others, we can contribute in ways that are distinctively geographical as well as precisely mission-oriented, and our experience can feed back into the literature of geography a richer base of concepts and evidence. This books says that we are ready. All we need to do is to persevere in our efforts to focus on the real problems, to establish our credentials as experts on energy as well as geography, to do consistently good work, and to communicate with those who will help to turn our findings into concrete actions.

4. FOOTNOTES

[1]One of our difficulties is that, in borrowing a body of principles and conceptual structures from economics, we have tended to adopt an artificial and simplistic view of human behavior (e.g., Aronson et al. 1984).

[2]The World Council of Churches has suggested that justice, sustainability, and participation are central to an "ethnic of ecological justice". To these, I add sufficiency, because one can imagine an

energy system which is just, sustainable, and participative--but so limited in the energy services it provides that it stunts social and economic development.

5. REFERENCES

Aronson, E., Stern, P., Wilbanks, T.J., et al. 1984. Energy use: The human dimension. San Francisco: W.H. Freeman & Co.

Berry, B.J.L. 1980. Creating future geographies. Annals of the Association of American Geographers 70: 449-58.

______. 1981. Applied geography at the public-private interface. Proceedings of Applied Geography Conferences 4: 3-5.

Gould, P. 1981. Letting the data speak for themselves. Annals of the Association of American Geographers 71: 166-76.

Hagerstrand, T. 1975. Comments at a Conference on Geography and Social Action, organized on the occasion of the Fiftieth Anniversary of the Department of Geography at the University of Michigan, Ann Arbor. October 9.

Harman, W. 1979. An incomplete guide to the future. New York: W.W. Norton.

Hillsman, E.L., and Alvic, D.R. 1985. Spatial relationships between steam electric power plants and utility service areas. Unpublished manuscript.

Kash, D.E., Fox, E.C., and Wilbanks, T.J. 1983. When will the next energy crisis occur? Oil and Gas Journal. December 19: 82-91.

King, L. 1976. Alternatives to a positive economic geography. Annals of the Association of American Geographers 66: 293-308.

Roy, R. 1979. A contemporary justification for the national R&D budget and process. Paper presented at the Annual Meeting of the American Association for the Advancement of Science, Houston, Texas. January.

Weiss, C.H., ed. 1977. Using social research in public policy making. Lexington, MA: Heath & Co./Lexington Books.

Wilbanks, T.J. 1979. Communications between hard and soft sciences. Review (Oak Ridge National Laboratory) 12: 24-29.

______. 1981a. Geography and jobs in the nonacademic world.

Geographical Perspectives 47: 1-5.

______. 1981b. Local energy initiatives and consensus in energy policy. Paper prepared for the Committee on Behavioral and Social Aspects of Energy Consumption and Production, National Research Council, National Academy of Sciences. March.

______. 1981c. Location and energy policy. In Applied geography: Selected perspectives, ed. J. Frazier, pp. 219-32. Englewood Cliffs, New Jersey: Prentice-Hall.

______. 1982a. Career public servants: What scientists worry about. Paper prepared for a Symposium on Science and the Career Public Servant, American Association for the Advancement of Science, Washington, D.C. January.

______. 1982b. Economic geography, relevance, and social change. Paper prepared for the annual meeting of the Southeastern Division, Association of American Geographers, Memphis, Tennessee. November.

______. 1983. Geography and our energy heritage. Materials and Society 7: 437-52.

______. 1984. Geography at ORNL. AAG Newsletter. April 1: 4-5.

______. 1985. Geography and national policy. Annals of the Association of American Geographers 75: in press.

Wilbanks, T.J. and Lee, R. 1985. Policy analysis in theory and practice. In Assessing regional consequences of large-scale energy projects, eds. B. Johansson and T.R. Lakshmanan. Amsterdam: North-Holland Publishing Company.

INDEX